RESEARCH HANDBOOK ON REMOTE WARFARE

RESEARCH HANDBOOKS IN INTERNATIONAL LAW

This highly original series offers a unique appraisal of the state-of-the-art of research and thinking in international law. Taking a thematic approach, each volume, edited by a prominent expert, covers a specific aspect of international law or examines the international legal dimension of a particular strand of the law. A wide range of sub-disciplines in the spheres of both public and private law are considered; from international environmental law to international criminal law, from international economic law to the law of international organisations, and from international commercial law to international human rights law. The *Research Handbooks* comprise carefully commissioned chapters from leading academics as well as those with an emerging reputation. Taking a genuinely international approach to the law, and addressing current and sometimes controversial legal issues, as well as affording a clear substantive analysis of the law, these *Handbooks* are designed to inform as well as to contribute to current debates.

Equally useful as reference tools or introductions to specific topics, issues and debates, the *Handbooks* will be used by academic researchers, post-graduate students, practicing lawyers and lawyers in policy circles.

Titles in this series include:

Research Handbook on International Law and Cyberspace
Edited by Nicholas Tsagourias and Russell Buchan

Research Handbook on Transnational Labour Law
Edited by Adelle Blackett and Anne Trebilcock

Research Handbook on Jurisdiction and Immunities in International Law
Edited by Alexander Orakhelashvili

Research Handbook on the Theory and Practice of International Lawmaking
Edited by Catherine Brölmann and Yannick Radi

Research Handbook on the International Penal System
Edited by Róisín Mulgrew and Denis Abels

Research Handbook on Disasters and International Law
Edited by Susan C. Breau and Katja L.H. Samuel

Research Handbook on International Law and Natural Resources
Edited by Elisa Morgera and Kati Kulovesi

Handbook on the Politics of International Law
Edited by Wayne Sandholtz and Christopher A. Whytock

Research Handbook on International Courts and Tribunals
Edited by William A. Schabas and Shannonbrooke Murphy

Research Handbook on Transitional Justice
Edited by Cheryl Lawther, Luke Moffett and Dov Jacobs

Research Handbook on UN Sanctions and International Law
Edited by Larissa van den Herik

Research Handbook on Remote Warfare
Edited by Jens David Ohlin

Research Handbook on Remote Warfare

Edited by

Jens David Ohlin

Vice Dean and Professor of Law, Cornell Law School, USA

RESEARCH HANDBOOKS IN INTERNATIONAL LAW

Edward Elgar
PUBLISHING

Cheltenham, UK • Northampton, MA, USA

Published by
Edward Elgar Publishing Limited
The Lypiatts
15 Lansdown Road
Cheltenham
Glos GL50 2JA
UK

Edward Elgar Publishing, Inc.
William Pratt House
9 Dewey Court
Northampton
Massachusetts 01060
USA

A catalogue record for this book
is available from the British Library

Library of Congress Control Number: 2017939814

This book is available electronically in the **Elgar**online
Law subject collection
DOI 10.4337/9781784716998

MIX
Paper from
responsible sources
FSC® C013056

ISBN 978 1 78471 698 1 (cased)
ISBN 978 1 78471 699 8 (eBook)

Typeset by Columns Design XML Ltd, Reading
Printed and bound in Great Britain by TJ International Ltd, Padstow

In Memoriam

Michael W Lewis

Professor of Law

6 December 1964–21 June 2015

Contents

Contributors

William C Banks is former Interim Dean of the College of Law; Board of Advisors Distinguished Professor of Law; Director of the Institute for National Security and Counterterrorism; Professor of Public Administration and International Affairs, Syracuse University.

Geoffrey Corn is Professor of Law, South Texas College of Law; formerly Special Assistant for Law of War Matters and Chief of the Law of War Branch, Office of the Judge Advocate General, United States Army; Chief of International Law for US Army Europe; Professor of International and National Security Law at the US Army Judge Advocate General's School.

Emily Crawford is senior lecturer and Director of the Sydney Centre for International Law (SCIL) at the University of Sydney.

Anthony Cullen is a Senior Lecturer in Law at the School of Law, Middlesex University, London.

Lydia Davies-Bright completed her LLM with distinction, and is a doctoral candidate at the University of Nottingham.

Gloria Gaggioli has a PhD in International Law and is Assistant Professor and Grantholder of Excellence at the University of Geneva, Law Faculty, Department of Public International Law and International Organization.

Robin Geiß is Professor of International Law and Security, University of Glasgow.

Terry D Gill is Professor of Military Law, University of Amsterdam and Netherlands Defence Academy.

Jelle van Haaster is an officer in the Royal Dutch Army and PhD candidate in cyber operations at the Amsterdam Centre for International Law and the Netherlands Defence Academy.

Robert Heinsch LLM is Associate Professor of Public International Law at the Grotius Centre for International Legal Studies, and the Director of the Kalshoven-Gieskes Forum on International Humanitarian Law and its IHL Clinic at Leiden Law School.

Ian S Henderson, PhD in Law, University of Melbourne, is a legal officer serving with the Royal Australian Air Force and is currently the Deputy Director of the Asia-Pacific Centre for Military Law.

Patrick Keane is a legal officer in the Royal Australian Air Force.

Mark Klamberg is Associate Professor at Stockholm University, Faculty of Law.

Henning Lahmann is Legal Adviser, iRights, Berlin.

Josh Liddy is a legal officer in the Royal Australian Air Force.

Peter Margulies is Professor of Law, Roger Williams University School of Law; BA, Colgate University, 1978; JD, Columbia Law School, 1981.

Michael W Meier is Attorney-Adviser at the Office of the Legal Adviser, Political-Military Affairs, United States Department of State.

Jens David Ohlin is Vice Dean and Professor of Law, Cornell Law School.

Mark Roorda is an officer in the Royal Netherlands Marine Corps and a PhD candidate in the operational and legal aspects of UAV targeting at the Amsterdam Centre for International Law and the Netherlands Defence Academy.

Nigel D White is Professor of Public International Law, University of Nottingham School of Law.

Preface

For this volume, I assembled an unsurpassed group of international law experts on the concept of remote warfare. Unfortunately, the volume is missing one key contributor. There is no chapter from Michael Lewis, professor of law at Ohio Northern University, and a former Top Gun F-14 navigator. Mike was scheduled to participate in the project but died, at the premature age of 50, at 10:47 p.m. on Sunday, 21 June 2015.

I met Mike Lewis during my first year of law teaching at Cornell Law School.[1] Mike was scheduled to give a lecture at the law school about torture and I was invited to give a commentary on his presentation. Mike had pre-circulated the paper that the presentation was based on. I disagreed with his thesis and pressed him sharply on its details during the event. His argument had the virtue of proposing a very workable standard for defining torture, but I felt it yielded counter-intuitive results for particular reasons that I articulated during the event. Afterwards, I was worried that I might have offended Mike, but it was not the case. Immediately after he got home, he wrote me a lovely note saying how much he appreciated our substantive exchange and was grateful that I had taken the time and energy to respond to his scholarship. He was a true scholar and intellectual.

In the ensuing five years, I spent much time reading and learning from Mike's other articles on IHL. This came at a crucial time for me as I was broadening my research agenda from exclusively ICL to include a wider range of IHL and law of war issues as well. I became heavily involved in debates about drones, targeted killings, targeting in general, and the relationship between IHL and human rights law. In all of these areas, I was heavily influenced by Mike's explanations and positions that he articulated in his many law review articles. And I should hasten to add that on most of these crucial questions I was in agreement with Mike. Although I disagree with the Obama Administration's legal positions on a few issues (definition of imminence, over-reliance on covert action and its consequences, use of the vague and indeterminate 'associated forces' moniker, etc.), the general tenor of my scholarship has been to

[1] Some of these remarks and thoughts were originally expressed in a post published on *Opinio Juris* on 25 June 2015.

recognize that the deep architecture of IHL continues to be fundamentally Lieberian. I came to this view of IHL by reading a great many sources, but I would rank Mike's articles near the top of that list. Simply put, I would not hold the views that I hold today if I had not been so richly educated by reading Mike's work.

I spent some time with Mike at the ethics and law of war conference at the University of Pennsylvania Law School. Mike was full of plans and we discussed the possibility of collaborating on future projects on the subject of the privilege of combatancy—a common interest for both of us. We hosted him at Cornell University last year as part of our university-wide Lund Critical debates series, where he debated Mary Ellen O'Connell on the use of drones.[2] Mike's presentation to the packed auditorium was both insightful and extremely clear. He had the ability to translate complex legal material to a wide audience, and Mary Ellen's thoughtful critique on U.S. policies made for a lively debate between the two of them.

As I set about working on a new collected volume on remote warfare, I emailed him in October to commission a chapter from him; he enthusiastically responded in the affirmative. Three weeks before he died, he informed me of his illness and said he could not definitively commit to the project anymore but was hopeful that he might still produce a chapter for it. Though he was still optimistic and making important plans for the future, I understood the nature of the diagnosis and prognosis because he gave me the name of his illness, but I labored under the illusion that we had more time. He was fighting cholangiocarcinoma, a very rare form of cancer that is especially lethal because it is often inoperable. But even so, when I heard the news about his death, I was shocked that the end had come so quickly. I was unprepared for the news even though in the back of my mind I inferred the seriousness of the situation.

I am devastated that we have been denied his voice for what should have been another 50 years. It highlights for me the fragility of life and our time on this earth and the ultimate unfairness by which some people are denied the privilege of a long life. But I take some comfort in knowing that he loved being a law professor and that we will be reading his work—unfortunately not in this volume but luckily in many widely read law review articles—in the years and decades to come.

JDO
Ithaca, NY
November 2016

[2] Video of the event can be found at http://www.cornell.edu/video/deaths-by-drone-are-they-illegal (accessed 25 April 2017).

Table of cases

Inter-American Court of Human Rights (IACHR)

International Criminal Court (ICC)

International Court of Justice (ICJ)

International Criminal Tribunal for the Former Yugoslavia (ICTY)

International Criminal Tribunal for Rwanda (ICTR)

NurembergTrials

Permanent Court of International Justice (PCIJ)

UN Human Rights Committee

National Cases

Australia

Colombia

Germany

Israel

Peru

Spain

United Kingdom

United States of America

Table of legislation

National Legislation

United Kingdom

United States of America

Introduction

Jens David Ohlin

1. THE FRAMEWORK OF REMOTENESS

The number of academic articles and books on drone warfare has increased exponentially in the last five years. Of course, this academic output is entirely justified, given the significance of drone technology—and the strategy of targeted killing—in modern warfare. Although there is much debate and little agreement in this literature, it is undeniable that the nature of warfare is changing, and the question addressed in this literature is how the current law, whether international humanitarian law (IHL), international human rights law, or jus ad bellum, should apply to drone strikes carried out in diverse operational situations.

In terms of sheer volume, the literature on cyber-war follows close behind. For some years, much of the literature offered legal analysis of hypothetical attacks. The coming age of cyber warfare was on the horizon, but had not yet appeared. The Stuxnet attack changed all of that. And now the cyber-intervention by Russia in the US election of 2016 has established that cyber activities (and cyber counter measures) will be ever-present in the strategic and diplomatic landscape. Whether one agrees that, on one end of the spectrum, the Russian hacking in that case constituted an 'act of war', as some US politicians declared, or on the other end of the spectrum perhaps simply a violation of standard 'norms' of international behavior, as President Obama indicated, the fact of the matter is that cyber interventions of all sorts will radically proliferate in the coming decade. With the advent of the *Tallinn Manual*, in its first and second manifestations, and other academic efforts, the invisible college of international lawyers is grappling with how to apply to cyberspace international legal norms whose foundation was laid many years ago when cyber-attacks were mostly the stuff of science fiction.[1]

Finally, the technological advances regarding Autonomous Weapons Systems (AWS) are perhaps less prevalent, though if one defines AWS sufficiently broadly, it is clear that current weapons development for the

[1] See M N Schmitt, *Tallinn Manual on the International Law Applicable to Cyber Warfare* (Cambridge University Press 2013).

United States, Russia and China already includes a substantial AWS component, so that the legal literature on AWS is arriving on the scene just in time. Although the concept of AWS is often misunderstood in the popular imagination as involving robotic infantry, the more typical application of the technology involves missile systems with autonomous targeting protocols—something that is not so hard to imagine. (The other typical application is cyber weapons with independent or quasi-independent targeting protocols.) The number of legal issues posed by these applications is immense.

The goal of the present volume is not to rehash the existing issues, which are already usefully articulated and debated in the literature. Rather, the point is to cut across the traditional categories and analyze these developments conceptually. Much of the literature atomistically considers drones, while another segment analyzes cyber, and a third literature addresses AWS. The prime motive of this volume is to both aggregate and disaggregate these subjects at the same time. The subjects are aggregated by considering drones, cyber, and AWS altogether, but then disaggregated by slicing off one aspect that is common to all three and investigating it in more detail.

Here is where the concept of remoteness enters the picture. One crucial aspect that unites drones, cyber and AWS is their remote capability—a potentially new form of warfare that is allowing operators to use ever more discriminating force while also receding further in time and space from the target of the military operation. This one aspect of the new technology—its remoteness—deserves its own investigation, and this volume provides it. Under this rubric, many questions can and should be asked. Is the remoteness of these technologies that much different than what came before? If it is different, is the difference one of degree or of kind? If these new technologies require additional regulation, or if existing regulation is already adequate, does their remoteness constitute an additional hurdle towards effective regulation? Do these remote technologies change the risk calculus for going to war in ways that are ethically or legally problematic? Are these technologies being used in ways that comply with jus ad bellum? In particular, are we witnessing a new category of warfare: remote self-defense?

The point here is not to defend a particular position with regard to these issues, but simply to defend the need for a legal volume that focuses exclusively on the notion of remoteness per se, as a field of serious academic inquiry. The nature of the remoteness is the Elephant in the Room, the obvious common characteristic of these technologies that has, nonetheless, moral and legal implications that are not so easy to trace or predict. Hopefully, the present volume is the beginning of that scholarly endeavor.

2. THIS VOLUME'S CONTRIBUTION TO THE DEBATE

In Chapter 1, 'Remoteness and reciprocal risk', I argue that reciprocal risk is not, and has never been, an essential component of IHL or the law of armed conflict. In canvassing the history of military technology, I argue that the strategic goal in all military engagements has been to maximize lethality to the target while minimizing risk to the operator. The most recent technologies, including drones, cyber-weapons, and AWS, are only the most recent (and extreme) versions of this familiar incentive that underlies all strategic warfare. In the chapter, I analyze the question of whether the radical *increase* in asymmetrical risk will make war too easy—and ultimately result in more warfare. I explore and respond to this objection. Along the way, the chapter also considers whether reciprocal risk is an essential component of ethical warfare—a possibility that is cautiously rejected. In the end, the chapter performs a debunking function by taking some of the most common complaints about these technologies—pertaining to their remoteness—in order to demystify them and provide the correct historical context for them. Seen in that light, drones, cyber-weapons, and even AWS start to look less like aberrations and more like the ultimate conclusion to the natural evolution of warfare.

In Chapter 2, 'The principle of distinction and remote warfare', Emily Crawford continues the discussion of remoteness by focusing on the fate of the 'intransgressible' principle of distinction in an era when military targeting technology is becoming ever more remote, with the operator of the weapon ever further in geographic distance from the target of the military strike. Crawford argues that this technological evolution carries both promises and perils for the principle of distinction. On the one hand, she takes seriously the argument that some remote technologies, including drones, allow for increased compliance with the principle of distinction, in part because of increased capacity to hover over a target and improved surveillance capacity: 'Drones, with the potential to spend days, even weeks observing a potential target, gathering copious data on the target, and allowing for complex and detailed assessments about the legality of a strike against such a target, without fear of discovery of such surveillance, thus may be exceptionally compliant with the principle of distinction.' On the other hand, she also warns that there are risks associated with remote attacks, whether by drone or cyber-weapons, because the 'clarity and precision offered by drones by way of their "real-time" video feed of targets is undermined by the very real technological problem of "latency"—the time delay between activities observed and videoed at the target site and the arrival of that video image via satellite to the pilots'. Crawford does more than simply trade in generalities, but instead gives specific examples of strikes where technology either helped or hindered compliance with the principle of distinction.

In Chapter 3, 'Modern drone warfare and the geographical scope of application of IHL: pushing the limits of territorial boundaries?', Robert Heinsch examines the way that remotely piloted vehicles have expanded the geographical scope of armed conflict. With this expansion has come uncertainty over the geographical scope of IHL as a regulator of lethal conduct. Specifically, the legal question Heinsch examines is whether drone strikes alone, if piloted remotely from the United States but deployed in a foreign state against a non-state actor, are sufficient to place the United States in a non-international armed conflict with that non-state actor. If the answer to that question is yes, then presumably the IHL rules applicable to non-international armed conflicts (NIACs) will govern and restrict those drone strikes. That being said, many human rights lawyers would resist this conclusion because they view IHL as being too permissive and insufficiently restrictive. Heinsch argues that the correct standard is whether there is 'protracted armed violence' between the state and the non-state actor, and there is nothing in that standard that would prevent a series of drone strikes from satisfying that standard. When applied to the facts of most drone strikes conducted by the United States, however, Heinsch argues that many of the strikes fall below the threshold for the existence of an armed conflict, because the violence is not sufficiently protracted. In that case, International Human Rights Law governs the killings.

In Chapter 4, 'The characterization of remote warfare under international humanitarian law', Anthony Cullen examines the concept of armed conflict as a legal term of art, and how it applies in the context of drones, cyber-attacks, and autonomous weapons. Cullen argues that a proper definition of 'armed conflict' cannot be constructed in the abstract, but instead requires consideration of the object and purpose of IHL as a field of legal regulation. Cullen identifies that object and purpose as the protection of civilians in armed conflict. Based on this simple cannon of interpretation, Cullen argues that IHL should be interpreted in such a way that it applies to armed conflicts pursued primarily with drones, cyber-weapons, or AWS—precisely because these modalities have the capacity to negatively impact civilians. As Cullen notes, 'If the law of armed conflict has a vanishing point in the 21st century, it is arguably that of remote warfare'. By this he means that the greatest risk of irrelevance for IHL is with regard to drones, cyber-weapons and AWS. This suggests that IHL as a discipline must pro-actively work to maintain its relevance as a source of constraint and regulation of the modalities of remote warfare. Since the future of armed conflict resides with these remote technologies, IHL risks irrelevance and obsoleteness if it does not rise to the challenge.

In Chapter 5, 'Remoteness and human rights law', Gloria Gaggioli asks whether, in contrast to Cullen's focus on IHL, human rights law has a distinct role to play in armed conflicts (or military force generally) that are characterized by the remote deployment of military assets, especially drones, but also to a lesser extent cyber weapons and AWS which also demonstrate degrees of remoteness. The author asks, critically, whether the human rights principles of legality, accountability, transparency, dignity and self-defense have a distinctive role to play in the regulation of remote force. Gaggioli moves away from the often-uncritical distrust of remote weaponry among human rights scholars and activists, and instead scrutinizes the very nature of remote force to determine whether—and to what degree—it offends these core principles of human rights law. The picture she paints is far subtler than that offered by some human rights activists, but it still recognizes a large conceptual space for human rights law to govern, and constrain, remote force.

In Chapter 6, 'Exploiting legal thresholds, fault-lines and gaps in the context of remote warfare', Mark Klamberg asks whether the current system of legal regulations regarding the use of force are adequate to the task of regulating military violence in a world of remote warfare. Specifically, Klamberg argues that the current scheme for regulating the use of force is full of gaps and thresholds that define what does—and does not—count as a use of force in the sense of the UN Charter or an armed conflict in the sense of IHL. It is precisely these gaps and fault lines that allow so-called gray conflicts to persist below the radar of particular legal regulations. Indeed, much of the rationale for states to deploy remote technologies, including drones, cyber-weapons and AWS, is that it might allow the state to project state interests while not triggering many of the legal restrictions that would attach to more conventional military interventions.

Part II of the Handbook focuses more specifically on the legal regulation of remotely piloted vehicles (drones) and cyber technology. In Chapter 7, 'Drone strikes: a remote form of self-defense?', Nigel D White and Lydia Davies-Bright argue that technological developments are changing the nature of international relations that give rise to the need for legal regulations of jus ad bellum. Specifically, White and Davies-Bright note that the advancement of remote technology is facilitating the deployment of military force in ways that place existing notions of sovereignty under intense pressure. States with remote technology can now attack far-flung territories with drones, causing those targeted states to worry about how to protect their own sovereignty against these intrusions. The authors connect this development to a resurgence in 'primordial understandings of sovereignty based on preservation of the nation state'. On the other hand, one might also view remote technology not just as a cause to this problem but as a *response* to it. Non-state actors are using the tactics of terrorism to inflict damage on state authorities in an attempt to undermine traditional notions of Westphalian sovereignty, while (some) states are pushing back with an aggressive campaign of extraterritorial military force designed to protect state authority. The authors conclude that 'the return to absolute forms of sovereignty by technologically advanced states is something more profound and alarming' because 'it represents a reversion to a very primitive view of the state whereby its promise to protect its citizens at all costs is used to circumvent the basic rights of individuals'. The chapter by White and Davies-Bright represents a powerful corrective to uncritical acceptance of the fate of self-defense arguments in an age of remote technology.

In Chapter 8, 'Drone warfare and the erosion of traditional limits on war powers', Geoffrey Corn continues the discussion of whether remote technology in general, and drones in particular, are transforming traditional avenues of legal regulation. Specifically, Corn asks whether the ubiquity of drones as a military platform has rendered obsolete the traditional constraints in US and international law for limiting the executive branch's deployment (or the sovereign's deployment) of military force. With regard to international law, even a limited number of drone strikes might qualify as an extraterritorial or transnational NIAC, as long as it is accepted that a NIAC may cross international borders and not remain exclusively *internal* to the state. If that is the case, the applicability of the label 'NIAC' encourages the use of military force because IHL sanctions a huge amount of violence, especially so in NIACs, which are under-regulated compared to their international cousins. But even without considering the classification dilemma of whether such conflicts are best described as IACS, NIACs, or something in between, the fact that they are armed conflicts at all means that a large amount of military force is permitted, at least with regard to jus in bello. (Jus ad bellum might be a different story, though other doctrines, including the 'unwilling or unable' test, perform similar justificatory work in that domain too.) With regard to national law, drone strikes might allow the executive branch to alleviate a national security threat while at the same time placing very few (or even no) service members in harm's way. This factor, inter alia, led Harold Koh and the Obama Administration to suggest that the War Powers Resolution 'clock' did not necessarily apply to military operations with few so-called 'boots on the ground'. For Corn, these factors, in separate legal domains, suggest a dangerous incentive to use drones and also suggest a weakening in the traditional resources of the legal system to constrain and discourage the use of military force. As Corn concludes, 'This advent of the transnational armed conflict theory, coupled with the capacity to conduct virtually risk-free attacks with decisive lethal force, has arguably incentivized an aggressive invocation of armed conflict'.

Chapter 9 switches gears and specifically covers the legal regulation of cyber conflicts as a form of cyber warfare. In 'Developing norms for cyber conflict', William C Banks provides a useful outline for how law of war doctrines will apply to cyber conflicts and the difficulty of translating, into the cyber domain, principles and norms that were originally developed for regulating traditional military activities. After outlining some of these interpretative difficulties, Banks notes that there are reasons for optimism. In particular, Banks notes that the US Department of Defense recently released its *Department of Defense Cyber Strategy*. In that document, which is admittedly a strategic rather than legal analysis, Banks sees the seeds of a burgeoning recognition that cyberwarfare will place immense pressure on the existing normative architecture and will require significant reconceptualization of classic IHL principles. For example, Banks believes that the traditional notions of 'use of force' and 'armed attack' are simply not useful for understanding cyber conflict, though they remain the central pillars of the existing jus ad bellum and jus in bello. As Banks concludes, the DoD *Cyber Strategy*, by moving away from such traditional notions, 'could provide a pillar of a normative architecture for cyber conflict' in the future.

In Chapter 10, 'Some legal and operational considerations regarding remote warfare: drones and cyber warfare revisited', Terry Gill, Jelle van Haaster and Mark Roorda argue that while drones advance the goals of range and covertness, cyber weapons advance the goals of anonymity, allowing the attacker to defeat the enemy's desire for attribution. Nonetheless, the authors paint a portrait of technological advances that can be, and should be, subject to the regular normative constraints of the law of war. In this sense, the authors disagree with those critics who argue that drones and/or cyber weapons require a radical re-working of jus ad bellum or jus in bello in order for these weapons to be effectively regulated. As the authors note, 'despite the new and to an extent revolutionary impact of these modes of remote warfare, it is clear they are and are, in principle, capable of being governed by the framework of international law, including the law relating to the use of force and the legal regimes which govern how and against whom or what force may and must be applied, in particular the humanitarian law of armed conflict and international human rights law where these are applicable'.

In the next chapter, 'Remote and autonomous warfare systems: precautions in attack and individual accountability', Ian S Henderson, Patrick Keane and Josh Liddy focus on the remoteness of AWS and the challenges that this poses for meeting the obligations, imposed by IHL, during the targeting process. The authors engage in an intensive analysis of the relevant requirements codified in Additional Protocol I (API), and in so doing, insert themselves into a long-running debate between activists who argue that AWS should be banned because they will necessarily violate the obligations of API and military analysts who argue that API may even impose an *obligation* to use AWS where doing so would improve targeting compliance. The authors take stock of the classic objections to AWS and cautiously and moderately support the conclusion that, in some limited circumstances, an AWS might better identify military targets, especially in contexts where military hardware is utterly unique and there are no civilian counterparts to the military equipment. That being said, it will likely be much more difficult to design and build an AWS that can reliably determine whether a human being is 'directly participating in hostilities' and therefore a lawful target on that basis. For proportionality assessments, an AWS might be trained and tested to determine whether its proportionality determinations are accurate. Although the determinations would need to keep a human being 'in the loop' in the strictest sense of that expression, 'it can be argued that human judgement and discretion is being applied by human decision makers electing to employ AWS at a certain location or time, or against certain kinds of targets, with knowledge of how the system will operate in that environment'.

In Chapter 12, 'Autonomous weapons systems: a paradigm shift for the law of armed conflict?', Robin Geiß and Henning Lahmann focus on the inevitable rise of autonomy in military operations, because 'the de-humanization of warfare is already well underway' and there is only a 'small step left towards fully autonomous weapons'. The authors therefore place the rise of AWS in historical context and suggest that the coming technological developments are just the last stage in a progression that has been long in the making. After developing this account, the authors turn their attention to the often-asserted problem of a responsibility gap for AWS—that is, the question of who will be held responsible when an AWS violates international criminal law or international humanitarian law. The authors concede that it is somewhat problematic to turn over responsibility for complying with IHL to a machine that operates according to an algorithm. On the other hand, the solution to this problem—insisting on 'meaningful human control' for all weapons—is inherently ambiguous because what counts as keeping a human 'in the loop' can mean many different things. However, the authors conclude that '[i]f an agreement on the concept of "meaningful human control" can be reached that comprises all of the three elements, then the risk of an allegedly insurmountable "accountability gap" becomes a non-issue'.

In Chapter 13, 'Making autonomous targeting accountable: command responsibility for computer-guided lethal force in armed conflicts', Peter Margulies argues that AWS will pose significant challenges for IHL compliance, though he sees the solution as active engagement rather than an outright ban on the new technology, which he equates with seeking to 'blink away the future of war'—a totally unrealistic proposition. Specifically, Margulies argues that the doctrine of command responsibility is the key to ensuring that deployment of AWS are IHL-compliant because command responsibility will ensure that those deploying the systems are held accountable for any violations that occur during the deployment. For Margulies, the correct standard will be 'dynamic diligence', which he defines as requiring ongoing: adjustments in the system's interface, assessments of its compliance with IHL, updates to its inputs, and flexibility in the parameters governing the system's operation. In this context, 'flexibility' means that the commander should not operate the system in situations where doing so would be inappropriate. The contribution of Margulies' intervention is that it provides concrete detail to explain how the doctrine of command responsibility could work in practice to ensure that military commanders do everything feasible to control and deploy AWS in a responsible manner.

The volume concludes with Chapter 14, 'The strategic implications of lethal autonomous weapons' by Michael W Meier. In the final chapter, Meier takes a more strategic perspective on AWS by combining strategic considerations with legal analysis. Meier focuses on four key issues: whether AWS make it easier for states to go to war, whether they promote 'unintended engagements', whether AWS will foster greater asymmetry in warfare and ironically spur more terrorist attacks in response, and whether AWS will proliferate and trigger a new arms race. Meier ends his chapter by suggesting diplomatic and regulatory actions that the United States might take in order to mitigate the potential instability that could be caused by the growing development of AWS. These include, inter alia, restrictions in some circumstances on the export of AWS technology to other states, similar to the existing UAV (drone) export policy that the US government has already implemented.

3. FUTURE DIRECTIONS

In the coming decades, the law regarding remote warfare will inevitably sharpen, as more and more state practice, in response to specific situations, will clarify the scope of existing customary law. Although the common refrain is that *we need a new treaty*, and that always remains a possibility, the vast majority of technological developments are either subsumed under existing treaty regulations or governed by rules of customary international law. Either way, the responses of the world community to these technologies over the coming decades will help form the legal framework for regulating them.

Although the future is bright for legal clarifications, the time for conceptual clarification is now. With the advancement of each new technology, 'armed conflict' as a category paradoxically seems to expand and contract at the same time. There is expansion because remote technologies allow for geographical distance between operator and target, or, in the case of AWS, between commander and target. This widens the geography of armed conflict—indeed globalizes it—scattering the 'participants' across the globe and even across virtual and cyber divides. However, at the same time, there is contraction of armed conflict because remote technologies, *in theory*, bring the promise of razor accurate targeting that will reduce civilian collateral damage. Indeed, even targeting against lawful combatants may be reduced. It is no surprise or coincidence that the advent of drones has allowed for the shift from status-based targeting to conduct-based targeting. The result is that individual fighters are targeted based on their level of threat or dangerousness, rather than the wholesale slaughter of hundreds or thousands of enemy fighters who are killed *en masse* simply because they belong to the same organization, whether it is a conventional army or a terrorist organization. This new era of status-based targeting, which is intimately wrapped up with remote technology, clearly represents a contraction of warfare in this sense.

With AWS, remoteness reaches into its heavenly extreme because the operator fades away into nothingness. Or, perhaps more extravagantly, the weapon *is* the operator. Either way, the distance of the individuals running the show has reached a level of conceptual remoteness perhaps unthinkable in past conflicts: a war without soldiers. If that sounds like an exaggeration, it certainly is. But perhaps it is more accurate to describe a discrete military *operation* launched without military personnel, who are remote in the sense of non-existent, because the AWS will execute the operation without significant human decision-making 'in the loop'. For some scholars, this remoteness is a recipe for a moral disaster because it will promote unrestrained killing without risk and atrocities committed without responsibility. For others, this remoteness holds the promise implicit in the humanitarian project itself: reducing the suffering on human beings by removing them from warfare as much as possible (at least on one side of the conflict, if not both). One could understand this classic dilemma in the AWS literature as a dispute about the value of remoteness in military engagements. Must humans remain intimately involved in each military operation, or conversely should we work as much as possible to get them removed from war, as long as we can maintain (or even improve) targeting accuracy? Boiled down to its guts, this is a dispute about *remoteness*.

PART I

THE CONCEPT OF REMOTENESS IN WARFARE

1. Remoteness and reciprocal risk

Jens David Ohlin

1. INTRODUCTION

The history of modern weaponry involves the construction of the technological capacity to produce lethal results while exposing the operator to the least amount of risk of death or injury. The most recent examples of this phenomenon are three new weapon categories: remotely piloted vehicles (drones), cyber-weapons, and Autonomous Weapons Systems (AWS). Each of these categories of weapons allows the attacking force to inflict military damage while the operators of the weapon remain safely shielded from the theater of operations.[1] The technological goal is therefore to generate an inverse proportionality between risk to the operator and lethality to the target.[2] By this standard then, the overall strategy is to create a system that grants the operator total immunity from risk but still inflicts maximum damage to the enemy.[3]

Recent increased technological capabilities have generated a divergent set of intuitive responses from different constituencies. For military planners in the United States, as well as coalition forces allied with them, the advent of remote killing by drone is a source of great pride, because it allows for greater force protection while still accomplishing the

[1] It is perhaps not correct that the operators remain completely shielded from attack. Rather, it is more correct to say that, as a comparative and relative matter, the operators of these weapons are *much less* subject to attack than the operators of more conventional weapons systems. But as the following chapter will demonstrate, the association between remoteness and risk is a difference of degree—not kind—from prior weapon systems.

[2] See Megan Braun, 'Predator Effect: A Phenomenon Unique to the War on Terror' in Peter L Bergen et al (eds), *Drone Wars: Transforming Conflict, Law, and Policy* (Cambridge University Press 2015) 253, 278 (describing the development of insect-sized drones that will operate in locations too dangerous for individual soldiers).

[3] Andy Dougan, *Through the Crosshairs: A History of Snipers* (Carrol & Graf 2005) 16 ('What primitive mankind sought was a weapon that could be used from a distance, leaving its user exposed to minimal risk of injury.').

mission.[4] Moreover, advocates for drone operations often insist that the parameters of the drone platform, including increased surveillance capacities as well as the ability to hover or circle over an intended target for an extensive period of time, allow drone operators to kill enemy targets with an unparalleled level of discrimination. In other words, the capacity to reduce civilian collateral damage is greatly enhanced. In contrast, critics are often harsh in their criticism of American and allied reliance on drone technology. Among the complaints are that drone attacks violate jus ad bellum in many instances, that the attacks produce collateral damage in the absence of risk, and that US military planners are all too willing to shift the risk of death from friendly military forces to enemy civilians who will be killed in the attack. The common assumption that resides beyond these diverse intuitions—both positive and negative—is that drone operations have transformed contemporary military engagements by changing the risk calculation.

For cyber-war, the calculation is slightly different. It is true that cyber-weapons involve remote operators who are physically distant from the scene of the ultimate attack. For example, the operators who designed and launched the Stuxnet computer virus against the Iranian nuclear power plant were nowhere near the centrifuges that were ultimately disabled and destroyed.[5] There is, however, a difference between drones and cyber-weapons: the United States and its allies view the latter not just as an opportunity for strategic force multiplication but also as an area of intense vulnerability and therefore a cause for grave concern. In particular, China, North Korea and Iran have remotely deployed, and threaten to deploy, cyber-weapons that the United States and other allied nations may only partially defend against.[6] Consequently, the risk is asymmetrical only in the following sense. China could launch a cyber-attack against US forces without endangering its Chinese cyber-soldiers, thus generating an asymmetry of risk. However, the United States could

---

[4] Ryan J Vogel, 'Drone Warfare and the Law of Armed Conflict' (2010) 39 Denv J Intl L & Policy 101, 102 ('Drone targeting has proven to be spectacularly successful—both in terms of finding and killing targeted enemies and in avoiding most of the challenges and controversies that accompany using traditional forces.').

[5] However, it should be noted that the virus might have entered the physical computer system through the connection of an infected USB device to a local computer. See Johann-Christoph Woltag, *Cyber Warfare: Military Cross-Border Computer Network Operations under International Law* (Intersentia 2014) 47–8.

[6] See David E Sangerjune, 'As Russian Hackers Probe, NATO Has No Clear Cyberwar Strategy' *New York Times* (16 June 2016).

do the same and launch a cyber-attack against Chinese forces without endangering American cyber-soldiers. The situation is therefore best described as 'reciprocal asymmetrical risk'.

It is unclear how the advent of AWS will transform reciprocal risk.[7] Just as in drone and cyber capabilities, some military nation-states will have the capacity to exploit the technology while others will not—at least not yet. The animating impulse behind AWS is to allow the 'operator' to remain further remote from the field of operation by transiting most of the operator tasks to the system itself.[8] One of the central reasons why the soldier—the human soldier—usually needs to be close to the kinetic effect is because the soldier needs to assess the situation and determine how to destroy the enemy. Air and naval power stretch that capability but they are only as effective as the intelligence regarding the target and they are only effective against some types of target. However, if a weapon system could—by itself—make strategic, legal, and even moral determinations about how to engage the target, the human soldier could remain safely out of harm's way in a remote location. Indeed, it would seem as if the whole strategic point of AWS is its promise of force protection combined with an argument that, like drones, they could reduce civilian collateral damage. Unlike drones, where the argument for lowering collateral damage is their ability to pin-point particular targets, the argument in favor of AWS is more specifically that their computerized algorithms would be immune from heuristic biases and other cognitive defects that infect human reasoning.[9] Like cyber-weapons, however, the asymmetrical risk would be balanced—it would be reciprocal in a deeper sense—in conflicts between military powers that both deploy AWS against each other.

The technological advances just described are novel, but the focus on reciprocal risk is not. The following section seeks to put these technological developments in historical context and will investigate the

---

[7] In using the phrase 'reciprocal risk', I follow George Fletcher's invocation of the term in a different context. See George P Fletcher, 'Fairness and Utility in Tort Theory' (1972) 85 Harv L Rev 537, 549 (strict liability and negligence as solutions for the unfairness of 'unexcused, nonreciprocal risk-taking').

[8] See Marco Sassoli, 'Autonomous Weapons—Potential Advantages for the Respect of International Humanitarian Law', Profiles in Humanitarian Assistance and Protection (2 March 2013); Christopher P Toscano, '"Friend of Humans": An Argument for Developing Autonomous Weapons Systems' (2015) 8 J Natl Security L & Policy 189, 213.

[9] See Gregory P Noone and Diana C Noone, 'The Debate over Autonomous Weapons Systems' (2015) 47 Case W Res J Intl L 25, 32 ('human error causes untold deaths—perhaps AWS can perform better').

moral and legal consequences of every belligerent's desire to reduce risk while maximizing lethality. In short, this chapter will investigate whether reciprocal risk is an essential component of ethical and lawful warfare, whether the technological capacity to produce asymmetrical risk through remoteness is historically novel or continuous, and whether recent advances on that front should be celebrated or criticized. For example, should we view the desire to create asymmetrical risk as fundamentally different from, or of a piece with, the shift from swords to guns? Are we witnessing a fundamental shift in warfare, or are we reading just the latest chapter in the same old story?

Specifically, this chapter will propose, explain and critically examine the concept of reciprocal risk. It will seek to determine whether there is, in fact, a historical norm in favor of reciprocal risk in warfare, and how the advent of drones, cyber-weapons and AWS have impacted this putative norm. After evaluating the alleged and often assumed rupture to reciprocal risk caused by technological innovation in weapons design, this chapter will then examine two familiar objections to these technologies. The first is whether the weapons will, by creating a severe asymmetry in risk, allow states to exercise force cavalierly, and remove an important check on warfare that helps limit the number of jus ad bellum violations across the globe. Having examined that anxiety, the final part of this chapter will ask whether reciprocal risk is an essential ethical component of basic norms of chivalry. This latter analysis will require an examination of legal principles under the Law of Armed Conflict (LOAC) and ethical principles embodied in just war theory.

2. A BRIEF HISTORY OF MILITARY RISK MANAGEMENT

Is it really true that the rapid development of drones, cyber-weapons and AWS has eroded reciprocal risk? A brief analysis of the history of weaponry suggests that the production of asymmetrical risk is not an outlier.[10] In fact, it has been the goal of weapon design ever since the abandonment of the club as an instrument of blunt-force killing.

Historians of weaponry generally regard the bow as the signature development in early weaponry. It allowed individuals to launch a deadly

---

[10] See Gabriella Blum, 'The Dispensable Lives of Soldiers' (2010) 2 J Legal Analysis 115, 132 ('governments try to protect their soldiers, partly by employing more aggressive force toward the enemy (including by "risk-transfer" from soldiers onto enemy forces and civilians)').

attack from a distance, perhaps even from a location hidden from view.[11] A target might be felled by an arrow without ever looking his killer in the eye. The attacker could therefore inflict a lethal result while minimizing, though not entirely erasing, his exposure to a lethal counter-attack. As one historian explains:

> At once safe and deadly, it was the ideal weapon of harassment. A combatant might spend an afternoon shooting away at long range with little fear of injury. Yet if the opportunity arose, he could move in closer and swiftly, silently dispatch an enemy with a single shot. It is not surprising, therefore, that the earliest actual image of combat, a Mesolithic cave painting at Morela la Vella in Spain, depicts men fighting with bows. Conceptually at least, the picture is a familiar one. The action is confused. The participants appear to be on the run, perhaps hoping to rip off a few quick shots before retreating. Moreover, all the combatants are armed symmetrically; only the bow is used.[12]

The advantage of bow and arrows over their predecessors—primarily fixed blades—was their ability to inflict damage at a distance.

It is important to recall that many of these early technological developments were motivated just as much by the requirements of hunting as they were by the requirements of warfare.[13] Action at a distance was crucial for catching prey. Managing risk was important in this endeavor as well, but not the primary motivating factor. While a few animals could harm a hunter during a close-quarters confrontation, most animals would simply flee. The real benefit of action at a distance was stealth and surprise. Regardless of its original motivation, the bow and arrow transformed warfare and effectively ended the paradigm of personal confrontations where soldiers were required to look the enemy in the eye before killing or injuring them.[14]

The introduction of guns accelerated the process that was first introduced with the bow and arrow.[15] Large-scale cannons used in the defense

---

[11] The javelin might be considered a forerunner of the bow and arrow and was used by the Mycenaeans for hunting but 'very rarely' in war. See A M Snodgrass, *Arms and Armour of the Greeks* (Cornell University Press 1967) 17.

[12] See Robert L O'Connell, *Of Arms and Men: A History of War, Weapons, and Aggression* (Oxford University Press 1989) 26.

[13] Ibid.

[14] The Mycenaeans probably used bows for warfare, even though the ancient Greeks who came later 'did not think highly of the bow for military use'. See Snodgrass (n 11) 17.

[15] See Malcolm Vale, *War and Chivalry* (University of Georgia Press 1981) 129 (discussing Don Quixote's complaint that war had become impersonal and

of fixed dwellings,[16] or used offensively in naval warfare, allowed armies to launch an explosive device over long distance with the artificial enhancement of explosive powder. The 'action' of action at a distance was no longer powered by human strength but by chemical ingenuity.[17]

Miniaturization of gun-power devices so that they could be carried by individual soldiers proved to be difficult to design and even more difficult to deploy. As the British historian John Keegan notes, hand-held firearms 'remained relatively ineffective' because they were 'fired by applying a burning match to an open touchhole, both prone to malfunction in wet weather, and they threw comparatively light balls only a short distance'.[18] The better option was the crossbow, a hybrid invention that had combined the propulsion of the firearm with the arrow, producing a deadly weapon: 'Armed with a crossbow a man might, without any of the long apprenticeship to arms necessary to make a knight, and equally without the moral effort required of a pike-wielding footman, kill either of them from a distance without putting himself in danger.'[19] In other words, crossbowmen were probably 'the first users of firearms'.[20] Consequently, the modern-day firearm was just the logical conclusion of a technological process first begun with the deployment of bows and arrows in battle.[21]

In conceptual terms, modern-day ordnances are basically improvements on the same paradigm. Artillery produces greater destruction, with greater accuracy, and at a greater distance than cannon fire. The strategic goal of artillery combat is to inflict damage from a range or location that

mechanical). See also J R Hale, 'Fifteenth and Sixteenth Century Public Opinion and War' (1962) 22 Past and Present 18, 21 ('by the beginning of the sixteenth century the gun had acquired a rich store of symbolic and associative overtones and was already rivalling the sword as the embracing symbol of war itself').

16 Vale (n 15) 130–31.

17 O'Connell (n 12) 162 ('Yet again, the great stabilizer was the imposition of the generic solid-firing smoothbore gun as the standard naval engine of destruction. So armed, all ships differed basically in degree rather than kind—the more guns, the more fighting power.').

18 John Keegan, *A History of Warfare* (Knopf 1993).

19 Ibid.

20 Ibid. See also Jean Liebel, *Springalds and Great Crossbows* (Royal Armouries 1998) 23 (dating crossbows to the 10th century); Blum (n 10) 75.

21 See Andrew Ayton, 'Arms, Armour, and Horses' in Maurice Keen (ed), *Medieval Warfare: A History* (Oxford University Press 1999) 186, 203–4 ('Massed archery by men able to unleash perhaps a dozen shafts a minute would produce an arrow storm, which at ranges of up to 200 yards left men clad in mail and early plate armour, and particularly horses, vulnerable to injury, while causing confusion and loss of order in attacking formations.').

is relatively immune from counter-attack by return fire. Cruise missiles launched from naval ships operate from the same strategic premise. Indeed, the logical culmination of this remoteness is Intercontinental Ballistic Missiles (ICBMs), which allow destruction of an enemy location without having to deploy mobile forces at all.

Before one gets to ICBMs, however, the game-changer for remote warfare came with the advent of air and naval power. One cause of World War II was a long-simmering dispute between Japan and the United States over naval power in the region. Japan needed naval dominance to transport ground troops to neighboring countries in pursuit of its imperial ambitions. The United States built a naval presence in the Pacific Ocean to counter this threat and check Japanese expansionism. The American capacity to deploy a massive naval presence was in one sense remote and in another sense not remote. It was not remote because the naval operators were located in the Pacific Ocean and subject to great risk. However, the use of naval power allowed the United States to project military force well beyond the American homeland, thus producing an element of remoteness that struck the Japanese as strategically intolerable, in turn leading to their decision to attack Pearl Harbor and start a conflict that they viewed, perhaps erroneously, as inevitable.

The specifics of naval warfare depend on remoteness. Aircraft carriers allow the major powers to deploy air assets—fighter jets—off the coast of the target state so that the strikers can move quickly to attack. At the same time, however, the aircraft carrier can in theory remain outside the direct radius of the target state's land-based defensive weapon systems, thus allowing the aircraft carrier to launch airstrikes from a safe and remote distance. Of course, if the target state has naval destroyers or air assets of its own, the aircraft carrier then becomes vulnerable, and must travel with an entire battle group of destroyers whose main purpose is to prevent enemy naval and air assets from getting within striking distance of the aircraft carrier. The carriers were originally conceived and designed at a time when they could produce an asymmetry of risk, although that advantage was quickly closed among major military powers; however, the asymmetry still persists in conflicts between major and lesser military powers.

3. MODERN TECHNOLOGICAL DEVELOPMENTS FOR REMOTE KILLING

Our next task is to determine whether the latest advancements in military technology are only the most recent examples of strategic asymmetries of risk, or whether new technology represents a fundamental breakdown in the paradigm of reciprocal risk.[22]

A. Drones

The deployment of remotely piloted vehicles produces a number of strategic advantages. First, it allows the projection of force without risking a human pilot. Second, since there is no human pilot, the drone can stay in the sky for an extensive period of time, allowing both increased surveillance and the ability to strike within minutes. Third, the increased surveillance allows the attacking force to ensure that they are hitting the right target. Fourth, the use of precision warheads with small yields serves the goal of discrimination, thus reducing the number of civilians potentially killed in the strike as collateral damage. Fifth, drones have a small footprint, therefore facilitating covert or clandestine action. These advantages are all contingent rather than intrinsic features of the drone platform, with the possible exception of the elimination of risk to the human operator.

Switching now to the legal criticisms mounted against targeted killing by drone, the vast majority of the objections have no direct connection to the remoteness of the pilot, which is the defining characteristic of drone operations, though some of the legal criticisms might be *indirectly* linked to remoteness. Consider the objections outlined below.

First, some critics argue that drone attacks violate the sovereignty of the territorial state where the strikes occur.[23] In this regard, strikes in Yemen, Pakistan and Somalia are sometimes referenced.[24] The United States has asserted that the strikes are justified by self-defense against al-Qaeda, a non-state actor, or associated forces that are co-belligerents of al-Qaeda in its non-international armed conflict against the United

---

22 For example, Paul Kahn argues that war without mutual risk produces 'an image of warfare without the possibility of chivalry'. See Paul Kahn, 'The Paradox of Riskless Warfare' (2002) 22 Phil & Pub Policy Quarterly 2, 4, cited in Blum (n 10) 137.

23 See Mary Ellen O'Connell, 'Unlawful Killing with Combat Drones: A Case Study of Pakistan, 2004–2009' in Simon Bronitt et al (eds), *Shooting to Kill: Socio-Legal Perspectives on the Use of Lethal Force* (Hart Publishing 2012) 263, 264 (discussing drone strikes in Pakistan).

24 For a discussion, see Sikander Ahmed Shah, *International Law and Drone Strikes in Pakistan: The Legal and Socio-Political Aspects* (Routledge 2015) 32.

States.[25] Of course, the drones cross the territorial borders of the territorial state, and if there is no self-defense argument against the territorial state, there is a non-trivial argument that the self-defense argument is inapplicable against the territorial state.[26] The United States has forcefully argued that self-defense is applicable in this context when the host state is unwilling or unable to redress the threat itself.[27] Formally speaking, the unwilling or unable test is a gloss on the necessity requirement for self-defense; if the host state is capable or willing to stop the threat, intervention by the foreign power is not necessary, and therefore self-defense is unavailable.[28] It is unclear, though, whether the US government views the infringement of the territorial state's sovereignty as a 'use of force' in violation of Article 2, thus requiring some Article 51 justification sounding in self-defense, or whether the infringement of sovereignty falls below the threshold of what would be considered an 'armed attack' under international law. Perhaps it is a mere counter-measure.[29] Without taking a view of the substance of this jus ad bellum debate, it is sufficient to note here that there is no *direct* connection between the remoteness of the pilot and the alleged jus ad bellum violations here. The issue would come up just as surely with a purportedly defensive attack carried out with manned aircraft or infantry

---

[25] See Harold Hongju Koh, Legal Adviser, US Department of State, Speech to the Annual Meeting of the American Society of International Law, Washington, DC (25 March 2010) ('As I have explained, as a matter of international law, the United States is in an armed conflict with al-Qaeda, as well as the Taliban and associated forces, in response to the horrific 9/11 attacks, and may use force consistent with its inherent right to self-defense under international law.').

[26] See Michael N Schmitt, 'Drone Law: A Reply to Un Special Rapporteur Emmerson' (2014) 55 Va J Intl L Dig 13, 16–17 ('Absent an "operational nexus" to the host State, the restrictive view would preclude an extraterritorial RPA strike, not because the individuals are unlawful targets, but rather because it is unlawful for the attacking State to cross the border in self-defense.').

[27] See Brian J Egan, Legal Adviser, US Department of State, Speech to the American Society of International Law, Washington, DC (1 April 2016) ('In particular, there will be cases in which there is a reasonable and objective basis for concluding that the territorial State is unwilling or unable to effectively confront the non-State actor in its territory so that it is necessary to act in self-defense against the non-State actor in that State's territory without the territorial State's consent.').

[28] See Ashley S Deeks, '"Unwilling or Unable": Toward a Normative Framework for Extraterritorial Self-Defense' (2012) 52 Va J Intl L 483, 522.

[29] For a discussion of countermeasures, see Sheng Li, 'When Does Internet Denial Trigger the Right of Armed Self-Defense?' (2013) 38 Yale J Intl L 179, 215.

(as was the case with the Bin Laden raid in Pakistan). To the extent that there is any connection at all, it is an *indirect* one. Perhaps the remoteness of the pilot removes an inherent constraint on using force cavalierly, thus giving the United States another incentive to violate jus ad bellum.[30] Since pilots will not be put in harm's way, there is less reason to be concerned about launching an international attack that could, according to some critics, violate jus ad bellum. This indirect argument will be evaluated in full in Section 4.

Second, drone attacks allegedly cause too much collateral damage to civilians.[31] Both journalists and legal scholars have reported on the vast human toll inflicted on the local population in areas where drone strikes have occurred.[32] Of course, collateral damage is not per se illegal unless it is disproportionate to the anticipated value of the military target. Moreover, even in cases where there are credible allegations of violations of the principle of proportionality, there is little to no evidence that the remoteness of the pilot is directly relevant to the production of the disproportionate collateral damage.[33] An air strike performed with a manned aircraft would have produced the same amount of collateral damage or, if one believes the US military, it is perhaps the case that the manned airplane might have produced *more* collateral damage. The same conclusions apply to criticisms that a particular drone strike violates international human rights law (IHRL). Under this argument, the strike is not governed by international humanitarian law (IHL) at all, and the more restrictive norms of IHRL apply, which prevent status-based targeting and would require capturing the target, using more traditional law enforcement methods, before lethal measures are employed against a

---

[30] See O'Connell (n 23) 267.

[31] Compare Shah (n 24) 28–9 ('A well-planned, targeted ground offensive with commando units would have been more effective in battling Al-Qaeda and sympathetic armed militias, and would have kept collateral damage, including civilian casualties, to a minimum.') with Jane Stack, 'Not Whether Machines Think, but Whether Men Do' (2015) 62 UCLA L Rev 760, 772 ('drones generate fewer civilian casualties than ground troops').

[32] See O'Connell (n 23) 271.

[33] See Stack (n 31) 772. One difficulty with assessing this question is the classified nature of the underlying data, which is rarely acknowledged or released by the US government. See Steven J Barela, 'Strategic Efficacy: The Opinion of Security and a Dearth of Data' in Steven J Barela (ed), *Legitimacy and Drones: Investigating the Legality, Morality and Efficacy of UCAVs* (Ashgate 2015) 271, 296 ('Secrecy of the drone program has rendered the necessary data for full assessment unattainable, and selective disclosures have only distorted an already fragmented picture.').

target that represents an imminent threat. But the same arguments could all be made—and have been made—against strikes carried out by manned vehicles.

There is one argument that asserts an indirect relationship between the remoteness of the killing and the alleged disproportionate collateral damage or the violation of human rights law. One could argue that the use of an unmanned aerial vehicle makes it more likely that civilians will be killed by collateral damage. The effect here would be to *transfer* risk from friendly forces *to* enemy civilians.[34] As noted above, it is empirically doubtful whether the move from manned aircraft to unmanned aircraft has this effect.[35] However, if one broadens the lens, a more defensible hypothesis emerges: that the selection of unmanned aircraft as compared to ground infantry forces, reduces the risk to friendly forces but dramatically increases the risk to enemy civilians.[36]

To answer this question, one must determine whether there is a legal obligation for attacking forces to use an alternative military tactic—such as ground forces—that would reduce collateral damage when compared to the default option, in this case a drone campaign. The question is whether this normative obligation is codified in existing law. Certainly, Additional Protocol I codifies a restrictive customary norm that requires attacking forces to take all reasonable precautions to reduce collateral damage as far as 'feasible'—an evaluative term that says little about how much attacking forces are permitted to prioritize friendly forces over enemy civilians.[37] Does this legal requirement *require* attacking forces to

---

[34] See O'Connell (n 23) 271 ('In the trailer in Nevada, the pilot knows she will not be attacked. She will go home to her family at the end of the day, coach a soccer game, make dinner, and help with homework.').

[35] For a discussion of this issue in the AWS context, see R Crootof, 'War, Responsibility, and Killer Robots' (2015) 40 North Carolina J Intl L & Commercial Reg 909, 923 (noting possibility for increased civilian harm but also concluding that this result is 'far from certain').

[36] See Stack (n 31) 772 ('The primary argument advanced to this end is that drones' significant reduction of the cost of war to the United States in terms of both "blood and treasure" will seduce policymakers into expanding the limits on what constitutes a "legitimate target," and engaging in more, longer, and less legitimate wars.').

[37] Protocol Additional to the Geneva Conventions of 12 August 1949, and relating to the Protection of Victims of International Armed Conflicts (Protocol I), 8 June 1977, Article 57. Although the United States is not a party to Additional Protocol I, many of its provisions are widely regarded as having ripened into customary international law, especially the provision requiring attacking forces to take all feasible precautions to reduce collateral damage.

forego unmanned aircraft in favor of ground forces where doing so would reduce collateral damage?[38]

The issue is contested. The US Law of War Manual recognizes an affirmative obligation to use a particular weapons platform to reduce collateral damage as far as possible, but 'the decision of which weapon to use will be subject to many practical considerations, including effectiveness, cost, and the need to preserve capabilities for other engagements'.[39] More importantly and convincingly, some IHL scholars who have considered this legal requirement believe that the obligation resides at the tactical level of the battlefield commander. In other words, a commander in a particular military engagement has an obligation, when presented with multiple weapons choices *at his or her disposal*, to select the option that will reduce collateral damage as far as possible, provided that the selection does not endanger friendly forces.[40] However, there is a question whether the normative obligation applies to strategic choices that are *not* within the purview of the battlefield commander. For example, if there are other weapons that *could* be deployed to the theater of operations, but remain warehoused far from the battlefield, is the nation in question responsible for its failure to bring the more discriminating weapons to the battle? No, argue several scholars.[41] Judged from this standard, a state cannot be faulted because it has decided, at the political level, to engage in an air-power campaign and has refused to

---

[38] In *Necessity in International Law* (Oxford University Press 2016), Larry May and I argue that attacking forces are under a moral obligation to subject themselves to a level of 'reasonable risk' in pursuit of their reduction of collateral damage.

[39] US Department of Defense, Law of War Manual (2015) § 5.11.3. The manual also notes that 'there would be few, if any, instances in which the use of a particular weapon system, such as precision-guided munitions or cyber tools, would be the only legally permissible weapon'. Ibid.

[40] See Michael N Schmitt and Eric W Widmar, '"On Target": Precision and Balance in the Contemporary Law of Targeting' (2014) 7 J Natl Security L & Policy 379, 402 ('Of course, attackers need only select less harmful means or methods that do not involve sacrificing military advantage and that are feasible. As an example, an attacker does not have to use a less powerful bomb against an insurgent leader in a building in order to avoid civilian casualties if doing so would significantly lower the likelihood of success (assuming all other IHL requirements are met).').

[41] Ibid ('For instance, although precision weapons may be available for an operation, they may be more useful at later stages of the campaign and thus need to be preserved, or the employment of a precision weapon may be infeasible because it would require increased risk to ground forces in order to designate a target.'). Compare with Shah (n 24) 28–9.

commit ground troops to the military operation, despite the fact that ground forces might do a better job of reducing civilian collateral damage. There is, of course, a strong moral argument for such an obligation, residing at the political level, but it is probably correct to conclude that mainstream international law is not yet prepared to recognize it as binding.

B. Cyber-weapons

Compared with remotely piloted vehicles, cyber-weapons involve an even further regression in the location of the combatant.[42] While the pilot of a drone might be far removed from the forward area of deployment, it is nonetheless the case that drones have a limited tactical range. Somewhere within a relatively close distance to the theatre of operations, the drone will need a base from which it can takeoff, land, refuel, and be repaired if necessary. These facilities are usually carefully sited to be out of harm's way, though they are still foreign military installations subject to discovery and attack in the event of an armed conflict. Though the pilot per se might remain in the United States, the actual deployment of the drone remains foreign.

In contrast, a cyber attack could, in theory, involve personnel who are all safely located in the United States and removed from the field of operations. Moreover, the target of the attack would have grave difficulty

---

[42] The application of the war paradigm to cyber warfare is contested, with some scholars believing that many cases of cyber attacks should be evaluated as 'actions short of armed conflict'. See Michael Newton and Larry May, *Proportionality in International Law* (Oxford University Press 2014) 280 (most cyber attacks are more analogous to embargoes than they are to traditional military attacks). Similarly, Mary Ellen O'Connell has argued that evaluating cyber attacks under the law of war paradigm only facilitates military control over cyber defenses. See Mary Ellen O'Connell, 'Cyber Security without Cyber War' (2012) 17 J Conflict Security L 187 ('The evidence shows that the USA, in particular, is building capacity and developing strategies that make the Department of Defense a major player in Internet use and protection. The concern with this development is that the Pentagon will conceive of cyber space as it does conventional space, with war fighting in mind. Yet, the international legal rules on the use of force, especially the rules on self-defence, raise important barriers to military solutions to cyber space problems. Indeed, the law of self-defence should have little bearing in discussions of cyber security. Even if some cyber incidents could fit a solid definition of what constitutes an armed attack, responding to such an attack will rarely be lawful or prudent if the response is a use of force. The emphasis, therefore, in terms of legal norms and commitment of resources should be in the non-military sphere.').

tracing and locating the source of the attack and the location of the hackers who launched the attack.[43] They might be located at the NSA headquarters (home of US Cyber Command), or they might be located in another installation. They might be located abroad, or they might even be civilian contractors (or hackers) hired by the attacking state to launch a particular cyber-attack.[44] In this sense, the offensive personnel might be located anywhere on the globe.[45]

Traditionally, scholars focus on this fact in the context of the attribution problem—the inability to determine which country launched the attack, which country is legally responsible for the infringement of the target state's sovereignty (assuming that the attack is not justified by principles of jus ad bellum articulated in the UN Charter), and by extension which state is subject to counter-attack as a right of response in legitimate defense.[46] However, it is less common for scholars to focus on remoteness as a risk-reducing strategy—though perhaps it was just too obvious to garner serious conceptual attention.[47] The risk-reducing nature of cyber-war is important, though, because it highlights the degree to which modern methods of attack are placing combatants at increasingly remote locations from the effects of their attacks.

One notable difference between drones and cyber is that for the former, the risk-reducing nature of the platform was the inspiration for its development, while for the latter, the risk-reducing nature of the platform

---

[43] See Scott J Shackelford and Richard B Andres, 'State Responsibility for Cyber Attacks: Competing Standards for a Growing Problem' (2011) 42 Geo J Intl L 971, 1016.

[44] See Nicolò Bussolati, 'The Rise of Non-State Actors in Cyberwarfare' in Jens David Ohlin et al (eds), *Cyberwar: Law and Ethics for Virtual Conflicts* (Oxford University Press 2015) 103.

[45] See Heather Harrison Dinniss, 'Participants in Conflict: Cyber Warriors, Patriotic Hackers and the Laws of War' in Dan Saxon (ed), *International Humanitarian Law and the Changing Technology of War* (Martinus Nijhoff 2013) 251, 258 (arguing that cyber warriors often work at great distances that alleviate them from the requirements of wearing uniforms or distinctive emblems); Mark Shulman, 'Discrimination in the Laws of Information Warfare' (1999) 37 Columbia J Transnational L 939, 956.

[46] See, eg, Nicholas Tsagourias, 'Cyber Attacks, Self-Defence and the Problem of Attribution' (2012) 17 J Conflict Security Law 229, 233; M C Waxman, 'Cyber-Attacks and the Use of Force: Back to the Future of Article 2 (4)' (2011) 26 Yale J Intl L 421, 445; T Rid and B Buchanan, 'Attributing Cyber Attacks' (2015) 38 J Strategic Stud 4, 5.

[47] But see Patrick Lin, George Bekey and Keith Abney, 'Robots in War: Issues of Risk and Ethics' in Rafael Capurro and Michael Nagenborg (eds), *Ethics and Robotics* (Akademische Verlagsgesellschaft 2009) 49.

is a mere by-product of the weapons system. The whole point of developing drones was to create a platform whereby the pilot would no longer be located on the aircraft. However, a cyber weapon is designed with the goal of launching an attack against a computer-based network in order to destroy or manipulate any civilian or military system that is partially controlled by, or connected to, a computer network or other device subject to manipulation. It just so happens that the deployment of such an attack almost never requires that the cyber attack be located in the general vicinity of the attack (although there might be some exceptions to this observation for so-called 'closed' systems). Generally speaking, though, cyber attackers can be located anywhere where there are sufficient network facilities to generate the attack. Usually, this location will be far removed from the location of the attack itself. But it would be wrong to describe this as a goal of cyber-warfare; rather, it is a happy coincidence (for the attacking force). The real goal is to attack computers and the systems that rely on them, and it just happens that the best way to do so is from the relatively safe confines of an office.

The target of a cyber attack might be a military installation, but the greatest utility of the cyber strategy is against dual-use (civilian and military) infrastructure as well as against revenue-enhancing operations that either directly or indirectly support the enemy's capacity to support its military. Both of these areas are controversial under existing law.[48] It might be objected that cyber will give attacking forces a greater incentive to attack such targets with little risk, and thus promote greater collateral damage on the part of civilians. The legitimacy of this argument depends, in part, on whether these attacks are lawful or not.

Consider first the question of attacking dual-use infrastructure targets.[49] Important examples include electrical grids, power plants, oil refineries, railroads, and bridges—all of which support both civilian and

---

[48] For a discussion of the legality of bombing dual-use infrastructure targets, see Henry Shue and David Wippman, 'Limiting Attacks on Dual-Use Facilities Performing Indispensable Civilian Functions' (2002) 35 Cornell Intl LJ 559, 575 ('Under our approach, if the advantage of attacking an indispensable object cannot be viewed as compelling in relation to the anticipated direct and indirect civilian harm, the military functions served by the indispensable object may still be thwarted, but must be thwarted by some other means. For example, if the electricity plant serves command and control, as well as water-purification, the attacks will need to target the command and control facilities directly rather than indirectly by way of the facilities' energy source.').

[49] See Eric Talbot Jensen, 'Unexpected Consequences from Knock-On Effects: A Different Standard from Computer Network Operations?' (2003) 18 Am Univ Intl L Rev 1160–68.

military operations. Under the governing standard announced in Additional Protocol I, objects are defined as military objects—and subject to attack—if 'their nature, location, purpose or use make an effective contribution to military action' and their 'total or partial destruction, capture or neutralization, in the circumstances ruling at the time, offers a definite military advantage'.[50] Under that definition, all of the dual-use infrastructure targets would seem to qualify as military targets, since it is clear that the destruction of a bridge, railroad, oil refinery, or power plant would certainly offer a 'definite military advantage'. It is therefore unlikely that such attacks are categorically prohibited by IHL. The most that can be said is that such attacks are subject to an additional constraint, the principle of proportionality, and that attacks are impermissible if the collateral consequences to the civilian population in destroying, say, the oil refinery or bridge, are disproportionate to the contemplated military advantage to be gained by their destruction.[51] However, even this is controversial.[52] The other position is that the principle of proportionality

---

[50] Additional Protocol I (n 37) Article 52(2).

[51] For a discussion of targeting dual-use infrastructure targets with cyber weapons, see Marco Roscini, *Cyber Operations and the Use of Force in International Law* (Oxford University Press 2014) 185 (concluding that '[t]he fact that an object is *also* used for civilian purposes does not affect its qualification under the principle of distinction: if the two requirements provided in Article 52(2) of Additional Protocol I are present, the object is a military objective but the neutralization of its civilian component needs to be taken into account when assessing the incidental damage on civilians and civilian property under the principle of proportionality').

[52] In 2013, an ICTY Trial Chamber held that the destruction of the old Stori Most Bridge used by military and civilian personnel was a violation of IHL because it violated the principle of proportionality. See *Prosecutor v Prlic*, Trial Chamber Judgment, ICTY Case No IT-04-74 (29 May 2013) para 1584 ('The Chamber therefore holds that although the destruction of the Old Bridge by the [Croatian armed forces] may have been justified by military necessity, the damage to the civilian population was indisputable and substantial. It therefore holds by a majority, with Judge Antonetti dissenting, that the impact on the Muslim civilian population of Mostar was disproportionate to the concrete and direct military advantage expected by the destruction of the Old Bridge.'). For a discussion of this case, see Martin Lederman, 'Is it Legal to Target ISIL's Oil Facilities and Cash Stockpiles?' *Just Security* (27 May 2016).

applies only to civilian collateral deaths and cannot transform what would otherwise be a lawful military *object*.[53]

The second example is the even more controversial question of targeting revenue-enhancing operations whose destruction would inhibit the *capacity* to launch and sustain a military campaign by denying a regime the capacity to develop the necessary financial and other resources to sustain the military.[54] Recent examples include the United States decision to target ISIS oil supplies and cash stockpiles.[55] In the case of the oil supplies, the oil was not necessarily destined for military vehicles but was instead to be sold in return for cash payments that might fund military operations. By destroying both oil and cash, the attacking forces denied to their enemy the ability to *fund* military operations.[56] This argument, if accepted, turns many proto-typically *economic* activities into legitimate military objectives. Again, this theory represents a slippery slope and is highly controversial, in part because so much of the citizenry's daily life is wrapped around economic activity, and at least one animating impulse of IHL is to protect the civilian population from the horrors of armed conflict. The targeting of objects involved in economic activity threatens that basic goal of the regulatory enterprise.

Both examples discussed above (dual-use and revenue-enhancing targets) are highly relevant for the case of cyber attacks, because it is plausible that a cyber-weapon might be ideally suited to go after either a dual-use infrastructure target such as an electrical grid, or to go after a war-sustaining economic activity, such as a financial system (for

---

[53] In *Prlic*, Judge Antonetti dissented and concluded, simply, that the Stori Most Bridge was a 'military objective' and 'there is no such thing as proportionate destruction'. *Prosecutor v Prlic*, Separate and Partially Dissenting Opinion of Presiding Judge Jean-Claude Antonetti, ICTY Case No IT-04-74 (29 May 2013) 325.

[54] In a recent article, Ryan Goodman argues that such attacks are, in principle, consistent with IHL, as long as they also meet other conditions, including the principle of proportionality. See Ryan Goodman, 'Targeting "War-Sustaining" Objects in Non-International Armed Conflict' (2016) 110 Am J Intl L (discussing precedent of Union destruction of cotton during the Civil War because it funded Confederate military operations).

[55] See Matthew Rosenberg, 'U.S. Drops Bombs Not Just on ISIS, but on Its Cash, Too' *New York Times* (20 January 2016).

[56] Goodman (n 54).

example, stock market operations).[57] If a cyber attack allows the attacking force to launch these questionable attacks, with limited risk to the attacking hackers, then the future of warfare might be waged with an ever-increasing impact on the civilian population who are more affected by the destruction of dual-use infrastructure targets and war-sustaining economic activity targets than by the destruction of traditional military targets.[58] This would be a worrisome result and perhaps reason to criticize, rather than celebrate, the increase in cyber attacks by virtue of their risk-free operation. In these situations, lowering risk to military personnel may come with a resulting, albeit indirect, increase in harm to the civilian population.[59]

So cyber attacks potentially allow a greater number of risk-free attacks against infrastructure targets that civilians rely on. The one thing holding militaries back, at this point, is the fear of cyber reprisals. If one country launches a cyber assault, one's enemies, provided that they are cyber-enabled, could potentially launch a brutal cyber-retaliation.[60] The result is a steady détente for the time being. However, that détente is only viable for states with adequate non-cyber military capabilities that can afford to forgo using cyber weapons. The real promise of cyber weapons is that they may represent a force multiplier—even a force equalizer—for states with the inability to spend their way to military dominance using conventional weapons. With no resources to win a typical arms race, a minor military power may view cyber weapons as its one shot to level the

---

[57] For example, a cyber attack could damage an irrigation system that is reprogrammed to flood crops that will feed the civilian population. See William Boothby, *The Law of Targeting* (Oxford University Press 2012) 398.

[58] See David Turns, 'Cyber War and the Concept of "Attack" in International Humanitarian Law' in David Saxon (ed), *International Humanitarian Law and the Changing Technology of War* (Martinus Nijhoff 2013) 209, 224–5 (noting risk of 'slippery slope towards a potential expansion of legitimate targets that would simultaneously expand the possibilities for indiscriminate attacks while curtailing the operation of the fundamental principle of distinction').

[59] See Turns (n 58) 224 n 77 (noting that an attack against an electricity generator may cause no initial physical damage but 'later down the line, such basics for the civilian population as water purification plants would shut down, leading to epidemics of disease from water contamination').

[60] However, it appears that some nations are better equipped than others to play this cat-and-mouse game. See David E Sangerjune, 'As Russian Hackers Probe, NATO Has No Clear Cyberwar Strategy' *New York Times* (16 June 2016) (reporting that while the US has substantial cyber offensive capabilities, NATO at the present moment has few options to engage in low-level cyber retaliation for attacks coming from Russia or China).

playing field with traditional military powers. Using this calculation, some states may not only utilize cyber-weapons but may wish to accelerate the transition to cyber conflict, a domain where they are more likely to succeed. This would represent a negative outcome for civilians who might be disproportionately harmed in a cyber conflict based on the likelihood that some civilian infrastructure targets may be tempting targets for a cyber offensive.

C. Autonomous Weapons Systems

Autonomous weapons involve the greatest attenuation of human operators; the level of risk to the operator could be reduced to zero by removing human beings from the 'loop' entirely.[61] While there are human beings involved in the design and construction of AWS, the systems could be deployed—at least in theory—without a human operator.[62] First, we should scrutinize whether this is an intended or unintended consequence of AWS. Like drones, it might be argued that AWS are designed to reduce risk to the human operator by turning over essential tasks to the weapon itself, thus allowing the relevant human beings to remain far from the field of battle.[63] However, this is probably an incorrect assumption. The animating impulse behind AWS is the promise that artificial intelligence is better equipped to make quick life-or-death decisions than a human operator, at least in certain operational contexts in which the AWS is designed to succeed.[64] For example, if an AWS could better distinguish between civilian and military aircraft, this would argue in favor of giving autonomous targeting protocols to an anti-aircraft

---

[61] See Daniel N Hammond, 'Autonomous Weapons and the Problem of State Accountability' (2015) 15 Chi J Intl L 652, 656 ('Unlike the activities of human-operated drones, an AWS's actions are not easily attributable to a particular person.').

[62] For a discussion, see David Akerson, 'The Illegality of Offensive Lethal Autonomy' in David Saxon (ed), *International Humanitarian Law and the Changing Technology of War* (Martinus Nijhoff 2013) 65, 86; Peter Asaro, 'Jus Nascendi, Robotic Weapons and the Martens Clause' in Ryan Calo et al (eds), *Robot Law* (Edward Elgar 2016) 367, 386 (concluding that a requirement of meaningful human control is suggested by ethical principles of humanity that were preserved by the Martens Clause).

[63] Crootof (n 35) 920–23.

[64] See Mark Klamberg, 'International Law in the Age of Asymmetrical Warfare, Virtual Cockpits and Autonomous Robots' in Jonas Ebbesson et al (eds), *International Law and Changing Perceptions of Security* (Brill Nijhoff 2014) 152, 167.

missile system.[65] The goal of this autonomous targeting would have nothing to do with reducing risk to human operators but rather would be inspired by the goal of reducing human error.[66] In this respect, it would appear that AWS should be grouped with cyber weapons where the reduction of risk to the combatants on the attacking side should be viewed as a collateral consequence rather than the animating impulse for the development of the weapon.

However, it is possible to imagine future applications of AWS where the goal is to reduce risk by achieving the ultimate degree of remoteness, that is, removal of the human operator entirely. This would, in fact, transcend remoteness entirely and transform it into pure disappearance.[67] This would be particularly desirable in the infantry context. That being said, for the moment the majority of contemplated AWS applications involve missile-targeting programs that would be deployed by land, sea or air. The other major area for AWS application is cyber. A cyber weapon could include an algorithm that identifies and assesses the nature of a particular computer system, whether it is military or civilian, for example, and then disables the computer system without requiring an executive command for an operator. In this respect, a cyber weapon might be more or less autonomous, thus suggesting that the most important development in the future of warfare might not be cyber or AWS but in fact the *combination* of the two strategies.

The infantry context, though a source of popular imagination, represents the future of AWS deployment, not its present. In the future, if AWS technology is sufficiently advanced, it could be deployed in the infantry context as a risk-management tool.[68] Robotic infantry devices, operating autonomously, could roam a city street and eliminate any and

---

65 Robert Sparrow, 'Building a Better Warbot: Ethical Issues in the Design of Unmanned Systems for Military Applications' (2009) 15 Science & Engineering Ethics 169.

66 Another benefit is the ability of an AWS to continue operating even when its communication link is severed. See Jeffrey S Thurnher, 'Examining Autonomous Weapon Systems from a Law of Armed Conflict Perspective' in Hitoshi Nasu and Robert McLaughlin (eds), *New Technologies and the Law of Armed Conflict* (Asser Press 2014) 213, 217. Compare with Peter Asaro, 'On Banning Autonomous Weapon Systems: Human Rights, Automation, and the Dehumanization of Lethal Decision-making' (2013) 94 Intl Rev of Red Cross 687, 691.

67 See Markus Wagner, 'The Dehumanization of International Humanitarian Law: Legal, Ethical, and Political Implications of Autonomous Weapon Systems' (2014) 47 Vand J Intl L 1371, 1373.

68 See Wagner (n 67) 1380 ('Moreover, the use of UMS reduces the risks to a military's own troops.').

all enemy combatants within a defined parameter.[69] The device could be programed to forego attacks that violate the principle of proportionality or other operational or legal (or moral) constraints. Advocates for AWS point out that such a system could increase IHL compliance since human beings during infantry deployments are subject to cognitive errors or weakness of the will. Using a robotic infantry device would avoid these difficulties. On the other hand, there is a grave risk that states might rush to deploy an AWS that is *less* capable of complying with IHL, when compared to its human counterparts, simply to avoid risk to its own troops.[70] In this situation, the lowering of risk to the attacking force would be accompanied by an increase of risk to the civilian population based on the possibility of errors committed by the system. Since infantry deployment of AWS is only a future possibility, and not a present reality, this problem should be classified as a hypothetical consequence of the AWS paradigm, and one that naturally flows from the incentives created by the risk reduction for the attacking force.[71]

Taken together, remotely piloted vehicles, cyber and autonomous weapons represent the logical culmination of a process that first began when human beings started using tools as weapons of warfare.[72] The goal

---

[69] See Heather M Roff, 'The Strategic Robot Problem: Lethal Autonomous Weapons in War' (2014) 13 J of Military Ethics 211, 212 ('Assuming that such machines will be able to identify correctly *combatants*, which is certainly questionable, we face an additional set of constraints: identifying legitimate objects/targets and setting military objectives.').

[70] Cf Jack M Beard, 'Law and War in the Virtual Era' (2009) 103 Am J Intl L 409, 423.

[71] I leave aside for the moment whether it is sufficient to generate an argument for banning AWS outright. See Human Rights Watch & Intl Human Rights Clinic, Harvard Law School, *Losing Humanity: The Case Against Killer Robots* (2012) 1–2. For a discussion, see Tyler D Evans, 'At War with the Robots: Autonomous Weapon Systems and the Martens Clause' (2013) 41 Hofstra L Rev 697, 730–31 ('However, HRW's "humanitarian" recommendation to preemptively ban AWS could actually result in a counter-humanitarian outcome: machines may "reduce risks to civilians by making targeting more precise and firing decisions more controlled [ ]especially compared to human-soldier failings that are so often exacerbated by fear, panic, vengeance, or other emotions ..."'); Michael N Schmitt, 'Autonomous Weapon Systems and International Humanitarian Law: A Reply to the Critics' (2013) 2 Harvard National Security J 1, 3 ('*Losing Humanity*'s recommendation to ban the systems is insupportable as a matter of law, policy, and operational good sense').

[72] D'Aspremont reads the current academic debate skeptically and suggests that international lawyers are predisposed to think of cyberspace as a 'problem' or 'gap' that can be resolved by 'intervening' with the tools of international law

of these technological advances consistently has been to increase lethality while reducing risk to operators. The major strategy for achieving this balance is physical distance: when the weapon can travel long distances, the operator can remain far from the target and out of harm's way. What recent technological advances demonstrate, however, is a capacity to lower risk asymmetrically—using methods that transcend mere physical or geographic remoteness. True, drones represent the logical culmination in risk reduction through physical remoteness, but cyber and autonomous weapons have achieved risk reduction in ways that have fundamentally transformed, or even eliminated, the significance of physical distance. In the case of AWS, the operator has not just moved from the front lines, but has been replaced entirely—the strategy of risk reduction taken to its logical conclusion.

4. OPTIMAL OR SUB-OPTIMAL LEVELS OF ASYMMETRIC RISK IN ARMED CONFLICT

Having established that the process of lowering asymmetrical risk through remoteness is a common but now accelerating aspect of modern warfare, our task is now to determine whether this process is normatively undesirable. In the prior section, we considered isolated arguments, within the context of each weapons platform, that lowering asymmetrical risk might have the perverse effect of increasing IHL violations. But now we consider a much larger and more powerful objection: that lowering asymmetrical risk will inevitably result in more jus ad bellum violations.[73] In other words, the advent of risk-free warfare will allow rogue states to launch attacks in violation of jus ad bellum.[74] In the past, what kept states in check—in addition to formal or informal international legal sanctions—was the possibility that states would not want to sacrifice a large percentage of their personnel in order to launch an attack. In a

(either existing frameworks or new rules). See Jean D'Aspremont, 'Cyber Operations and International Law: An Interventionist Legal Thought' (2016) 21 J of Conflict & Security L.

[73] The most sophisticated analysis (and critique) of this objection was offered in Kenneth Anderson, 'Efficiency *in Bello* and *ad Bellum*: Making the Use of Force too Easy?' in C Finkelstein, J D Ohlin and A Altman (eds), *Targeted Killings: Law and Morality in an Asymmetrical World* (Oxford University Press 2012) 374, 389 ('Resort to force is "too easy" if it results in unjust interventions; otherwise not.').

[74] For an example of this argument, see M Shane Riza, *Killing without Heart* (Potomac 2013) 77–80.

sense, the personal costs of war were the greatest check on jus ad bellum violations. With risk-free warfare possible, what is to stop rogue nations from unleashing a continuous stream of jus ad bellum violations?[75]

This objection presupposes that states care about their own citizens. If a despotic tyrant rules a state, the authority to decide whether to resort to force may be concentrated in a particular individual who is indifferent to the suffering his decisions impose on his subjects. Consequently, the reduction of asymmetrical risk would have little impact on the tyrant's calculation regarding the costs of potentially violating jus ad bellum. The same argument might apply even when the state is relatively democratic, as long as membership of the armed forces is not distributed equally across the population. If the decision-making authority is concentrated among political elites who are less likely to serve—or have family members who serve—in the military, they may be insensitive to the costs associated with warfare.[76] Again, in this situation, the reduction in asymmetric risk may have limited impact on the decision-making process of the policy elites. The elites may already be predisposed to the use of force and the risks on the domestic population may not be decisive for their selfish calculus.

For the sake of evaluating the jus ad bellum argument against remote warfare, I will assume that there are a non-trivial number of states where the risk to the military is important to the ruling class, and where the reduction of asymmetrical risk will remove an important barrier to the exercise of force. This rhetorical assumption is important for purposes of evaluating the objection on its own terms. Does it succeed? Does remote warfare risk increasing jus ad bellum violations?

Consider the following hypothetical. Assume for the sake of the argument that State A is deciding whether to launch an invasion against State B. Let us also presuppose that State A is aware that the majority of the world community will criticize the attack as a violation of core principles of jus ad bellum under public international law generally and the UN Charter specifically. Although State A is concerned about the reputational costs associated with a perceived violation of the UN Charter, the officials leading State A are more concerned about a report from their own military generals that the military campaign will result in an initial loss of life estimated between 500 and 2,000 military personnel, during the first two weeks of the military campaign. This news may not

---

75 See Crootof (n 35) 923–6.

76 This has led some to suggest that the US should reinstate a universal draft. See Kathleen Frydl, 'Why America Needs the Draft' *The American Interest* (16 January 2014).

necessarily change the decision of State A, but it is enough to give its leaders pause before deciding to proceed. It makes it less likely, ceteris paribus, that State A will decide to go ahead with the invasion. Now, let us change the hypothetical and assume that the military deaths are reduced to near zero because of the remote technologies that will be used during the two-week military campaign. It is not controversial to assume that this decision will make it more likely that State A will launch the military campaign. It is important not to exaggerate the point. This does not mean that this one factor is the sole criterion that will influence the outcome of the decision. Rather, it simply means that, ceteris paribus, State A will be more likely to launch the invasion with the remote technology than they would if they had no access to the remote technology. This provides ample illustration of the oft-heard objection that access to technologies that lower asymmetrical risk will inevitably make it easier for states to go to war in violation of the UN Charter.

In evaluating the legitimacy of this objection to remote weapons, it is important to separate out the two elements of the argument. The first element is the claim that lowering asymmetrical risk will encourage more states to launch more attacks. The second element is the claim that lowering asymmetrical risk will mean more attacks in *violation of the UN Charter*. I concede the first element of the argument but wish to question the second element. Simply put, there is strong reason to think that remote technology will increase the number of attacks, but there is no reason to think—absent other information—that remote technology will increase the number of jus ad bellum violations.[77]

How can we put a wedge between these two different elements of the argument? The answer depends on recognizing the exact opposite scenario: some states forego the use of military force, even in situations when principles of jus ad bellum suggest that they are entitled to launch an attack, simply because the costs of vindicating their jus ad bellum rights are inappropriately high. Consider the following situation: State B is situated close to a belligerent and aggressive neighbor, State C. The government officials in State C are in an expansionist mood and seek control over their neighbor, State B, even though State C has no viable claim under international law to the territory of State B. Consequently, State B is in a position of having to decide whether it will expend military resources in fighting off State C's aggression. Now let us also assume that State C has substantial military assets, and with it, the capacity to undertake an unambiguous injustice against State B.

---

[77] See Anderson (n 73).

In deciding whether to resist, State B will be guided by, inter alia, two considerations. The first is whether it can actually succeed in repelling State C's aggression, that is, whether it has sufficient military strength to push back State C's military advance. The second consideration is the cost that it must bear, in terms of the human lives of its own military personnel, in repelling the aggression (assuming that its defensive effort will succeed). This last consideration is especially relevant in the context of our discussion. One could well imagine State B deciding that so many of its personnel will be killed during the operation that it is not worth fighting back. Consequently, it will forego violence and effectively accede to State C's wishes. The result is that State B puts down its weapons and decides to let its country be annexed by State C.

Assuming that defensive force is permitted against politically motivated military aggression, the result here is troubling.[78] We have constructed the hypothetical in such a way that it is assumed that State B has a jus ad bellum right to defend itself against an unlawful aggression. However, because the costs of exercising that right are too high, State B declines to exercise the right. From the perspective of jus ad bellum, this has to be counted as a poor outcome.

Now imagine, for the moment, that State B has significant remote capabilities as part of its arsenal, and that it can defend itself using technology that reduces asymmetrical risk, thus giving it either strategic parity or even strategic advantage over State C. Now, the disincentive to exercising its jus ad bellum right is removed. State B may now decide that the cost of exercising its inherent right of legitimate defense is bearable. This is only made possible because of the advent of remote technology: drones, cyber weapons or AWS. In this situation, this would be an enhancement to jus ad bellum, not a negative.

How realistic is this hypothetical? Would it ever happen in reality? My own view is that it happens more often than we are comfortable recognizing. In Crimea, the Ukrainian government basically gave up

---

[78] Not everyone follows this assumption. For example, the philosopher David Rodin has argued that military force is not morally justified when used against political aggression—one country's invasion of a sovereign state in order to rule it rather than destroy it. See David Rodin, 'The Myth of National Self-Defense' in Cécile Fabre and Seth Lazar (eds), *The Morality of Defensive War* (Oxford University Press 2014) 69, 70–75.

control over Crimea, while Russian troops, engaged initially in unacknowledged force, were already on the territory of Crimea.[79] The position of the Ukrainian government, as well as at least some international law scholars, was that Russian attempts to annex Crimea were illegal, and that the Ukrainian government would have been justified under the UN Charter in using military force to protect its sovereign borders, which included Crimea. However, the Ukrainian government did not launch a counter-attack and surrendered the territory to Russia. At least part of the concern was the loss of life to individual Ukrainian soldiers who would have died during the conflict.[80] If the Ukrainian government could have entertained an armed conflict with a lower risk to its own troops, it might have been more willing to exercise its jus ad bellum rights. In this case, lowering asymmetrical risk would have promoted—rather than obstructed—Charter values.

A similar story could be told regarding Nazi aggression in Europe during World War II. It was clear to almost everyone involved that Nazi aggression in Europe was illegal. First, Hitler moved into the Rhineland, despite prior treaty obligations not to do so; his forces then pursued the Anschluss with Austria. After that, Hitler annexed the Sudetenland, which had been under the control of Czechoslovakia. At least one of the reasons Hitler was not stopped earlier was because the human toll of resistance was too high. Sometimes, jus ad bellum principles allow war, rather than condemn it, so any military advantage that promotes the recourse to war could, in these scenarios, actually maximize the principles embedded within jus ad bellum.

Again, a similar story could be told regarding humanitarian intervention, though this point is subtle because arguably unilateral humanitarian intervention violates the UN Charter and its promotion of state sovereignty and territorial integrity. However, many international lawyers support humanitarian intervention in limited cases,[81] even though the weight of today's legal doctrine is against such intervention. There may be cases where humanitarian interventions are not pursued because the risk to military personnel performing the intervention is just too high.

[79] David M Herszenhorn, Patrick Reevell and Noah Snieder, 'Russian Forces Take Over One of the Last Ukrainian Bases in Crimea', *The New York Times* (22 March 2014).

[80] For a discussion of the capitulation, see Patrick Reevell and Noah Sneidermarch, 'For Ukraine Military in Crimea, Glum Capitulation and an Uncertain Future' *The New York Times* (22 March 2014).

[81] See Jens David Ohlin, 'The Doctrine of Legitimate Defense' (2015) 91 Intl L Stud 119, 121.

And if remote technology could reduce that risk, the reduction in risk would increase the number of humanitarian interventions. Whether this is a good thing or not depends on the legality of humanitarian intervention.

This is no mere theoretical point. Neither the United States nor the world's other major powers engaged in any significant military intervention in Rwanda during its genocide. This was so despite the fact that many individuals, at the United Nations and elsewhere, were aware of the coming slaughter.[82] Why was nothing done? A military intervention large enough to stop the genocide would have put American lives at risk, and there was little appetite in the United States for more war casualties. The military intervention in Somalia had been particularly difficult for the Clinton Administration, leading to both military disaster and a sense that American troops were being sacrificed for an ill-conceived mission with limited connection to American interests.[83] The Clinton Administration was not interested in replicating this result and consequently failed to intervene in Rwanda. Had there been an opportunity to intervene in Rwanda with lower asymmetrical risk, through drones or AWS, perhaps the United States could have averted the genocide. Of course, these military technologies did not exist then, but the next time a 'Rwanda' occurs, the available military options will include options to lower risk. A similar story can be told with regard to Serbia. Although in that case the United States did intervene, the intervention came late and was hampered by various concerns over risk to US personnel. With more technological options to reduce risk, the United States might have intervened earlier, and more aggressively, against Serbia (or other actors) during the Balkan conflicts.

Expressed abstractly, the point here is simple: remote technologies reduce the costs associated with going to war. However, they reduce the costs associated with *all* wars, both good and bad. And unless one is a pacifist who believes that all wars are bad, then we should be concerned about a situation where we inhibit morally necessary, or legally authorized wars, by over-regulated warfare.[84] The goal of international law should be to reduce war to zero, but in the absence of that utopia, it

---

[82] For a first-hand account, see Roméo Dallaire, *Shake Hands with the Devil: The Failure of Humanity in Rwanda* (Random House Canada 2003) 240.

[83] The specifics of the Somalia intervention are detailed in Mark Bowden, *Black Hawk Down* (Grove Atlantic 1999).

[84] Pacifism is poorly understood as a philosophical doctrine. The best and most recent account of the theory is found in Larry May, *Contingent Pacifism: Revisiting Just War Theory* (Cambridge University Press 2015).

should reduce all illegal wars but not discourage lawful wars—or at the very least an important subset of them.

Is there any reason to think that enforcing a system of reciprocal risk would somehow discourage illegal wars but encourage lawful wars? I see no intuitive reason why this would be the case. It would seem, rather, that tinkering with reciprocal risk is a blunt tool in the regulator's toolbox—it would simultaneously discourage morally odious and morally commendable conflicts. To correctly determine the soundness of this proposal, one ought to tally up the benefits of preventing the odious conflicts and then subtract the harm associated with preventing morally beneficial counterattacks. Whether this would be an improvement in the status quo would be anyone's guess—this hypothetical calculation is incredibly difficult to envision. From this vantage point, the discouragement of morally and legally appropriate wars ought to count as a serious deficit in the proposal and would be an example of the perverse effects of overregulating warfare. At the very least, this problem ought to sound a cautionary note into the argument that remote technologies ought to be restricted because they make the resort to armed force too easy.[85] As the preceding analysis has shown, sometimes making the resort to force easier is, all things considered, a better result—depending on whether the conditions for jus ad bellum are satisfied or not.

5. RECIPROCAL RISK AS AN ETHICAL REQUIREMENT FOR CHIVALRIC KILLING

The final objection to remote warfare is that its reduction of asymmetrical risk is fundamentally incompatible with deeper norms of chivalric warfare—norms that are either codified in existing provisions of IHL, embedded ethical norms stemming from the moral requirements for belligerency, or 'professional' norms that attach to soldiers.[86] In the following section, I canvass each of these domains but conclude that the reduction of asymmetrical risk, while intuitively troubling, does not violate norms of chivalric warfare.[87]

---

[85] See Anderson (n 73) 389–90.

[86] Riza (n 74) 88. See also Martin L Cook, 'Drone Warfare and Military Ethics' in David Cortright et al (eds), *Drones and the Future of Armed Conflict: Ethical, Legal, and Strategic Implications* (University of Chicago 2015) 46, 60–61.

[87] The concept of chivalry, as a social institution, dates back to the year 1000. For a discussion, see Georges Duby, *William Marshal: The Flower of*

First, consider the codified requirements of IHL. Although chivalry is clearly a background principle that helps explain the historical development of IHL, few scholars view it as a guiding principle equal in significance to the principles of humanity or necessity. Those principles have a determinate legal status that structure and inform the analysis of particular rules in IHL. The principle of necessity was explicitly codified in Articles 12, 13, and 14 of the Lieber Code, and forms the foundation for the rule that allows privileged combatants to kill enemy combatants. Similarly, the principle of humanity also emanates from the structure and penumbra of existing IHL. Just as the principle of necessity was explicitly referred to in the Lieber Code, so too the principle of humanity was referred to in the Martens Clause, and provides the analytical foundation for the rules that regulate unnecessary suffering of soldiers, the protection of POWs and the prohibition against killing them, and in general the rules that seek to insulate the civilian population, as much as possible, from the horrors of warfare.

Can the same thing be said of the principle of chivalry? Probably not, though it may still be a *background* legal principle in IHL.[88] Soldiers meet each other on the battlefield as equals. Indeed, this is the foundation for the moral equality of combatants—the rule that extends immunity to all privileged combatants for their acts of battlefield killing, regardless of whether they are parties to the 'right' or 'wrong' side of the conflict as far as jus ad bellum is concerned. If this is what is meant by the principle of chivalry, then yes, combatants meet each other as equals on the battlefield. But if 'chivalric warfare' means an equality of arms, a

---

Chivalry (Pantheon 1987); Antonio Santosuosso, *Barbarians, Marauders, and Infidels: The Ways of Medieval Warfare* (Westview 2004) 172.

[88] For example, see Terry Gill, 'Chivalry: A Principle of the Law of Armed Conflict?' in Marcel Brus and Brigit Toebes (eds), *Armed Conflict and International Law: In search of the Human Face: Liber Amicorum in memory of Avril McDonald* (Asser Press 2013). Gill argues that chivalry retains its significance for contemporary IHL and provides the foundation for many particular rules codified in existing treaties. However, Gill does not identify killing at a distance as a violation of chivalry: '(Long range) missile warfare, whether by means of ballista, long or crossbow, musket, artillery, machinegun or sniper rifle (or for that matter helicopter gunships or missile armed unmanned aerial vehicles or "drones"), has been part of warfare for centuries and is the "great leveler", making no distinction between rank, class, skill or bravery of the recipients … In short, chivalry and martial honour have never precluded maximizing one's advantages and neutralizing those of the opponent …' Ibid 47.

personal connection between killer and killed,[89] or a symmetrical assumption of risk, there is simply no way to logically deduce this more extensive notion of chivalry from the moral equality of combatants.

It is true, however, that chivalry may help explain other rules in international law that remain viable, notably the extensive protections that IHL extends to civilian populations.[90] Some of these requirements may owe their existence to requirements of chivalry.[91] For example, the law requires combatants not to engage in attacks that will cause disproportionate collateral damage—a requirement that places soldiers in some degree of risk at the expense of protecting the civilian population. In some cases, soldiers may even be under an obligation to take all feasible measures to reduce civilian casualties as far as possible—again a requirement that imposes demands on soldiers that may cause them harm. Similarly, in IHL, a soldier may not assert duress or necessity as an excuse to the war crime of killing innocent civilians. Traditionally, these rules are justified by reference to the inherent dignity or moral worth of civilians. But for each of these rules, the law's prioritization of the civilian (even an enemy one) over the soldier could be seen as a partial outgrowth of the principle of chivalry; it is dishonorable for a soldier to put himself or herself above the civilian, and sacrifices may be required. Again, though, there is nothing in this notion of chivalry that requires soldiers to bear additional risk simply because lowering asymmetrical risk is *inherently* dishonorable.[92]

Second, we should consider whether established norms of warfare, stemming either from moral philosophy or professional standards,[93]

---

[89] See O'Connell (n 23) 271; see also David Grossman, *On Killing: The Psychological Cost of Learning to Kill in War and Society* (Back Bay 1996) 187; Jennifer M Welsh, 'The Morality of "Drone Warfare"' in Cortright et al (n 86) 24, 43–4 (concluding that drone pilots are simultaneously far removed and 'quite close' to their targets at the same time).

[90] Gill (n 88) 47.

[91] See Thomas Wingfield, 'Chivalry in the Use of Force' (2001) 32 U Tol L Rev 111, 136 ('This quest for honor—the desire to fight with swords, both literally and metaphorically—counterbalances the dark side of armed conflict, the base desire to destroy, simply because one can. In this struggle, the law of chivalry is an elegant weapon, for a more civilized age.').

[92] Another example is the prohibition against perfidy. See Walter G Sharp, 'The Effective Deterrence of Environmental Damage During Armed Conflict: A Case Analysis of the Persian Gulf War' (1992) 137 Mil L Rev 1, 31.

[93] Some have argued that the principle of necessity could be constrained by interpreting it through the lens of professional standards and discipline. See Yishai Beer, 'Humanity Considerations Cannot Reduce War's Hazards Alone:

require reciprocal risk. This argument implies that killing from a distance while remaining out of danger is cowardly and that 'real men' put themselves in harm's way when they engage in lethal attacks during combat.[94] For example, after the 9/11 attacks, author and cultural critic Susan Sontag offered the following observations in *The New Yorker*:

> Where is the acknowledgment that this was not a 'cowardly' attack on 'civilization' or 'liberty' or 'humanity' or 'the free world' but an attack on the world's self-proclaimed superpower, undertaken as a consequence of specific American alliances and actions? How many citizens are aware of the ongoing American bombing of Iraq? And if the word 'cowardly' is to be used, it might be more aptly applied to those who kill from beyond the range of retaliation, high in the sky, than to those willing to die themselves in order to kill others. In the matter of courage (a morally neutral virtue): whatever may be said of the perpetrators of Tuesday's slaughter, they were not cowards.[95]

Although Sontag did not come out and say explicitly that American pilots are cowards, she did say that they are more appropriately considered cowards than are suicide bombers. Why? The relevant difference between the two is that the suicide bomber not only risks his life but in fact *sacrifices* it, while the pilot risks little or nothing. And Sontag said this before the great proliferation of drone strikes. With those technological advancements, Sontag might have been even more categorical in her condemnation of the 'cowardly' actions of drone pilots, who basically risk nothing of their own personal safety.

Sontag's tone-deaf assessment was met with predictable outrage. Coming on the heels of the worst terrorist attack on American soil, it is no surprise that the public had little interest in her apparent lionization of suicide bombers and her criticism of American military personnel. However, the public reaction is neither here nor there. The more important question is whether Sontag was on to something or not from the perspective of moral or political theory. And in this regard, it does

---

Revitalizing the Concept of Military Necessity' (2015) 26 Eur J Intl L 801, 805 ('Constraining the brute force of a military is in its own self-interest and enhances its operational effectiveness.'). This example is an attempt to make professional standards morally and legally relevant. The question is whether that methodology is generalizable to other contexts.

94 For a contrasting view, see Drone Pilot, 'It is War at a Very Intimate Level' in Peter L Bergen et al (eds), *Drone Wars: Transforming Conflict, Law, and Policy* (Cambridge University Press 2015) 113, 116 ('Just because you are separated by technology does not mean you are separated emotionally.').

95 Susan Sontag, 'Tuesday, and After' *The New Yorker* (New York, 24 September 2001).

seem odd to suggest that there is a 'morally neutral virtue' that would require personal risk in the fighting of war. The most that can be said in favor of the position is that: fighting in a war while risking one's life is particularly courageous, but failure to do so should not be considered a moral deficit. In other words, courage is supererogatory; its presence should be celebrated but its absence is no reason to condemn an action as immoral or illegal. Indeed, there are many situations, such as duress, where we celebrate those who demonstrate moral heroism but refuse to condemn those who fail to live up to that unreasonably high standard. The second insight one might draw from Sontag's exposition is that the United States is quick (perhaps too quick) to act as the world's 'self-proclaimed superpower' simply because it can project military force without risk—an argument that we already considered above and mostly rejected.

Finally, consider reciprocal risk from the perspective of professional standards of conduct that reign among soldiers. When the technology of guns advanced enough, in both range and precision, to allow for a practice that could fairly be called sniping, it immediately drew attention as a questionable practice.[96] Over time, the unease remained because some soldiers consider it a 'dirty practice' or 'cold-blooded'.[97] But for others, snipers are virtuous because they are more capable of respecting the principle of discrimination.[98]

The use of snipers was specifically designed to be a force equalizer—to counter the enemy's strategic advances in other domains.[99] Also, the use of the sniper was intimately wrapped up with a new

---

[96] See Michael E Haskew, *The Sniper at War: From the American Revolutionary War to the Present Day* (Macmillan 2005) 8; Andy Dougan, *The Hunting of Man: A History of the Sniper* (Fourth Estate 2004).

[97] See Joanna Bourke, *An Intimate History of Killing: Face-to-face Killing in Twentieth-century* (Basic 2000) 54. See also Robert Graves, *Good-Bye to All That* (1957) 132 ('While sniping from a knoll in the support line, where we had a concealed loop-hole, I saw a German, about seven hundred yards away, through my telescopic sights. He was taking a bath in the German third line. I disliked the idea of shooting a naked man, so I handed the rifle to the sergeant with me. "Here, take this. You're a better shot than I am." He got him; but I had not stayed to watch.'), quoted in Michael Walzer, *Just and Unjust Wars* (Basic Books 2000) 140.

[98] Dougan (n 3) 295 ('The sniper is the smartest of "smart" weapons. Through the fog of war he is able to distinguish friend from foe, target from survivor, in confused surroundings.').

[99] For example, the British used snipers to counter Napoleon's army. See Brandon Webb, *The 21st Century Sniper: A Complete Practical Guide* (Skyhorse

development that saw a change in practice regarding who was targetable in warfare. In the past, officers and other military personnel who held support positions were usually not targeted in battle; the advent of sniping coincided with a breakdown of that cultural norm, leading some to suggest that snipers had 'foresworn the chivalrous code of battle from the old days'.[100] For example, during the US military engagement in Somalia, a young woman was killed by an American sniper, which provoked a public outcry. One soldier, responding to the event, wrote:

> Although I am a proponent of the employment of snipers, I believe that most people feel there is something unclean—unsporting or not chivalrous—about their use. Some see sniping as a dishonorable method of prosecuting war and expect more from our military. The alleged killing of the young Somali woman only reinforced the skeptics' view and illustrated how a misplaced round can instantly result in human tragedy.[101]

However, what was problematic about that event was uncertainty over the status of the target, not the manner in which the target was defeated. If there is a chivalric code of warfare that requires reciprocal risk, and prohibits killing at a distance, it is an element of the chivalric code that has long since been abandoned,[102] a development that was already well underway by the time of the US Civil War.[103] Sniping was rampant

---

2010) 11. See also Dougan (n 3) 9 (discussing snipers used during the siege of Lichfield in 1643).

[100] Webb (n 99) 11. See also Blum (n 10) 75 (noting that crossbows allowed commoners to strike knights from a distance—an earlier example of a similar problem).

[101] Lawrence E Casper, *Falcon Brigade: Combat and Command in Somalia and Haiti* (Lynne Rienner 2001) 134.

[102] See Allen J Frantzen, *Bloody Good: Chivalry, Sacrifice, and the Great War* (University of Chicago Press 2004) 1–3 (arguing that chivalry continued as an important concept during World War I, despite the popular assumption that when 'young men filled with illusions of chivalry were ordered to walk into machine-gun fire, an ancient brotherhood fell before the weapons of a new age').

[103] See Adrian Gilbert, *Stalk and Kill: The Sniper Experience* (St Martin's Press 1997) 27 (describing the American Civil War as a 'golden age for the military rifle' but noting that 'this simple yet profound transformation in the conduct of war was not realized at the time, and massed columns of brightly uniformed soldiers were regularly slaughtered by devastating fire from a single line of riflemen'); Matthew J Grow, *'Liberty to the Downtrodden': Thomas L. Kane, Romantic Reformer* (Yale 2009) 223 (noting that Thomas Kane believed that it was fundamentally dishonorable for General Ashby to have been felled by a sniper's bullet). Grow writes that 'The Incident not only demonstrates how chivalry defined appropriate action in combat for Kane, but also how the chaos

during World War I.[104] If the chivalric code defined honorable killing as personal, that was the chivalric code of a prior era.[105]

6. CONCLUSION

Reduction of asymmetrical risk is, in many ways, the goal of strategic warfare.[106] The preceding analysis has demonstrated a historical continuity in every army's attempt to project military power but remain at arm's length from the enemy's strategic capabilities. Drones, cyber weapons and autonomous weapons are just the latest instantiation of an ancient imperative of strategic warfare. The goal of combat is to exploit the gap between one's own zone of lethality and the enemy's—projecting power while reducing or eliminating risk. Indeed, military experts recently warned the US government that China's development of a new missile would allow the Chinese military to strike an American base in Guam that was previously beyond the range of Chinese missiles—negating the point of locating the base in Guam.[107] The desire to reduce asymmetrical risk is ever-present both in new technological platforms but also in more conventional assets, such as ballistic missiles.

Reduction of asymmetrical risk has consequences, though the preceding analysis has demonstrated that, contrary to received intuitions, none of these consequences run afoul of either jus in bello or jus ad bellum considerations. While risk-free weapon platforms might raise the risk of more armed conflict, this fact applies equally to good and bad wars, thus

---

of battle and the bitterness of the war rendered such concern for chivalry increasingly antiquated and impractical for most soldiers as the war wore on.' Ibid.

[104] Gilbert (n 103) 53. See also Adrian Gilbert, *Sniper: The Skills, the Weapons, and the Experiences* (St Martin's Press 1994) 37–68.

[105] See Matthew Strickland, *War and Chivalry: The Conduct and Perception of War in England and Normandy, 1066–1217* (Cambridge University Press 1996) 17, 30 (discussing the chivalric code among captured knights which broke down with regard to the treatment of external enemies who were slaughtered upon capture); John Gillingham, '1066 and the Introduction of Chivalry into England' in George Garnett and John Hudson (eds), *Law and Government in Medieval England and Normandy* (Cambridge University Press 1994) 31, 32–3.

[106] Gill (n 88) 47.

[107] See Brad Lendon, 'U.S. must beware China's "Guam killer" missile', CNN.com (15 May 2016) ('Guam, home to Andersen Air Force Base and Apra Naval Base, has been as a place from where the U.S. could project power across the Pacific while having its forces at relatively safe distance from possible threats, including North Korea and China.')

leading to the inescapable conclusion that in some circumstances, remote technology has salutary benefits. As for jus in bello, reduction of risk is only problematic if it comes with an increase in risk to the civilian population that is greater than what the law allows. If, however, the level of risk to civilians complies with the law, the lowering of risk to friendly troops is not, itself, a legal vice.

2. The principle of distinction and remote warfare

Emily Crawford

1. INTRODUCTION

The notion of 'remote warfare' is, arguably, not a new phenomenon. Since the invention of the crossbow, a belligerent involved in an armed conflict was no longer limited to engaging in hand-to-hand combat, and could attack and kill an enemy at a distance. The invention of the pistol and cannon only widened the gap between belligerents on the battlefield. However, even with such weapons, one needed to be in proximity to one's adversary—one needed to be able to visually identify one's enemy before being able to target them. Truly remote warfare—warfare conducted thousands of miles from active hostilities—was only made possible with the technological breakthroughs of the 20th century, such as aerial bombardment, inter-continental ballistic weaponry, and unmanned armed aerial vehicles.

These new weapons and new modes of delivery were, and are still, subject to the extant laws of armed conflict (also known as international humanitarian law or IHL).[1] The 'newness' of the weaponry, and the

[1] The terms IHL, international humanitarian law, and the law of armed conflict will be used interchangeably throughout this chapter. The main treaties of IHL examined in this chapter are the Geneva Conventions of 1949 (Comprising Geneva Convention for the Amelioration of the Condition of the Wounded and Sick in Armed Forces in the Field of 12 August 1949 (hereinafter Geneva Convention I or GCI) 75 UNTS 31; Geneva Convention for the Amelioration of the Condition of Wounded, Sick and Shipwrecked Members of Armed Forces at Sea of 12 August 1949 (hereinafter Geneva Convention II or GCII) 75 UNTS 85; Geneva Convention Relative to the Treatment of Prisoners of War of 12 August 1949 (hereinafter Geneva Convention III, GCIII or the POW Convention) 75 UNTS 135; and Geneva Convention Relative to the Protection of Civilian Persons in Time of War of 12 August 1949 (hereinafter Geneva Convention IV, GCIV, or the Civilians Convention) 75 UNTS 287) and the Additional Protocols of 1977 (The Additional Protocols comprise Protocol Additional to the Geneva Conventions of 12 August 1949, and relating to the Protection of Victims of International Armed Conflicts of 8 June 1977 (hereinafter Protocol I or API),

absence of any specific treaty obligation that regulates its use, does not relieve parties to the conflict from observing the fundamental principles of IHL, such as the principle of distinction.[2] The principle of distinction provides that participants in an armed conflict must at all times distinguish between civilians and civilian objects, and military personnel and military objectives.[3] Only military personnel and military objects may be made the subject of direct attacks; civilians and civilian objects are immune from direct targeting.[4] The principle of distinction has been called one of the 'fundamental and intransgressible principle[s]'[5] of IHL, and one of the 'cardinal principles ... constituting the fabric of humanitarian law'.[6] The principle of distinction has formed part of most of the IHL treaties and instruments since the beginning of modern treaty IHL.[7]

Giving effect to the principle of distinction can be achieved in a number of ways; on a 'macro' level, it includes not intentionally targeting installations or objects that are purely civilian in nature, purpose, or use—such as schools or hospitals.[8] It is also at the core of provisions that, for example, obligate parties to a conflict to locate military

---

1125 UNTS 3, and Protocol Additional to the Geneva Conventions of 12 August 1949, and relating to the Protection of Victims of Non-International Armed Conflicts of 8 June 1977 (hereinafter Protocol II or APII), 1125 UNTS 609).

[2] See for instance Article 35 of API, which provides that 'In any armed conflict, the right of the Parties to the conflict to choose methods or means of warfare is not unlimited'; and Article 36 of API, which provides that 'In the study, development, acquisition or adoption of a new weapon, means or method of warfare, a High Contracting Party is under an obligation to determine whether its employment would, in some or all circumstances, be prohibited by this Protocol or by any other rule of international law applicable to the High Contracting Party.'

[3] Article 48, API.

[4] Civilians and civilian objects will lose their immunity from targeting if they are deemed as having a military purpose. For civilian objects, immunity from targeting is lost if the object is considered as having a military purpose or use. For civilians, immunity is lost if the civilian takes direct part in the hostilities. See further Article 51(3), API for the targeting of civilians taking direct part in hostilities, and Article 52(2), API for the general rules regarding the targeting of objects.

[5] *Legality of the Threat or Use of Nuclear Weapons* (Advisory Opinion), [1996] ICJ Rep 226, 257 (hereinafter *Nuclear Weapons*).

[6] Ibid.

[7] Discussed in further detail in the next section of this chapter.

[8] See Article 12, API (on medical units) and Article 52, API (on civilian installations).

objectives away from civilian populations,[9] and to ensure that any targeting of military objectives is carried out proportionately[10] and in a discriminate manner.[11] On a 'micro' level, the principle of distinction can be respected through the wearing of uniforms and the open carriage of arms by members of the armed forces participating in the hostilities,[12] or the marking of buildings or installations in ways that clearly indicate its prima facie immunity.[13] Determining whether an object or person is civilian or military in nature can also be achieved through intelligence gathering, which can assist in making the determination as to whether a prima facie civilian object is being used for military purposes, or whether a person who might appear to be a combatant (because they are, say, armed with weaponry) is actually a civilian.

In theory, certain forms of remote warfare are ideal for compliance with the principle of distinction. Technologically advanced weaponry, such as unmanned aerial vehicles (also known as UAVs, or drones), are able to conduct precision attacks, eliminating targets with a degree of exactness and surety unmatched by previous technologies such as missiles or bombs. In the realm of cyber-hostilities, precisely engineered software or computer code can target and disable very specific objectives, ensuring that only specific objectives are affected by the attack, leaving other systems untouched.[14] Equally, however, the remoteness of such warfare can make distinction assessments and distinction-compliant targeting a harder task. This chapter therefore examines certain questions that arise regarding the principle of distinction and remote warfare. What impact does the remoteness of these means and methods of warfare have on the principle of distinction? Does the fundamental 'remoteness' of these kinds of attacks—drone attacks and cyber-attacks—mean that compliance with the principle of distinction is made easier or harder? That is to say, does the physical removal of the attacker from the immediate or proximate vicinity of the target make respecting the principle of distinction more or less achievable? And if compliance with the principle of distinction is facilitated by these remote means and methods of war, how much of that is due to the 'remoteness' of the

---

[9] Article 58(b), API.

[10] Articles 51(5)(b) and 57(2)(a)(iii), API.

[11] Article 51(4), API.

[12] Article 4A(2), GCIII and Article 44(3), API.

[13] See for example Article 18, API on the marking of persons and property involved in the care of the wounded, sick and shipwrecked.

[14] Examples of such precision warfare are explored in more detail below.

weapons? Is the remoteness of the drone pilot or the cyber-attacker fundamentally linked to distinction-compliant warfare?

This chapter will examine these questions, and interrogate how the 'fundamental and intransgressible'[15] principle of distinction interacts with new forms of remote warfare. The first part of the chapter provides a brief historical overview of the development of the principle of distinction and its place in the law of armed conflict. The second part of the chapter explores how the principle has been operationalized in practice. The third part of the chapter then looks at the unique problems and issues that arise regarding remote warfare and compliance with the principle of distinction, focussing on two specific types of remote warfare: targeted killing with armed UAVs and the emergent field of cyber warfare. Finally, the last part of the chapter examines the degree to which the 'remoteness' of these kinds of remote warfare is connected to their potential for compliance with the principle of distinction.

2. THE PRINCIPLE OF DISTINCTION IN THE LAW OF ARMED CONFLICT: HISTORICAL DEVELOPMENT OF THE PRINCIPLE AND THE CURRENT INTERNATIONAL LAW OF DISTINCTION

The principle of distinction provides that parties to an armed conflict must, at all times, distinguish between civilians and civilian objects (which are not to be made subject to attack), and military personnel and military objects (which may be directly targeted). The principle of distinction has been a cornerstone of the modern treaty law of armed conflict, first included in General Orders No. 100, better known as the Lieber Code—the instructions given to the Union Armies during the American Civil War.[16] Article 22 of the Code stated that:

> as civilization has advanced during the last centuries, so has likewise steadily advanced, especially in war on land, the distinction between the private individual belonging to a hostile country and the hostile country itself, with its men in arms. The principle has been more and more acknowledged that the unarmed citizen is to be spared in person, property, and honor as much as the exigencies of war will admit.

---

[15] *Nuclear Weapons* (n 5), 257.

[16] Instructions for the Government of Armies of the United States in the Field, Prepared By Francis Lieber, Promulgated as General Orders No. 100 by President Lincoln, 24 April 1863.

This provision was reaffirmed in the St Petersburg Declaration of 1868, which, in its preamble, stated that: 'the progress of civilization should have the effect of alleviating as much as possible the calamities of war; [t]hat the only legitimate object which States should endeavour to accomplish during war is to weaken the military forces of the enemy.'[17]

The Lieber Code and the St Petersburg Declaration codified a theory of civilian immunity that had for some time been present in the writings of leading jurists and theorists on the law of armed conflict. In his 1762 work *The Social Contract*, Jean Jacques Rousseau wrote that: 'the object of the war being the destruction of the hostile State, the other side has a right to kill its defenders, while they are bearing arms; but as soon as they lay them down and surrender, they cease to be enemies or instruments of the enemy, and become once more merely men, whose life no one has any right to take.'[18] That belligerent forces should direct attacks solely against military targets was also commented on by Lassa Oppenheim, who stated that while 'during antiquity and the greater part of the Middle Ages, war was a contention between the whole of the populations of the belligerent state',[19] the situation had notably changed, and that:

> 'gradually a milder and more discriminative practice grew up … war nowadays is a contention of States through their armed forces. Those private subjects of the belligerents who do not directly or indirectly belong to the armed forces do not take part in it; they do not attack and defence; and no attack ought therefore to be made upon them.'[20]

The principle of distinction continued to form part of the law of armed conflict in all its iterations throughout the late 19th and early 20th centuries. While the principle of distinction was not explicitly defined in

---

[17] Preamble, Declaration Renouncing the Use, in Time of War, of Explosive Projectiles under 400 Grammes Weight (St Petersburg Declaration) 1868.

[18] Jean Jacques Rousseau, *The Social Contract, Or Principles of Political Right* (1762, translated by G D H Cole, public domain). However, see comments by Geoffrey Best, who dismissed Rousseau's comments as 'pretentious and imprudent', 'defective and disadvantageous', and a 'well-meaning but practically useless maxim [which] merely encouraged self-deception among the French' (in *Humanity in Warfare: The Modern History of the International Law of Armed Conflicts* (Weidenfeld and Nicholson 1980), 55–9).

[19] Lassa Oppenheim, *International Law: A Treatise* (2 Volumes, Longmans Green 1912), Vol 2, 63.

[20] Ibid.

either the Hague Regulations of 1899 and 1907,[21] or the Geneva Conventions of 1949, the principle is nonetheless at the core of articles such as Article 25 of the 1899 and 1907 Hague Regulations[22] which prohibited 'the attack or bombardment of towns, villages, habitations or buildings which are not defended'.[23] Distinction is also at the heart of provisions in the Geneva Conventions that prohibit acts of violence against persons who do not, or no longer, take 'active part in the hostilities, including members of armed forces who have laid down their arms and those placed *hors de combat* by sickness, wounds, detention, or any other cause'.[24] Indeed, as noted in the Commentary to the Additional Protocols of 1977, the 'entire system established in The Hague in 1899 and 1907 and in Geneva from 1864 to 1977 is founded' on the principle of distinction.[25]

It was not until the adoption of the Additional Protocols in 1977 that the principle of distinction was given express codification, in Article 48 of Additional Protocol I, which states that:

> In order to ensure respect for and protection of the civilian population and civilian objects, the Parties to the conflict shall at all times distinguish between the civilian population and combatants and between civilian objects and military objectives and accordingly shall direct their operations only against military objectives.[26]

The principle of distinction is reiterated or reaffirmed in a number of other articles in the Additional Protocols, forming the basis for provisions such as Article 51 of Protocol I, which states that:

---

21 The relevant Hague instruments for the purposes of this chapter are Hague Convention II with Respect to the Laws and Customs of War on Land 1899, 187 *CTS* 429 and Hague Convention IV Respecting the Laws and Customs of War on Land 1907, 205 *CTS* 227.

22 Article 25 of the Hague Regulations of 1899 (annexed to Convention (II) with Respect to the Laws and Customs of War on Land (187 *CTS* 429)) and the Hague Regulations of 1907 (annexed to Convention IV Respecting the Laws and Customs of War on Land (205 *CTS* 277)).

23 Though the Regulations do not expressly prohibit attacking civilians, the principle can be inferred from the instruments.

24 Common Article 3, GCIII.

25 Yves Sandoz, Christophe Swinarski and Bruno Zimmerman (eds), *Commentary on the Additional Protocols of 8 June 1977 to the Geneva Conventions of 12 August 1949* (ICRC, Geneva, 1987 hereinafter *AP Commentary*) 598.

26 Article 48, API.

1. The civilian population and individual civilians shall enjoy general protection against dangers arising from military operations.
2. The civilian population as such, as well as individual civilians, shall not be the object of attack. Acts or threats of violence the primary purpose of which is to spread terror among the civilian population are prohibited.
3. Civilians shall enjoy the protection afforded by this Section, unless and for such time as they take a direct part in hostilities.[27]

The principle of distinction is also at the core of international instruments on landmines[28] and cluster munitions,[29] and is applicable in both international and non-international armed conflicts.[30] It is considered customary international law[31] and violation of the principle of distinction is a war crime in both international[32] and non-international armed conflicts.[33] Domestic and international courts have affirmed the centrality of the principle of distinction to the law of armed conflict.[34]

---

[27] Article 51(1)–(3), API. See also Article 13 of APII, which enshrines the principle of distinction in situations of non-international armed conflict.

[28] Preamble, Convention on the Prohibition of the Use, Stockpiling, Production and Transfer of Anti-Personnel Mines and on their Destruction, 2056 UNTS 211.

[29] Preamble, Convention on Cluster Munitions, 2688 UNTS 39. See also the Convention on Prohibitions or Restrictions on the Use of Certain Conventional Weapons which may be deemed to be Excessively Injurious or to have Indiscriminate Effects 1342 UNTS 137, Protocol II, Amended Protocol II, and Protocol III.

[30] Jean-Marie Henckaerts and Louise Doswald-Beck (eds), *Customary International Humanitarian Law* (3 volumes, Cambridge University Press 2005; hereinafter *ICRC CIHL Study*) Rule 1.

[31] Ibid.

[32] Article 8(2)(b)(i)–(ii), Rome Statute of the International Criminal Court, 2187 UNTS 90 (hereinafter Rome Statute).

[33] Article 8(2)(e)(i), Rome Statute.

[34] See the decisions in the International Criminal Tribunal for the Former Yugoslavia of *Blaškić* (Case No. IT-95-14, Judgment, 3 March 2000, §180); *Martić* (Case No. IT-95-11, Review of the Indictment under Rule 61, 8 March 1996, § 10); *Kordić and Čerkez* (Case No. IT-95-14/2-A, Appeal Judgment, 7 December 2004, § 54); *Kupreškić et al* (Case No. IT-95-16-T, Judgment, 14 January 2000, § 521); *Strugar* (Case No. IT-01-42-T, Judgment, 31 January 2005, §§ 220-221); *Tadić* (Case No. IT-94-1-AR72, Decision on the Defence Motion for Interlocutory Appeal on Jurisdiction, 2 October 1995, § 127), and *Galić* (Case No. IT-98-29-T, Judgment, 5 December 2003, §§ 27, 45); the Inter-American Commission on Human Rights in *Juan Carlos Abella v Argentina* (*Case 11.137*, Report No. 55/97, Inter-Am. C.H.R., OEA/Ser.L/V/II.95 Doc. 7 rev. at 271 (1997), § 177); the Israeli courts in *Military Prosecutor v Omar Mahmud Kassem et al* (Israel Military Court, Ramallah, 13 April 1969, 42 *ILR*

3. OPERATIONALIZING THE PRINCIPLE OF DISTINCTION: HOW IS IT OBSERVED AND RESPECTED IN PRACTICE?

In order to give effect to the principle of distinction, parties to the conflict must take certain steps to ensure compliance. In its simplest permutation, the principle of distinction can be respected by limiting targeting to only military objects or objectives. In practice, this means that the persons entrusted with assessing and selecting objects to be targeted as part of the military campaign must undertake to acquire and review relevant intelligence and information about the nature, purpose, location or use[35] of proposed targets, to ensure that they are military in character, and not civilian.[36] For example, during the 1990 war with Iraq, US commanders would review available intelligence to ensure that selected targets were military and not civilian in nature.[37] Furthermore, individual members of the armed forces serving in the field were each given pocket-sized 'rules of engagement cards',[38] which reaffirmed the

---

470 (1971), 480), The Public Committee against Torture in *Israel et al v the Government of Israel et al* (HCJ 769/02, 13 December 2006 (hereinafter *Targeted Killings*), §§ 23, 26); and *Physicians for Human Rights v Prime Minister of Israel* (HCJ 201/09, 19 January 2009, § 21); the decision in the Federal Court of Australia of *SZAOG v Minister for Immigration & Multicultural & Indigenous Affairs* ([2004] FCAFC 316, 26 November 2004, para 17); the Colombian Constitutional Court, Constitutional Case No. C-037/04 (§§ 35–6), Constitutional Case No. T-165/06 (at §§ 7–8); and Constitutional Case No. C-291/07 (§ 78); the Peruvian Constitutional Court, *Gabriel Orlando Vera Navarrete* (Case No. 2798-04-HC/TC, 9 December 2004, § 15); and the Spanish Supreme Court in the case of *Couso* (13 July 2010, Section II(II), Sexto, § 2, 13).

35 Article 52(2), API. The exact scope of the nature/location/purpose/use test is discussed in more detail below.

36 Article 57 of API outlines the obligations to take precautions in attack, including the obligation to 'do everything feasible to verify that the objectives to be attacked are neither civilians nor civilian objects and are not subject to special protection but are military objectives within the meaning of paragraph 2 of Article 52'.

37 See the US Department of Defense report, The Conduct of the Persian Gulf War: Final Report to Congress Pursuant to Title V of The Persian Gulf Conflict Supplemental Authorization and Personnel Benefits Act of 1991 (Public Law 102-25) (GPO 1991), Appendix O.

38 *Desert Storm – Rules of Engagement*, Pocket Card, US Central Command, January 1991 (reprinted in Gary Solis, *The Law of Armed Conflict: International Humanitarian Law in War* (Oxford University Press 2010) 517–8).

principle of distinction in its summarized instructions: 'fight only combatants; attack only military targets; spare civilian persons and objects'.[39]

Respecting the principle of distinction also requires parties to the conflict to take precautions against the effects of attacks. Thus, parties to a conflict must:

> endeavour to remove the civilian population, individual civilians and civilian objects under their control from the vicinity of military objectives; ... avoid locating military objectives within or near densely populated areas; [and] ... take the other necessary precautions to protect the civilian population, individual civilians and civilian objects under their control against the dangers resulting from military operations.[40]

The principle of distinction is thus a two-fold obligation, a 'negative obligation not to attack or harm ... [and an] affirmative obligation to protect actively'.[41] Applying the principle of distinction is dependent on making an assessment as to an object's status, and whether that status renders the object targetable or immune from targeting. To that end, one must know the rules on targeting and the law on defining a military objective, which is codified in Additional Protocol I.

(a) The Rules on Targeting Objects

For objects, such as buildings or vehicles, Article 52(2) of Protocol I defines military objectives as 'those objects which by their nature, location, purpose or use make an effective contribution to military action and whose total or partial destruction, capture or neutralization, in the circumstances ruling at the time, offers a definite military advantage.' While by no means a straightforward or uncontested definition,[42] an object may be a 'military objective' in one of four ways. The first, and most obvious, is that an object could be a legitimate target because of its

---

39 Ibid.

40 Article 58, API.

41 Michael Bothe, Karl Joseph Partsch and Waldemar Solf (eds), *New Rules for Victims of Armed Conflicts: Commentary on the Two 1977 Protocols Additional to the Geneva Conventions of 1949* (Martinus Nijhoff 1982) (hereinafter *New Rules*) 322–3.

42 See Stefan Oeter, 'Means and Methods of Combat' in Dieter Fleck (ed), *The Handbook of Humanitarian Law in Armed Conflicts* (3rd edn, Oxford University Press 2013) 169. See also Hays Parks, 'Air War and the Law of War' (1990) 32 Air Force L Rev 1, 137–44; Yoram Dinstein, 'Legitimate Military Objectives under the Current Jus in Bello' in Andru Wall (ed), 'Legal and Ethical Lessons of NATO's Kosovo Campaign' (2002) 78 Intl L Stud 139, 144–5.

military *nature*: the object is a military command centre, a military base, or a military vehicle, like a tank.[43] Second, an object's *location* may render it a target: For example, a narrow mountain pass that provides the only means of egress for an armed group to launch attacks could be considered a military objective because of location.[44] Thus, objectives rendered targetable due to location 'includes areas that are militarily important because they must be captured or denied the enemy'.[45] An object's *use* renders it targetable if that use is military—a school or place of religious worship being used to plan and launch attacks would lose its protected status and become a lawful target. Thus, for example, in Iraq in 2003, it was discovered that mosques, hospitals and schools were being used as military bases; targeting such installations would thus be permissible under the law of targeting.[46] Finally, and connected to an object's use, is its *purpose*, which may render the object a legitimate target, purpose being 'concerned with the intended future use of an object'.[47]

Operationalizing the principle of distinction in relation to the law on targeting thus requires military objects to be accurately assessed *as military objectives* under the nature/location/use/purpose test. Furthermore, in order to comply with the principle of distinction, parties to a conflict must not attempt to make a military object appear to have a protected civilian function; a military base must not be disguised to look like a hospital or any other internationally protected installation.[48] The four possible bases for targeting—nature, location, purpose or use—mean

---

43 Solis (n 38) 525; Dinstein (n 42) 146–7.

44 Dinstein (n 42) 150.

45 Solis (n 38) 525.

46 Gregory Fontenot, EJ Degen and David Tohn, *On Point: The United States Army in Operation Iraqi Freedom* (Combat Studies Institute Press 2004) 214. See also Human Rights Watch, *Off Target: The Conduct of the War and Civilian Casualties in Iraq* ((accessed 26 April 2017 at https://www.hrw.org/report/2003/12/11/target/conduct-war-and-civilian-casualties-iraq), 66–79) for a discussion of IHL violations by the Iraqis during the conflict, including the use of schools, mosques and hospitals as military sites.

47 *AP Commentary* (n 25), 636.

48 Article 38, API states that 'it is prohibited to make improper use of the distinctive emblem of the red cross, red crescent or red lion and sun or of other emblems, signs or signals provided for by the Conventions or by this Protocol. It is also prohibited to misuse deliberately in an armed conflict other internationally recognized protective emblems, signs or signals, including the flag of truce, and the protective emblem of cultural property'.

that nearly any object, even prima facie civilian objects, may be subject to direct attack if they are being used for military purposes.[49]

(b) The Rules on Targeting Persons

With regard to targeting persons, the law takes an essentially similar approach to that taken regarding objects—a person may be targeted if they are directly contributing to the hostilities because of their status or function. Thus, a person who is a member of the armed forces of a party to the conflict can be lawfully targeted due to their status as a member of the military or associated forces (and thus, by their presumed continued direct contribution to the armed conflict).[50] As Sassòli and Olson note,

---

[49] Additionally, for the attack to be lawful, it must be shown that targeting or attacking the object, with a view to its destruction, capture, or neutralization, offers a definite military advantage. Determining what amounts to a definite military advantage is difficult. Some kinds of attacks which might seem to offer a military advantage to an attacking party are not allowable, because the advantage offered may be of a political character, or the advantage is too abstract to quantify. Thus, an attack on an enemy for the sole purpose of demoralising the enemy population is not legitimate—as Michael Bothe notes 'air attacks have a definite impact on the morale of the entire population and, thus, on political and military decision-makers ... [but] this type of "advantage" is political, not military. The morale of the population and of political decision-makers is not a contribution to "military action"'. See Michael Bothe, 'Targeting' in Wall (n 42) 180. Economic targets are likewise generally considered too removed from the armed conflict to offer a definite military advantage. See Dinstein (n 42) 146, quoting the *San Remo Manual*, 161. See also APV Rogers, *Law on the Battlefield* (3rd edn, Manchester University Press 2012) 70–71. The US approach is that war-sustaining industries are legitimate targets—this is based on practice in the US Civil War, where cotton fields in Confederate territory were often targeted and destroyed by the Union forces, on the grounds that cotton was almost the only source of income for the Confederate war effort. See Solis (n 38) 523. However, State practice suggests that the US is alone in taking this broad view.

[50] These are defined in Article 4(A), Geneva Convention III as: '(1) Members of the armed forces of a Party to the conflict as well as members of militias or volunteer corps forming part of such armed forces; (2) Members of other militias and members of other volunteer corps, including those of organized resistance movements, belonging to a Party to the conflict and operating in or outside their own territory, even if this territory is occupied, provided that such militias or volunteer corps, including such organized resistance movements, fulfil the following conditions: (a) that of being commanded by a person responsible for his subordinates; (b) that of having a fixed distinctive sign recognizable at a distance; (c) that of carrying arms openly; (d) that of conducting their operations

'combatants are part of the military potential of the enemy and it is therefore always lawful to attack them for the purpose of weakening that potential'.[51]

Persons can also be targeted because they are carrying out military functions. While civilians are, prima facie, immune from being directly targeted in situations of armed conflict,[52] this immunity can be lost if the civilian takes a direct or active part in the hostilities.[53] Thus, civilians become targetable because they are fulfilling a military function. While the exact scope of the term 'direct participation in hostilities' is a contentious and much debated concept,[54] generally speaking, direct participation in hostilities can be taken to mean commission of any hostile act, 'understood to be acts which by their nature and purpose are intended to cause actual harm to the personnel and equipment of the armed forces'.[55] One of the Commentaries to the Additional Protocols expands on direct participation in hostilities:

> it is clear that civilians who personally try to kill, injure or capture enemy persons or to damage material are directly participating in hostilities. This is also the case of a person acting as a member of a weapons crew, or one

---

in accordance with the laws and customs of war; (3) Members of regular armed forces who profess allegiance to a government or an authority not recognized by the Detaining Power; [and] (6) Inhabitants of a non-occupied territory, who on the approach of the enemy spontaneously take up arms to resist the invading forces, without having had time to form themselves into regular armed units, provided they carry arms openly and respect the laws and customs of war.'

[51] Marco Sassòli and Olson, 'The Relationship Between International Humanitarian and Human Rights Law Where it Matters: Admissible Killing and Internment of Fighters in Non-International Armed Conflicts' (2008) 90 *IRRC* 599, 606.

[52] Article 51(1)–(2), API.

[53] In Protocol I, DPH is outlined in Article 51(3) providing that 'civilians shall enjoy the protection afforded by this Section, unless and for such time as they take a direct part in hostilities'.

[54] For an overview of the differing approaches to the notion of direct participation in hostilities, see Emily Crawford, *Identifying the Enemy: Civilian Participation in Armed Conflict* (Oxford University Press 2015), specifically, Chapter 3.

[55] *AP Commentary* (n 25), 618. The commentary on the DPH provision in Article 13(3) of APII is essentially identical to that of Article 51(3) in API and is based on the notion of 'acts of war that by their nature or purpose struck at the personnel and *matériel* of enemy armed forces', and would likely include preparation for and return from combat activities.

> providing target information for weapons systems intended for immediate use against the enemy such as artillery spotters or members of ground observer teams.[56]

Bothe et al also include 'preparation for combat … [for example] direct logistic support for units engaged directly in battle such as the delivery of ammunition to a firing position'[57] within the scope of direct participation in hostilities.[58] Operationalizing the principle of distinction with regard to individuals can be achieved for example by active participants in the hostilities wearing a uniform or a fixed distinctive emblem and carrying their arms openly.[59]

4. REMOTE WARFARE AND THE PRINCIPLE OF DISTINCTION: THE UNIQUE BENEFITS AND PROBLEMS FOR RESPECTING THE PRINCIPLE OF DISTINCTION IN THE PRACTICE OF REMOTE WARFARE

Given the primacy of the principle of distinction to IHL, and its centrality to targeting decisions, what happens when the principle of distinction becomes difficult, even impossible, to observe? Remote warfare—warfare conducted at a significant remove from active hostilities—can often complicate distinction assessments. The common indicia used to assess something as having a military purpose—for instance, the open

---

[56] *New Rules* (n 41) 303.

[57] Ibid.

[58] However, the Bothe Commentary (n 41) excludes from DPH 'civilians providing only indirect support to the armed forces, such as workers in defence plants or those engaged in distribution or storage of military supplies in rear areas' as such persons 'do not pose an immediate threat to the adversary' (303).

[59] Hans-Peter Gasser and Knut Dörmann, 'Protection of the Civilian Population' in Fleck (n 42) 233. Note however that neither the Conventions specifically, and the law of war more generally, prescribe the wearing of a uniform as a requirement for combatant status; the need to wear a fixed, distinctive sign visible at a distance (under Article 4A(4), Convention III) does not mean one must wear a 'complete head-to-toe outfit that one normally associates with regular armed forces … states are free to choose their armed forces uniform, so long as it is readily distinguishable from the enemy and civilians'. See William Ferrell, 'No Shirt, No Shoes, No Status: Uniforms, Distinction, and Special Operations in International Armed Conflict' (2003) 178 Mil L Rev 94, 106.

carriage of arms or the wearing of a uniform—can sometimes be difficult to ascertain from a live video feed being captured by a drone hovering at 30,000 feet over a target. Likewise, a computer network attack against a military base might have originated from a civilian hacker looking to cause mischief, rather than a direct participant in an armed conflict aiming to inflict military damage.

The difficulty in deciphering the exact character and nature of the source of the attack (or the subject of the attack) can make the appropriate, lawful response difficult. By the same token, however, the technological precision that often accompanies remote warfare can provide parties to a conflict the opportunity to more carefully observe the principle of distinction. The next section of this chapter will thus examine these different kinds of remote warfare, and specifically analyze the benefits and drawbacks that come with remote means and methods of warfare and the obligation to distinguish between the civilian and the military in armed conflict.

(a) Drone Warfare

(i) What is drone warfare?

At this stage of the chapter, it is useful to explain, even in brief detail, something of the modes of remote warfare examined herein, for the purposes of exploring their relationship to distinction assessments, and how remote warfare and distinction assessments interact. Drones, also known as unmanned aerial vehicles (UAVs), are unmanned aircraft capable of sustained flight, either autonomously or otherwise under remote-control by a pilot.[60] Drones and drone warfare have come to especial prominence in the last 15 years due to their usage in the conflicts in Afghanistan and throughout the world against al-Qaeda and associated forces. The use of drones is commonplace in modern armed

---

[60] The US Department of Defense defines a drone as a 'powered, aerial vehicle that does not carry a human operator, uses aerodynamic forces to provide vehicle lift, can fly autonomously or be piloted remotely [and carries] a lethal or nonlethal payload'. Department of Defence, *Dictionary of Military and Associated Terms*, 331 Joint Publication 1-02 (2011), 577. The Harvard Manual on International Law Applicable to Air and Missile Warfare makes the distinction between drones that are equipped with weapons, dubbed unmanned combat aerial vehicles, and those drones do not and are not capable of carrying weapons, known as unmanned aerial vehicles. Article 1, Harvard Manual.

conflict: most reports place the current US drone fleet at over 7,000;[61] another 40 States have either acquired or are planning to acquire UAVs (both unarmed and armed) for their own militaries.[62]

Drones can range from palm-sized[63] to the size of small passenger aircraft[64] and larger.[65] The larger, armed drones, have a flight range of up to 1,000 nautical miles, and can stay aloft at a maximum ceiling of 50,000 feet for over 24 hours before needing to refuel. Armed drones can be equipped with payloads such as laser-guided air-to-surface missiles, with current research investigating the possibility of air-to-air weapons capabilities.[66] Drones are also equipped with high-resolution cameras that enable the pilots and sensor operators to observe proceedings on the ground, and to relay the visuals captured back to analysts for assessment. Thus, a pilot and sensor operator, located at an air force base outside of Las Vegas,[67] can pilot a UAV from an airfield in Saudi Arabia[68] or

---

[61] Stanford Law School International Human Rights and Conflict Resolution Clinic and New York University School of Law Global Justice Clinic, *Living Under Drones: Death, Injury and Trauma to Civilians from US Drone Practices in Pakistan* (2012), accessed 26 April 2017 at http://www.livingunderdrones.org/report/, 8; Spencer Ackerman and Noah Shachtman, 'Almost 1 in 3 US Warplanes is a Robot', *Wired*, 9 January 2012, accessed 26 April 2017 at http://www.wired.com/2012/01/drone-report/.

[62] Philip Alston, *Study on Targeted Killings*, Report of the Special Rapporteur on Extrajudicial, Summary or Arbitrary Executions, UN Doc A/HRC/14/24/Add.6, 28 May 2010, 9.

[63] The Black Hornet, billed as the world's smallest military-grade spy drone, weighs 16 grams and is four inches in length. It is unarmed, and is in current use with British soldiers stationed in Afghanistan (see Spencer Ackerman, 'Palm-Sized Nano-Copter is the Afghanistan War's Latest Spy Drone', *Wired*, 4 February 2013, accessed 26 April 2017 at http://www.wired.com/2013/02/black-hornet-nano/).

[64] The Predator drone measures 27 feet in length – see further specifications at the US Air Force website, accessed 26 April 2017 at http://www.af.mil/AboutUs/FactSheets/Display/tabid/224/Article/104469/mq-1b-predator.aspx.

[65] The Reaper drone measures 36 feet in length – further specifications are listed at the US Air Force website, accessed 26 April 2017 at http://www.af.mil/AboutUs/FactSheets/Display/tabid/224/Article/104470/mq-9-reaper.aspx.

[66] 'MQ-9 Reaper', Global Security, accessed 26 April 2017 at http://www.globalsecurity.org/military/systems/aircraft/mq-9.htm.

[67] Reports indicate that Creech Air Force Base in Nevada, USA, support a number of remote drone sorties – see further Chris Woods, 'CIA's Pakistan Drone Strikes Carried Out By Regular US Air Force Personnel', *The Guardian*, 14 April 2014, accessed 26 April 2017 at http://www.theguardian.com/world/2014/apr/14/cia-drones-pakistan-us-air-force-documentary/print.

Turkey,[69] and observe and attack sites or persons in Yemen,[70] Pakistan[71] or Afghanistan.[72]

(ii) The benefits of drones for observing the principle of distinction
It must first be acknowledged that there has been widespread criticism levelled at States in their use of drones as a weapon of war, with the United States the primary target of such criticism, mainly in relation to their broad targeting policy.[73] A report in *The New York Times* in May 2012 revealed that the criteria for the selection of targets for drone strikes included 'all military-age males in a strike zone'.[74] An administration source justified this broad approach (which classifies all 'people in an area of known terrorist activity, or found with a top al-Qaeda operative'[75] as lawful targets), by stating that 'Al Qaeda is an insular, paranoid organization—innocent neighbours don't hitchhike rides in the back of trucks headed for the border with guns and bombs'.[76] Such a broad

---

[68] Micah Zenko and Emma Welch, 'Where the Drones Are: Mapping the Launch Pads for Obama's Secret Wars', *Foreign Policy*, 29 May 2012, accessed 26 April 2017 at http://foreignpolicy.com/2012/05/29/where-the-drones-are/.

[69] Ibid.

[70] See reports on drone strikes in Yemen at The Bureau of Investigative Journalism, accessed 11 May 2017 at https://www.thebureauinvestigates.com/projects/drone-war/charts?show_casualties=1&show_injuries=1&show_strikes=1&location=yemen&from=2004-1-1&to=now.

[71] See reports on drone strikes in Pakistan at The Bureau of Investigative Journalism, available at https://www.thebureauinvestigates.com/projects/drone-war/charts?show_casualties=1&show_injuries=1&show_strikes=1&location=pakistan&from=2004-1-1&to=now.

[72] See reports on drone strikes in Afghanistan at The Bureau of Investigative Journalism, available at https://www.thebureauinvestigates.com/projects/drone-war/charts?show_casualties=1&show_injuries=1&show_strikes=1&location=afghanistan&from=2015-1-1&to=now.

[73] See *Living Under Drones* (n 61) 31; Columbia Law School Human Rights Clinic and the Centre for Civilians in Conflict, *The Civilian Impact of Drones: Unexamined Costs, Unanswered Questions* (2012), accessed 26 April 2017 at http://web.law.columbia.edu/sites/default/files/microsites/human-rights-institute/files/The%20Civilian%20Impact%20of%20Drones.pdf; and Kevin Jon Heller, '"One Hell of a Killing Machine": Signature Strikes and International Law' (2013) 11 J Intl Crim Just 89.

[74] Jo Becker and Scott Shane, 'Secret "Kill List" Proves a Test of Obama's Principles and Will', *The New York Times*, 29 May 2012.

[75] Ibid.

[76] Ibid. The targeting criteria used by the US also seems to change depending on who is carrying out the strikes—the military or the CIA—with one US Congressional aide stating that, for the CIA, 'their standards of who is a

targeting policy is indeed problematic, in that such persons are being targeted for reasons of geographical location or who they are found in the company of, rather than concrete evidence of direct participation in hostilities.

However, that the *policy* underpinning the use of drones is problematic does not automatically mean that the *use* of drones per se is problematic. Indeed, drones theoretically offer the potential for better compliance with the principle of distinction. As noted by Blank and Noone, drones can 'loiter over a target for hours, even days, providing a much more refined assessment of who is in the area and at what times—information critical to minimizing collateral damage'.[77] Drones, equipped with sophisticated high-resolution cameras and sensor technologies, allow for those engaging in surveillance and targeting to spend a considerable amount of time assessing and evaluating the conditions prevailing in situ prior to engaging in any attack.[78] Such a luxury of time to analyse and assess, in real time, intelligence on the ground is arguably immensely conducive to ensuring that all targeting decisions are compliant with the principle of distinction.

Furthermore, the remoteness of the pilots arguably adds an additional safeguard to the proceedings. As the pilots and sensor operators are thousands of miles from the hostilities and in no danger of being targeted themselves, they are relieved of the threat and pressure of an active combat situation. As Peter Singer has noted:

> The unmanning of the operation also means that the robot can take risks that a human wouldn't otherwise, risks that might mean fewer mistakes … [the] removal of risk … allows decisions to be made in a more deliberate manner than normally possible. Soldiers describe how one of the toughest aspects of fighting in cities is how you have to burst into a building and, in a matter of milliseconds, figure out who is an enemy and who is a civilian and shoot the ones that are a threat before they shoot you, all the while avoiding hitting any civilians … unmanned systems can remove the anger and emotion from the

---

combatant are different' to the standards of the Department of Defence and the US military (see Ken Dilanian, 'Debate Grows Over Proposal For CIA To Turn Over Drones To Pentagon', *The Los Angeles Times*, 11 May 2014, accessed 26 April 2017 at http://www.latimes.com/world/middleeast/la-fg-yemen-drones-20140511-story.html).

[77] Laurie Blank and Gregory Noone, *International Law and Armed Conflict: Fundamental Principles and Contemporary Challenges in the Law of War* (Wolters Kluwer 2013) 533.

[78] See Michael Lewis and Emily Crawford, 'Drones and Distinction: How IHL Encouraged the Rise of Drones' (2012–2013) 44 Georgetown J Intl L 1127, 1153–4.

> humans behind them. A remote operator isn't in the midst of combat and isn't watching his buddies die around him as his adrenaline spikes; he can take his time and act deliberately in ways that can lessen the likelihood of civilians being killed.[79]

The luxury of time and safety for those engaging in the hostilities may thus create an environment better conducive to making distinction assessments. As such, remote warfare as conducted by drones offers the potential for greater compliance with the principle of distinction, due to the unique qualities both of the drone itself, and to the remoteness from active hostilities of the operators and decision-makers.[80]

(iii) The drawbacks of drones for observing the principle of distinction

However, despite these potential advantages offered by drones in conducting lawful, distinction-compliant warfare, it has often been the case that drone attacks have *not* been compliant with the principle of distinction. Indeed, the purported benefits of drones—their precision and the clarity that their remove from active hostilities brings—have often proven to be the factors that contribute to their violation of the principle of distinction. That is to say, the remote nature of drone attacks can be (and has been) directly connected to indiscriminate attacks against both military and civilian targets. This has been manifested in a number of ways.

First, the clarity and precision offered by drones by way of their 'real-time' video feed of targets is undermined by the very real technological problem of 'latency'—the time delay between activities observed and videoed at the target site and the arrival of that video image via satellite to the pilots.[81] The gap between the action as observed by the drone and the action as seen by pilots and sensor operators can be enough to result in the target having fled to a civilian-heavy area. Indeed, as noted in a *New York Times* report in 2012, 'senior operatives with al-Qaeda in the Arabian Peninsula told a Yemeni report that if they hear

---

[79] Peter Singer, 'Military Robots and the Laws of War', *The New Atlantis*, Winter 2009, accessed 26 April 2017 at http://www.thenewatlantis.com/publications/military-robots-and-the-laws-of-war.

[80] See Michael Schmitt, 'Drone Attacks under the Jus ad Bellum, and Jus in Bello: Clearing the "Fog of Law"' (2013) 13 YBIHL 311, 320–21.

[81] See *Living Under Drones* (n 61) 9; Rob Blackhurst, 'The Air Force Men Who Fly Drones In Afghanistan By Remote Control', *The Telegraph*, 24 September 2014.

an American drone overhead, they move around as much as possible'.[82] The time-delay between identifying the target to be attacked and actually attacking that target is more pronounced in remote warfare, and thus opens up the possibility for the eventual attack to be non-compliant with the principle of distinction.[83]

Furthermore, an additional problem arises regarding remoteness and drone strikes. While the video feed of the object or person under surveillance is clearly of a high quality, it may not be high quality enough for the purposes of an accurate distinction assessment. That is to say, even with a high quality video feed, there have been examples of misidentification of persons, who have been targeted because they *appear*, wrongfully it turns out, to be engaged in hostilities. This has been seen in a number of instances in the last 15 years. In the first of these instances, in February 2002, three men were killed in a drone attack in Khost, in Afghanistan's Paktia province. These men were spotted by Predator drone operators, who noticed that two of the men were 'acting with reverence'[84] towards a third 'tall man'.[85] The operators, and their superiors, came to the conclusion that the tall man was Osama bin Laden.[86] Moreover, the three men were found at a former *mujahedeen* base known as Zhawar Kili.[87] The order to attack was given, and the men were killed by Hellfire missiles fired from the Predator. In reporting the attack, US government officials stated that they could not be sure who precisely they killed, but they were 'convinced it was an appropriate

---

82 Mark Mazetti, 'The Drone Zone', *The New York Times*, 6 July 2012.

83 Furthermore, the precision of the munitions used has also been disputed, with reports stating that the blast radius of a Hellfire missile—the kinds of munitions usually attached to Reaper drones, can be anywhere from 15 to 20 metres, not including the amount of shrapnel that is also discharged following an attack. *Living Under Drones* (n 61) 10. This shortcoming in drone warfare is not inherently connected to its remote nature however, but rather a shortcoming of the weaponry itself, regardless of its method of deployment.

84 Quoted in John Sifton, 'A Brief History of Drones', *The Nation*, 27 February 2012, accessed 26 April 2017 at http://www.thenation.com/article/166124/brief-history-drones.

85 Ibid.

86 FBI reports had calculated bin Laden's height at 195.6 cms (6 ft 5 in). Federal Bureau of Investigation, 'Most Wanted Terrorists: Usama bin Laden', accessed 11 May 2017 at https://vault.fbi.gov/osama-bin-laden/Osama%20Bin%20Laden%20Part%2001%20of%2002/view.

87 Zhawar Kili was known to US authorities; in 1998, President Bill Clinton had authorized missile attacks against Zhawar Kili. See Brian Glyn Williams, *Afghanistan Declassified: A Guide to America's Longest War* (University of Pennsylvania Press 2012) 182.

target'[88] and that 'the initial indications afterwards would seem to say that these are not peasant people up there farming'.[89] However, it was later uncovered that the 'tall man' and his associates were local villagers, who had gone to the long-abandoned[90] campsite to forage for scrap metal. One of the men, Daraz Khan, was tall (at 5ft 11in, or 180cms), but was at least five to seven inches (approximately 12 to 17 cms) shorter than bin Laden. No connection between Daraz Khan and his companions and al-Qaeda could be made out post mortem.[91]

A similar incident, in this instance a remote attack conducted by Apache helicopters, took place in Iraq, on the first day of Operation Ilaaj in Baghdad in 2007. US troops on the ground in an area known as New Baghdad had been under fire all morning from rocket-propelled grenades (RPGs) and small arms;[92] Apaches were hovering over the area, providing air cover and reconnaissance for ground troops, using video feeds taken from cameras mounted on the helicopters.[93] The crews observed what they thought were around a dozen insurgents, some armed with machine guns, including a man they believed was carrying an RPG launcher. The crews requested and were given permission to fire on the people; in a series of attacks, over a dozen people were killed or wounded.[94] In the aftermath of the attack, it was discovered that the RPG launcher was actually a telephoto lens carried by a Reuters journalist; while the men in his company were armed, it did not appear that any of them were insurgents, or had been participating in the hostilities. Certainly none were aiming or firing at anyone.[95]

---

88 Quoted in John Burns, 'A Nation Challenged: The Manhunt; US Leapt Before Looking, Angry Villagers Say', *The New York Times*, 17 February 2002.

89 Ibid.

90 Ibid.

91 Ibid.

92 Tom Cohen, 'Leaked Video Reveals the Chaos of Baghdad Attack', *CNN*, 7 April 2010, accessed 26 April 2017 at http://edition.cnn.com/2010/WORLD/meast/04/06/iraq.journalists.killed/.

93 Reports suggest that the helicopters could have been around 800 metres, or over 2,500 feet, away from the men attacked. See Tom Cohen, ibid.

94 See further the report on the attack, issued by the Pentagon, *Investigation Into Civilian Casualties Resulting From An Engagement On 12 July 2007 In The New Baghdad District Of Baghdad, Iraq*, accessed 26 April 2017 at http://i2.cdn.turner.com/cnn/2010/images/04/06/6–2nd.brigade.combat.team.15-6.investigation.pdf.

95 Cohen (n 92). The men being observed were not seen aiming weapons, nor was there any *post facto* evidence to suggest that the men on the ground were actually linked to the insurgency.

This example, and that of the drone attack in Zhawar Kili, is demonstrative of one of the major issues regarding remote warfare and respecting the principle of distinction. Making the distinction between civilian and military, between targetable and non-targetable, still to some degree depends on a visual assessment; if someone looks like they are taking a direct part in hostilities, then they can be targeted. However, if one is unable to accurately assess whether a person is taking direct part—in this case, because one is too far removed to accurately identify the precise nature of the target—how can a distinction assessment be made? The scope for misidentification seems considerable—indeed, this was noted in the Pentagon report into the Baghdad attack:

> It must be noted that details which are readily apparent when viewed on a large video monitor are not necessarily apparent to the Apache pilots during a live-fire engagement. First of all, the pilots are viewing the scene on a much smaller screen than I had for my review ... two individuals can be seen carrying cameras with large telephoto lenses slung from their right shoulders ... the cameras could easily be mistaken for ... rifles, especially as neither cameraman is wearing anything that identifies him as media or press.[96]

The very remoteness of the targeting decision-makers thus seems a hindrance to making accurate distinction assessments. Arguably, a soldier in the field in close proximity to the hostilities would not mistake a camera for a rifle.

However, even if there seem to be strong indicia of direct participation in hostilities, and thus lawfulness of targeting, the distinction assessment can still be incorrect. That is to say, if a person is carrying a weapon—as was seen in the Baghdad incident—does this necessarily mean that they can be targeted? The law of armed conflict would suggest not: the mere fact that a person is carrying arms is insufficient grounds to support the contention that such a person is directly participating in hostilities, and thus liable for targeting. Indeed, the ICTY in *Simić* refuted the argument that 'the possession of weapons, in itself, creates a reasonable doubt as to the civilian status'[97] of an individual. This very point was made also by a Yemeni intelligence official who noted that carrying weapons was commonplace in Yemen: 'Every Yemeni is armed ... how can they differentiate between suspected militants and armed Yemenis?'[98] A

---

[96] Pentagon Report (n 94) 2.

[97] *Prosecutor v Simić et al*, Case No. IT-95-9-T, Judgment, 17 October 2003, § 659.

[98] Adam Entous, Siobhan Gorman and Julian E Barnes, 'US Relaxes Drone Rules', *The Wall Street Journal*, 26 April 2012.

person carrying a rifle may seem to be targetable at 50,000 feet; but at 10 feet, they may simply be a civilian hunter or farmer. Thus, the remoteness of the means and methods of warfare can be problematic to a distinction-compliant targeting assessment.

(b) Cyber Warfare

(i) What is cyber warfare?

As with the above section on drones, this next section of this chapter benefits from having a very brief backgrounder on cyber warfare, and how it can be conducted. Academics and practitioners[99] in the field have categorized cyber warfare in two broad groupings: computer network attack (CNA) and computer network exploitation (CNE). A computer network attack is defined as 'operations to disrupt, deny, degrade, or destroy information resident in computers and computer networks, or the computer and networks themselves',[100] while computer network exploitation is defined as 'the ability to gain access to information hosted on information systems and the ability to make use of the system itself'.[101] Cyber warfare is a form of information warfare—the use of technology and other means and methods of warfare 'to influence, disrupt, corrupt, or usurp the decision-making of adversaries and potential adversaries', while using those same means and methods to protect one's own decision-making and war-fighting capabilities.[102]

---

[99] See Heather Harrison Dinniss, *Cyber Warfare and the Laws of War* (Cambridge University Press 2012) 4; Michael Schmitt, 'Computer Network Attack and the Use of Force in International Law: Thoughts on a Normative Framework' (1999) 37 Colum J Transnatl L 885; Myriam Dunn Cavelty, 'Cyberwar' in George Kassimeris and John Buckley (eds), *The Ashgate Research Companion to Modern Warfare* (Ashgate 2010); US Joint Chiefs of Staff, Joint Publication 3-13, *Information Operations*, 13 February 2006, accessed 11 May 2017 at http://www.information-retrieval.info/docs/jp3_13.pdf.

[100] Background Document, *Expert Meeting on Direct Participation in Hostilities under International Humanitarian Law*, 2 June 2003 (hereinafter DPH Background Document 2003), 15.

[101] Ibid 15.

[102] US Joint Chiefs of Staff, Joint Publication 1-02, *Department of Defense Dictionary of Military and Associated Terms*, 8 November 2010 (as Amended Through 15 March 2014), accessed 26 April 2017 at http://www.dtic.mil/doctrine/new_pubs/jp1_02.pdf, 127. See also Michael Schmitt, 'Computer Network Attack: The Normative Software' (2001) 4 *YBIHL* 53, 54.

Cyber warfare involves its own unique 'weapons', and, like any weapon, they can be deployed discriminately or indiscriminately.[103] Cyber weapons can target specific computer systems, render ineffective satellite or other communications systems, destroy or manipulate codes and other software in operating systems and otherwise hamper, hinder, or eliminate outright an adverse party's ability to actively conduct hostilities.[104] The key element in cyber warfare is that the weapons are cyber-based; they are electronic weapons, software-based weapons. While they can produce kinetic 'real-world' effects, they are weapons that exist solely in the form of computer code.

Cyber warfare is, perhaps, the quintessential example of remote warfare—the attacking party need only have a computer and internet access in order to launch their attack, which can be undertaken from anywhere in the world. There are, theoretically, no geographical limitations on the conduct of cyber warfare; the attacker could be within 50 feet of the target, or on the other side of the planet. This remoteness brings with it its own benefits and drawbacks for compliance with the principle of distinction.

(ii) The benefits of cyber warfare for respecting the principle of distinction

That the principle of distinction applies to cyber war is generally undisputed; the *Tallinn Manual on the International Law Applicable to Cyber Warfare* affirms, in Rule 31, that the principle of distinction applies to cyber-attacks.[105] The Tallinn Manual, which defines a cyber attack as a 'cyber operation, whether offensive or defensive, that is reasonably expected to cause injury or death to persons or damage or destruction to objects',[106] thus prohibits any cyber attack against civilians

---

[103] For a more detailed description and analysis of cyber weaponry, see generally Jeffrey Carr, *Inside Cyber Warfare* (2nd edn, O'Reilly 2012) 141–60; Jason Andress and Steve Winterfeld, *Cyber Warfare: Techniques, Tactics and Tools for Security Practitioners* (Elsevier 2011) 104–68; Sean Watts, 'Combatant Status and Computer Network Attack' (2010) 50 Vir J Intl L 391, 397–410; Arie Schaap, 'Cyber Warfare Operations: Development and Use under International Law' (2009) 64 Air Force L Rev 121, 134–9.

[104] See, for example, so-called 'logical weapons', which scan for vulnerabilities in networks and systems of adversaries, access such systems, and either retrieve information from them or otherwise disrupt such systems, to render them inoperative or defective. See Andress and Winterfeld (n 103) 83.

[105] Michael Schmitt (ed), *The Tallinn Manual on the International Law Applicable to Cyber Warfare* (Cambridge University Press 2013) Rule 31.

[106] Tallinn Manual, Rule 30.

or civilian objects,[107] and reaffirms that the civilian population as such, as well as individual civilians, shall not be the object of cyber attack.[108]

Cyber warfare certainly has the potential to be remarkably compliant with the principle of distinction. The scientific precision of cyber weapons can mean that parties to a conflict engaging in cyber attacks can attack only military computers or computer-dependent systems, while leaving civilian systems untouched. An example of this kind of precision can be seen in the 2010 Stuxnet attack on the Iranian nuclear facility in Natanz. The Stuxnet worm was a 'logical weapon'—a piece of malware designed to disrupt a network.[109] Stuxnet was very carefully constructed to target only the computers operating within the Natanz facility, and only in very specific ways: the worm caused the IR-1 centrifuges used for enriching uranium to spin at higher or lower rates than was within optimal operations limits; the result was reported mechanical damage to some centrifuges, and sub-optimal performance of other centrifuges (which prevented the enrichment of uranium).[110] Examination of the Stuxnet virus revealed that 'though the virus had been designed to propagate and spread fairly indiscriminately within a network, it had been coded to only execute its payload where specific conditions were fulfilled that would indicate it was on the targeted system'.[111] Cyber weapons can thus be very specifically engineered, and therefore exceptionally compliant with the principle of distinction: 'an attacking coder could easily set the conditions of the payload deployment to ensure that

---

107 Tallinn Manual, Commentary to Rule 31, 112.

108 Tallinn Manual, Rule 32—unless such civilians are DPH (rule 35).

109 See further Paulo Shakarian, Jana Shakarian and Andrew Ruef, *Introduction to Cyber-Warfare: A Multi-Disciplinary Approach* (Elsevier/Syngress 2013), specifically Chapter 13, which examines the Stuxnet attack in detail. 'Stuxnet' is the name that was given to the worm by hackers who had discovered the worm once it infected computers and networks outside of Natanz; the codename given to the worm by its purported creators—the US and Israel—was 'Olympic Games'. A detailed analysis of the Olympic Games program can be found in David Sanger, *Confront and Conceal: Obama's Secret Wars and Surprising Use of American Power* (Crown 2012) 188–225.

110 See Shakarian et al (n 109) 224–35, for an overview of the Stuxnet attack; see also Paulo Shakarian, 'Stuxnet: Cyberwar Revolution in Military Affairs', *Small Wars Journal*, 14 April 2011, accessed 26 April 2017 at http://smallwarsjournal.com/jrnl/art/stuxnet-cyberwar-revolution-in-military-affairs.

111 Harrison Dinniss (n 99) 203–4. That Stuxnet got out to the web outside Natanz was due to human error, and not a 'fault' per se in the virus.

the system being attacked included a piece of software or a system file that is unique to the military systems targeted'.[112]

(iii) The drawbacks of cyber warfare for respecting the principle of distinction

That cyber weapons can be engineered to be compliant with distinction does not, however, necessarily overcome an additional problem regarding cyber warfare and distinction. Due to the interconnected nature of the internet and cyberspace, differentiating between civilian and military objects becomes more complicated. As Geiß and Lahmann point out, 'because of the systemic technological set-up of cyberspace in times of armed conflict, basically every cyber installation—possibly even cyberspace as such—potentially qualifies as a military objective'.[113] The internet and World Wide Web, and the hardware and software that create the web, are used by the military in the same way—indeed, at the same time—as civilians; the 'systemic dual nature'[114] of cyberspace makes it difficult, perhaps even 'impossible'[115] to separate out the military from the civilian. In targeting the military aspect of the Web, one would be also targeting the civilian; the destruction or incapacitation of the military element would necessarily mean the destruction or incapacitation of the civilian element of the Web. As noted by Geiß and Lahmann, military software and codes:

> ... travelling in cyberspace are split up into various data packages, all of which may travel via different (civilian) channels and typically traverse various civilian systems when travelling through cyberspace ... even in a single cyber attacks, a wide range of physical cyber infrastructures—namely servers, routers, cables or satellites, as well as software—are used to make effective contributions to military action and would thus qualify as legitimate military targets.[116]

Indeed, estimates from 2010 indicated that 98 per cent of US government communications, including classified communications, were transmitted

---

[112] Ibid 257.

[113] Robin Geiß and Henning Lahmann, 'Cyber Warfare: Applying the Principle of Distinction in an Interconnected Space' (2012) 45 Israel L Rev 381, 383; see also Eric Talbot Jensen, 'Cyber Warfare and Precautions Against the Effects of Attacks' (2010) 88 Texas L Rev 1522, 1542.

[114] Geiß and Lahmann, ibid.

[115] Ibid.

[116] Ibid 385–6.

over civilian owned and civilian operated networks and systems.[117] The profound interconnectivity of military and civilian in cyberspace makes respecting the principle of distinction noticeably difficult.

Furthermore, the very nature of participation in cyber warfare seems at odds with one of the ways in which the principle of distinction is operationalized—that is, the act of a member of the armed forces (or other direct participant) distinguishing himself or herself from civilians and the civilian population. As noted above, one of the ways in which such a distinction is made is by the direct participant wearing a fixed distinctive sign recognizable at a distance, or carrying their arms openly. These requirements were introduced into the law of armed conflict: 'in an era when warfare involved a certain amount of physical proximity between opposing forces. For the most part, combatants could see one another and hence distinguish between combatant and non-combatant.'[118] However, cyber warfare does not require such proximity between participants—one does not physically see one's adversary—but the principle of distinction remains paramount. How does one respect the principle of distinction in such instances? How does the military know that the attack comes from another military source, and is not merely the work of a criminal or a civilian malcontent, seeking to attack a military network for notoriety?[119] A cyber-attack launched by a criminal or civilian 'hacktivist' against the military would be most appropriately dealt with by domestic law enforcement rather than the law of armed conflict; an attack against such a person would likely be a violation of IHL. For those in the armed forces responding to a cyber-attack, respecting and observing the principle of distinction is thus no easy task.[120]

Indeed, civilian computers and computer software may be unwittingly involved in cyber attacks in armed conflict. For example, botnets are

---

[117] Michael McConnell, Former Director of National Intelligence, Keynote Address at the Texas Law Review Symposium, *Law at the Intersection of National Security, Privacy, and Technology*, 4 February 2010.

[118] Harrison Dinniss (n 99) 145.

[119] The US DoD, as well as US government networks, are constantly subject to hacking attempts—for example, 2012 reports from the Pentagon stated that it was subject to near ten million cyber attacks per day. Zachary Fryer-Biggs, 'U.S. Military Goes on Cyber Offensive', *Defence News*, 24 March 2012, accessed 26 April 2017 at http://www.space4peace.org/articles/us_cyber_offensive.htm.

[120] See Harrison Dinniss (n 99) 145–9, for an analysis of the issues raised by the principle of distinction and cyber war, specifically in the context of the distinction between direct participants (civilian or military) and non-direct participants.

networks of hijacked or hacked computers, which can have been accessed and exploited by a hacker, without the knowledge of the owner.[121] These 'zombie' computers are then controlled by the hacker to perform tasks such as distribution of spam emails or the launching of denial of service attacks.[122] Civilians could thus find that their computers have been used to conduct cyber attacks against military cyber-infrastructure or material, thus unwittingly directly participating in an armed conflict. Indeed, botnets were used in the cyber attacks on Estonia in 2007 and Georgia in 2008.[123] As such, civilians may find themselves liable for attack without ever having intentionally participated in the hostilities. As with drone warfare, the remoteness of cyber warfare may actually make distinction-compliant conduct in armed conflict harder, rather than easier, to observe.

5. DISTINCTION AND REMOTE WARFARE: IS THE 'REMOTENESS' OF THE WARFARE ESSENTIAL TO COMPLIANCE WITH THE PRINCIPLE OF DISTINCTION?

Finally, it is interesting to reflect on the degree to which the 'remoteness' of remote warfare is fundamental to the ability (or not) to respect the principle of distinction. Is there anything *inherent* to these forms of remote warfare that makes them fundamentally more discriminate means or methods of warfare, capable of respecting the principle of distinction better than other means or methods?

To some extent, there is nothing unique to drones and cyber warfare from the perspective of IHL. They are just newer delivery methods for attacks, updated versions of the projectiles, bombs and other munitions that have always been a part of warfare. That these weapons use computers and code does not necessarily make them any different to those weapons which use gunpowder or bullets—all of the weapons share the fundamental component of being delivery systems for attacks which cause injury or death to persons, and destruction to property.

---

[121] Heli Tiirmaa-Klaar, 'Botnets, Cybercrime and National Security' in Heli Tiirmaa-Klaar, Jan Gassen, and Elmar Gerhards-Padilla, *Botnets* (SpringerBriefs in Cybersecurity, 2013) 3.

[122] Denial of service attacks aim is to 'flood the target of the attack with an abnormally large amount of legitimate traffic to the effect of rendering it inaccessible to other users'. Shakarian, Shakarian and Ruef (n 109) 12–13.

[123] See Gadi Evron, 'Battling Botnets and Online Mobs' (2008) 9 GJIA 121, 124; Heli Tiirmaa-Klaar (n 121) 16–21.

Furthermore, the 'remoteness' of these weapons is not especially unique or unusual; the advance of technology over the last 150 years has meant that adversaries have been able to attack one another from farther away than previously imaginable. A drone strike from 50,000 feet in the air is simply a different version of a bomb dropped from 30,000 feet from a WW2 B-29 aircraft, or indeed a projectile fired from a WW1 tank or cannon several miles from its target. The remoteness of these new technologies of war is simply a continuation of an ongoing evolution of remote warfare.

That being said, both drones and cyber warfare share a common quality to their remoteness that is unique to the process of remote warfare, and one that is particularly relevant for respecting the principle of distinction. Drones and cyber warfare are unique in allowing the attacker to conduct their attack from anywhere in the world, far removed from the location of the attack—indeed, far removed from any active hostilities. Drone pilots and cyber attackers can engage in hostilities and launch attacks without the fear that they may be subject to injury or death due to their participation in hostilities. The nature of drone and cyber warfare—as discussed above—also means that these 'cubicle warriors'[124] have a great deal of time in which to assess their targets and make targeting decisions. As such, these two factors—time and safety—mean that attacking parties can be exceptionally faithful to the principle of distinction.

In this respect, the remoteness of those conducting the attacks is arguably pivotal to their ability to respect the principle of distinction. Drones, with the potential to spend days, even weeks observing a potential target, gathering copious data on the target, and allowing for complex and detailed assessments about the legality of a strike against such a target, without fear of discovery of such surveillance, thus may be exceptionally compliant with the principle of distinction. Indeed, as Daniel Rothenburg notes:

> the link between broad surveillance and complex, integrated data analysis might allow far more accurate determinations of who is and is not a member of a non-state armed group while identifying which civilians are directly participating in hostilities ... where such information is linked with precision killing, one could imagine drone deployment as embodying the promise of

---

[124] Lambèr Royakkers and Rinie van Est, 'The Cubicle Warrior: The Marionette of Digitalized Warfare' (2010) 12 Ethics Inf Technol 289.

> legal war, to direct lethal force only against legitimate targets—combatants and military installations—while protecting civilians.[125]

That being said, neither drones nor cyber warfare exist in a vacuum—while the weapons themselves may offer the potential for strict compliance with the principle of distinction, they still may be used in indiscriminate ways.

6. CONCLUDING THOUGHTS

The principle of distinction places considerable obligations on parties to a conflict; parties to a conflict must ensure that they do not intentionally target civilians or civilian objects which serve no meaningful military purpose or function. Furthermore, parties to a conflict must ensure that they protect civilians and civilian objects within their own control, and as far as possible protect such civilians and civilian objects from being impacted by attacks from the adversary. In using remote means and methods of warfare, parties to a conflict must comply with the principle of distinction. As explored in this chapter, remotely-operated weapons and remotely-conducted methods of warfare have the potential to be notably compliant with the principle of distinction. However, such means and methods are not always perfectly compliant with the principle of distinction—shortcomings in the technology of the weaponry and methods of warfare exist, which can result in violations of the principle of distinction. As such, it remains incumbent on parties to armed conflicts to ensure that the principle of distinction retains its paramount status in targeting decisions, in choice of weaponry, and in choice of methods of warfare.

---

[125] Daniel Rothenburg, 'Drones and the Emergence of Data-Driven Warfare' in Peter Bergen and Daniel Rothenburg (eds), *Drone Wars: Transforming Conflict, Law, and Policy* (Cambridge University Press 2014) 449.

3. Modern drone warfare and the geographical scope of application of IHL: pushing the limits of territorial boundaries?[1]

Robert Heinsch

1. INTRODUCTION

The 'war' against international terrorism, but also the rapid development of new weapons technologies, has led to a change in the way that hostilities are conducted, especially in the last 10 to 20 years. The times where wars were fought as an inter-state conflict, with man-to-man combat on a clearly defined battleground with the object to obtain territory, seem to be over. Nowadays, more and more fighting activities take place by using remote controlled drones and other comparable weapons systems. As a result, we now witness an increasing distance between the initiator and the target of an attack, for example, the targeted killing of potential terrorists in the mountains of Pakistan and Afghanistan or in the outskirts of Somalia and Yemen by drones which are controlled from an operation center far removed from the targets, such as in the United States, the Sahel-region, or in the United Arab Emirates. The questions that arise for international humanitarian law (IHL) are whether these scenarios have to be seen (1) within the scope of IHL, and if so, whether (2) the existing rules are still able to deal with this type of weapon, or whether we need a reform of the current IHL regime. This chapter will focus primarily on the first question, and examine whether

[1] This chapter is a revised and amended version of a previously published article of the author: Robert Heinsch, 'Unmanned Aerial Vehicles and the Scope of the "Combat Zone": The Geographical Scope of Application under International Humanitarian Law in Modern Warfare' (2012) 25 J of Intl L of Peace & Armed Conflict 184–92. With agreement of the previous publisher, the article has been expanded and updated in order to include new developments between 2012 and 2016. The author would like to thank his researcher, Mr Abdulahi Abdulrahman Abdalla, LLM, for his invaluable assistance in updating the current chapter and collecting the respective material.

we have to think about expanding the concept of the geographical scope of IHL, or whether the current system is sufficient to cover all situations which are connected with remote warfare in armed conflict situations. Furthermore, it will look briefly at how this issue is related to the application of international human rights law, especially in cases of so-called 'targeted killings', and what the possible future developments in this area could be.

Since the increase in use of unmanned aerial vehicles (UAVs) in the aftermath of 11 September 2001, commentators and NGOs have raised the question whether the use of UAVs is legal under international law.[2] In order to answer this question, it is decisive to know which legal regime is actually applicable, that is whether it is IHL, international human rights law (HRL), or maybe national criminal law that needs to be taken as the appropriate standard. Because drones can be deployed far away from the operator of the weapon, and also in areas which are not directly connected with the actual area of fighting, this is where the question of the geographical scope of application of the law of armed conflict becomes pertinent. The application of IHL offers the possibility for operators of drone attacks to aim at military targets and persons legally if the target is to be seen as a legitimate military objective. This unique opportunity offered by IHL feeds a certain desire—especially from the military side—to extend the geographical scope of the application of IHL

---

[2] See for example B Emmerson, 'Report of the Special Rapporteur on the Promotion and Protection of Human Rights and Fundamental Freedoms while Countering Terrorism', UN Doc A/HRC/25/59, 11 March 2014, para 70; Amnesty International, 'Will I be Next? Drone Strikes In Pakistan', accessed 2 May 2017 at http://www.amnestyusa.org/sites/default/files/asa330132013en.pdf (22 October 2013), International Human Rights and Conflict Resolution Clinic, Stanford Law School & Global Justice Clinic, NYU Law School, 'Living under Drones: Death, Injury, and Trauma to Civilians From US Drone Practices in Pakistan', September 2012; M E O'Connell, 'Unlawful Killing with Combat Drones—A Case Study of Pakistan, 2004–2009' in Simon Bronitt et al (eds), *Shooting to Kill: Socio-Legal Perspectives on the Use of Lethal Force* (Hart Publishing 2012); B Boothby, 'The Law Relating to Unmanned Aerial Vehicles, Unmanned Combat Aerial Vehicles and Intelligence Gathering from the Air' (2011) 24 Humanitäres Völkerrecht—Informationsschriften 81; M W Lewis, 'Drones and the Boundaries of the Battlefield' (2012) 47 Texas Intl L J; R J Vogel, 'Drone Warfare and the Law of Armed Conflict' (2010–2011) 39 Denver J of Intl L & Poly 101; P Stroh, 'Der Einsatz von Drohnen im nicht-internationalen bewaffneten Konflikt' (2011) 24 Humanitäres Völkerrecht—Informationsschriften; D Fleck, 'Unbemannte Flugkörper in bewaffneten Konflikten: neue und alte Rechtsfragen—Kommentar' (2011) 24 Humanitäres Völkerrecht—Informationsschriften 78.

in order to legitimately target objects and persons which might be the orchestrators of, for example, terrorist attacks. The motivation for wanting to extend the geographical scope in these situations—and by this having a legal justification for killing a terrorist suspect, if that person is to be seen as directly participating in hostilities—seems to come from the general uncertainty whether one (or even a couple) of drone strikes against non-state actors alone already qualifies as either an international or non-international armed conflict (NIAC) under IHL.

This is especially obvious because the current requirements for establishing the existence of a NIAC requires 'protracted armed violence' between government authorities and organized armed groups, or between such groups, while an international armed conflict (IAC) traditionally requires the 'use of force in war-like manner' by two States. Consequently, there is the chance that one or two isolated drone strikes would not reach the threshold of a NIAC. Therefore, it would be easier to argue for the operators of drone strikes, who want to use the IHL privilege and therefore being able to legally target other fighters and military objectives, if the geographical scope of an existing armed conflict could be extended to cover also operations outside of the actual 'hot' combat zone.[3] Otherwise, the alternative is that these actions have rather to be seen as 'targeted killings' outside of an armed conflict, which would have to be evaluated according to human rights law.[4]

---

[3] As will be explained below in section 2.2, the current author sees the term 'combat zone' merely as a factual term which does not have any consequences for the legal application of IHL. The correct question one needs to ask is the question whether we have an 'armed conflict' and whether the respective activities are covered by the geographical scope of the respective armed conflict. On the geographical scope of application of IHL, see for example Lewis (n 2) and L R Blank, 'Defining the Battlefield in Contemporary Conflict and Counterterrorism: Understanding the Parameters of the Zone of Combat' (2011) 39 Georgia J Intl & Comp L 1; see also L Arimatsu, 'Territory, Boundaries and the Law of Armed Conflict' (2009) 12 Yearbook of Intl Humanitarian L 157 (2009).

[4] On the topic of targeted killing, cf D Kretzmer, 'Targeted Killing of Suspected Terrorists: Extra-Judicial Executions or Legitimate Means of Defence' (2005) 16 Eur J Intl L 171; N Melzer, *Targeted Killing in International Law* (Oxford University Press 2008); A Margalit, 'Did LOAC Take the Lead? Reassessing Israel's Targeted Killing of Salah Shehadeh and the Subsequent Calls for Criminal Accountability' (2012) 17 J Conflict & Sec L 147.

1.1 The Use of Modern Technology in Recent Conflicts

The phenomenon that we have witnessed since the turn of the century is a growing use of modern technologies in armed conflicts and especially unmanned aerial vehicles.[5] While the invention of remote controlled weapons dates back to the time during and after World War II,[6] and examples of their use as an intelligence gathering tool could also be witnessed during the Vietnam War and the 1991 Gulf War,[7] the real importance in warfare of this special form of weaponry was show-cased shortly after the terrorist attack on the World Trade Center and the ensuing 'War against Terror'.[8]

During the last 20 years the number of states possessing, trading and using military drones has steadily increased.[9] While at the beginning it was mainly the US who made use of this technology, nowadays a handful of states have used drones during an armed conflict, a list which includes Turkey, Iraq and Nigeria, for example.[10] More significantly, the number of states with drones as part of their military arsenal is higher, and includes states such as China, France and India.[11] The use of military drones is not limited to states only; there are several non-state actors, such as Hamas and Hezbollah, who have been involved in developing

[5] On the topic of modern technologies in armed conflict, see D Saxon, *International Humanitarian Law and the Changing Technology of War* (Brill 2013).

[6] M E O'Connell (n 2) 2.

[7] Ibid 3.

[8] Chris Woods, 'The Story of America's Very First Drone Strike' *The Atlantic*, 30 May 2015.

[9] See Elisabeth Bumiller, 'A Day Job Waiting for a Kill Shot a World Away' *New York Times*, 29 July 2012. ('By 2015, the Pentagon projects that the Air Force will need more than 2,000 drone pilots for combat air patrols operating 24 hours a day worldwide. The Air Force is already training more drone pilots—350 last year—than fighter and bomber pilots combined. Until this year, drone pilots went through traditional flight training before learning how to operate Predators, Reapers and unarmed Global Hawks. Now the pilots are on a fast track and spend only 40 hours in a basic Cessna-type plane before starting their drone training'.)

[10] New America, *International Security Database—Worlds of Drones: Military*, accessed 2 May 2017 at http://securitydata.newamerica.net/world-drones.html.

[11] For an overview of the states that are using drones, see Drone Wars UK, Who has Drones?, accessed 2 May 2017 at https://dronewars.net/6-who-has-drones/.

and deploying military drones—albeit of a more rudimentary type.[12] Correlating to this development is the increase of the trade in armed drones. Although the trade of drones in the period between 2010 and 2014 corresponds to 0.3 per cent of the total worldwide trades in arms, the states importing drones increased from 26 to 35 states between 2010 and 2014,[13] with several other states indicating the intention to acquire drones in the near future.[14] Furthermore, the amount of drone bases has seen an increase, particularly in Central Africa, the MENA-region, and in the Horn of Africa, indicating a further entrenching of the use of drones during military and counter-terrorism operations.[15]

Drones have various practical advantages for the military; for example the fact that since they are unmanned the risk of losing a pilot is reduced to literally zero (although of course there is still the possibility that the drone controllers are targeted at their control center). Furthermore, they are much cheaper in comparison to a regular fighter plane, and they usually have a much greater range. They can stay in the air for up to 336 hours gathering information about possible targets or waiting for the right moment to attack the desired object or person.[16] Although they are much slower than normal airplanes and in that regard it is much easier for them to be intercepted by an efficient air defense, this disadvantage seems to be outweighed by the strategic advantages listed above.

1.2 Change of Conflict Type

One reason for the increasing use of unmanned aerial vehicles probably lies in the fact that the respective technology has now reached a sophistication which seems to offer armed forces a distinction and

---

[12] Peter Bergen and Emily Schneider, 'Hezbollah armed drone? Militants' new weapon' CNN, 22 September 2014; William Booth, 'Israel accepts truce plan; Hamas balks' *Washington Post*, 15 July 2014.

[13] George Arnett, 'The numbers behind the worldwide trade in drones' *The Guardian*, 16 March 2015.

[14] Clay Dillow, 'All These Countries Have Armed Drones Now' *Fortune*, 12 February 2016.

[15] Nick Turse, 'US Military Is Building A $100 Million Drone Base In Africa' *Intercept*, 29 September 2016; Craig Whitlock, 'U.S. Drone Base in Ethiopia is Operational' *Washington Post*, 27 October 2011; Craig Whitlock and Greg Miller, 'U.S. Building Secret Drone Bases in Africa, Arabian Peninsula, Officials Say' *Washington Post*, 20 September 2011.

[16] The QinetiQ Zephyr, a British drone currently holds the record for the longest flight time by a drone with 336 hours and 22 minutes. See https://en.wikipedia.org/wiki/Unmanned_aerial_vehicle (accessed 2 May 2017).

unprecedented military advantage and efficiency. However, one cannot deny that this technological development goes hand in hand with a change in the kind of conflict situations that we witness today.[17] International armed conflicts with states fighting against states have become almost the exception, while conflicts between government forces and organized rebel groups and/or terrorist movements have become the overwhelming majority.[18] And especially the phenomenon of a globalized terrorist movement which acts worldwide and beyond national borders has posed challenges to governments which can partly be countered in a better way by the use of these modern technologies.[19]

The rise in the use of drones has brought the possibility for states—such as the United States—to expand the ways of targeting terrorist suspects. The United States is now able to target possible terrorist suspects and objectives far away from American territory, in States such as Somalia, Pakistan or Yemen.[20] All this while the operator of the drone sits safely in a control center in Texas in front of a computer screen, and therefore offering an unprecedented opportunity to limit the risk of danger and casualties of their armed forces.[21] This distance between the initiator and the object of the attack is not completely new. The above-mentioned development fits into a pattern within the development of weapons and military history; namely the ever-increasing distance between the attacker and its target. In past armed conflicts in the 20th century, due to the use of aerial bombardment and the use of missiles and artillery, a certain dislocation between one side and the other side has not

---

[17] See L Reydams, 'A la guerre comme à la guerre: patterns of armed conflict, humanitarian law responses and new challenges' (2006) 88 Intl Rev Red Cross 729; see also R Geiß, 'Asymmetric Conflict Structures' (2006) 88 Intl Rev Red Cross 757.

[18] See on this L Blank and A Guiora, 'Teaching an Old Dog New Tricks: Operationalizing the Law of Armed Conflict in New Warfare' (2010) 1 Harv Natl Sec J 45, 53.

[19] For an overview of the rise and reach of global terrorism since 1970, see, for example: http://terror.periscopic.com/ (accessed 2 May 2017).

[20] For an overview of the drone strikes in these respective states, see https://www.thebureauinvestigates.com/projects/drone-war (accessed 2 May 2017).

[21] For a detailed description of the American drone programme, see International Human Rights and Conflict Resolution Clinic, Stanford Law School & Global Justice Clinic, NYU Law School, *Living under Drones: Death, Injury, and Trauma to Civilians from US Drone Practices in Pakistan* (2012) ch 1; see also M E O'Connell (n 2) 2–11.

been completely uncommon.[22] However, the kind of warfare nevertheless has changed. While traditionally the forces of two sovereign states would face each other, now we often see that government agencies are targeting non-state actors and vice versa, and in contrast to the past internal conflicts, the non-state actor usually is not present on the territory of the respective state.

1.3 Two Main Scenarios—Which Law is Applicable?

As a consequence, we are today faced with two main scenarios: (a) attacks from non-state actors against a state which can either originate from the state itself (for example, 11 September 2001) or from a third, maybe neighboring state (for example, Hezbollah troops attacking Israel from Lebanese soil), or (b) attacks from government agencies against non-state actors being present on the territory of a third state (for example, the United States targeting suspected terrorists hiding in the mountains of Western Pakistan). The alternatives in which it comes to armed violence between government forces and non-state actors with a cross border dimension especially raise a crucial question: which legal regime is applicable, and—to be more precise—what is the geographical scope of IHL and/or human rights law? The question we have to ask ourselves in these situations is: are we dealing with an IAC, a NIAC, or a mere enforcement action which has to comply with human rights law (if the scope of application for the respective human rights treaty is actually given)? The kind of conflicts which are described above are often coined with the term 'transnational armed conflict',[23] and the

---

[22] B Boothby, 'Some Legal Challenges Posed by Remote Attack' (2012) 94 Intl Rev Red Cross 579, 593; see also Philip Alston, 'Report of the Special Rapporteur on Extrajudicial Killings or Summary Executions' (2010) para 84: 'Furthermore, because operators are based thousands of miles away from the battlefield, and undertake operations entirely through computer screens and remote audiofeed, there is a risk of developing a "Playstation" mentality to killing'.

[23] Instructive on the phenomenon of transnational armed conflicts, see C Kress, 'Some Reflections on the International Legal Framework Governing Transnational Armed Conflicts' (2010) 15 J Conflict & Sec L 245; G S Corn and E T Jensen, 'Transnational Armed Conflict: A "Principled" Approach to the Regulation of Counter-Terror Combat Operations' (2009) 42 Israel L Rev 46; G S Corn, 'Hamdan, Lebanon, and the Regulation of Hostilities: The Need to Recognize a Hybrid Category of Armed Conflict' (2007) 40 Vanderbilt J of Transnational L 295; see also 2016 ICRC Commentary on the First Geneva Convention, para 472 with further references in fn 194.

respective debate surrounding this terminology circles around the question of what kind of IHL is actually applicable to these kinds of conflicts, namely the regime governing IACs, the one dealing with NIACs, or neither of those two.[24] However, although the term 'transnational armed conflict' accurately describes the cross-border dimension of these situations, it already assumes that there is actually an armed conflict taking place. This is understandable if one has scenarios in mind like the Hezbollah-Israel conflict on Lebanese territory.[25] But if we look at situations of drone attacks in Pakistan, Somalia and Yemen, one should not exclude the possibility that the situation remains *below* the threshold of an armed conflict. For the purpose of this chapter we will focus mainly on the second category of conflicts.

2. THE RELATIONSHIP BETWEEN THE GEOGRAPHICAL SCOPE OF APPLICATION AND THE TRADITIONAL IHL CONCEPT OF 'ARMED CONFLICT'

2.1 Why It Is Important to Define the Geographical Scope of Application

In order to decide whether the use of armed violence in these situations has to be judged by IHL, we need to find standards in order to define the meaning and ambit of the geographical scope of the application of IHL. This is especially important since a single drone attack as such usually does not fulfil the requirements of an 'armed conflict' which is required for IHL to be applied. The definition of the geographical scope of application[26] shows us when and especially *where* it is permitted to use,

---

[24] See Corn and Jensen (n 23) 4–6; Kress (n 23) 255–7.

[25] See, on the Hezbollah conflict, S Mahmoudi, 'The Second Lebanon War: Reflections on the 2006 Israeli Military Operations against Hezbollah' in O Engdahl and P Wrange (eds), *Law at War: The Law as it was and the Law as it should be, Liber Amicorum Ove Bring* (Martinus Nijhoff 2008).

[26] On the geographical application of IHL, see C Greenwood, 'Scope of Application of Humanitarian Law' in D Fleck (ed), *Handbook of International Humanitarian Law* (Oxford University Press 2008) paras 216–221; see also K Schöberl, 'Konfliktpartei und Kriegsgebiet in bewaffneten Auseinandersetzungen—zur Debatte um den Anwendungsbereich des Rechts internationaler und nicht-internationaler bewaffneter Konflikte' (2012) 25 Humanitäres Völkerrecht—Informationsschriften.

for example, drones against combatants (in an IAC) or 'fighters' (in a NIAC) within the realms of IHL. If the attack falls within the geographical scope, the use of drones to kill targeted persons would be legal in a case where the targeted person is to be seen as a combatant (in IACs) or as a 'fighter' in NIACs.[27] As has been put forward in the traditional view on this matter: 'military operations may *not* be carried out beyond the area of war'.[28] At face value, the term 'geographical scope' sufficiently indicates where the legal framework of IHL would be applicable. The term has a penumbra of uncertainty[29] surrounding it, both at a legal level—how does the geographical scope apply in the contemporary content of warfare, and at a factual level—how to determine when the geographical scope of application of IHL is applicable? This last problem is enhanced by the use of synonyms in everyday language which do not fully reflect the correct meaning of the legal concept. Different terms have been used to indicate the geographical scope of application, such as 'combat zone'[30] 'conflict zone', 'theatre of war', or 'area of military operations'.[31] The reason for the fact that there are so many different connotations for the same concept will be examined in the further process of this analysis.

In any case, the problem which is usually discussed especially from the perspective of the military and politicians in context with the evaluation of the legality of drone strikes is the question whether the target is located within the 'combat zone',[32] 'zone of conflict'/'zone of active hostilities',[33] 'hot battlefield'[34] or just the 'battlefield'.[35] The reason for

---

[27] On the concept of combatants, see K Ipsen, 'Combatants and Non-Combatants' in D Fleck (ed), *Handbook of International Humanitarian Law* (Oxford University Press 2009) paras 301–331.

[28] Greenwood (n 26) para 216 (italics by author).

[29] H L A Hart, *The Concept of Law* (Oxford University Press 1961).

[30] L Blank (n 3) 4.

[31] K Anderson, 'Targeted Killing and Drone Warfare: How We Came to Debate Whether There Is a "Legal Geography of War"' Hoover Institution Stanford University 14.

[32] L Blank (n 3) 4.

[33] J Daskal, 'The Geography of the Battlefield: A Framework for Detention and Targeting Outside the 'Hot' Conflict Zone' (2013) 161 U Penn L Rev 1203; J Pejic, 'Extraterritorial targeting by means of armed drones: Some legal implications' (2014) 96 Intl Rev Red Cross 94.

[34] K Schöberl, 'Boundaries of the Battlefield: The Geographical of the Laws of War' in A Clapham, P Gaeta and M Sassoli (eds), *The 1949 Geneva Conventions: A Commentary* (Oxford University Press 2015) 73.

[35] See Lewis (n 3) 299.

this is, of course, that from a traditional viewpoint, military operations may not be carried out beyond the area of war; that area—which at least in traditional inter-state conflicts—was to be seen as already rather extensive. In the context of an IAC, it has been stated that it includes 'all the territory of the parties to the conflict, and the high seas, and exclusive economic zones, including the exclusive economic zones of neutral states'.[36] Furthermore, it has been put forward 'that when two states engage in armed conflict, the conflict extends to all of the states' territory, regardless of the actual incidence of intense fighting'.[37] Indeed, it has been argued that 'it [also] extends to ships and *planes far from the state's territory or actual combat zone*: "The combat zone on land is likely to be quite limited in geographic scope, yet naval and air units may attack targets in distant areas".'[38]

2.2 Geographical Scope and Its Different Wordings

At the same time, it is important to stress that probably nobody promulgates that hostilities must be conducted throughout the whole area of war in order for IHL to be applied. As one commentator has formulated:

> Military operations will not normally be conducted throughout the area of war. The area in which operations are actually taking place at any given time is known as the "area of operations" or "theatre of war". The extent to which a belligerent today is justified in expanding the area of operations will depend upon whether it is necessary for him to do so in order to exercise his right of self-defense. While a state cannot be expected always to defend itself solely on ground of the aggressor's choosing, any expansion of the area of operations may not go beyond what constitutes a necessary and proportionate measure of self-defense. In particular, it cannot be assumed—as in the past—that a state engaged in armed conflict is free to attack its adversary anywhere in the area of war.[39]

Which brings us to the question of how the above-mentioned terms such as 'combat zone' or 'hot battlefield' actually relate to the main legal regime in question, the law of armed conflict or IHL. One could also ask

---

[36] Greenwood (n 26) para 216.

[37] M E O'Connell, 'Combatants and the Combat Zone' (2009) 43 U Richmond L Rev 845.

[38] Y Dinstein, *War, Aggression and Self-Defence* (4th edn, Cambridge University Press 2005) 19–20, cited by O'Connell (n 37).

[39] Greenwood (n 26) para 221.

more precisely: Are these terms actually terms used in IHL? Terms such as 'combat zone',[40] 'zone of combat'[41] and 'battlefield areas',[42] which are used regularly in the context of armed conflicts, and which can also be found at various points at least in the commentaries to the Geneva Conventions, are, however, not legal terms as such, which would determine the scope of applicability of the law of armed conflict; something which—inaccurately—seems to be implied by some commentators and/or certain politicians.[43] For example, and probably as a 'manifestation of the blending of armed conflict and operational counterterrorism, terms such as "zone of combat" have been characterized by some commentators as broadly as anywhere where terrorist attacks are taking place, or perhaps even being planned and financed'.[44] Moreover, when political and military 'leaders invoke the battlefield or the zone of combat, they seek to harness the authority to use force as a first resort against those identified as the enemy (terrorists, insurgents)'.[45] In this regard, it is clearly the desire to make sure that the deployment of drones in the combat zone is in accordance with the standards of international law. Therefore, it is crucial to acknowledge that terms such as 'combat zone', 'area of war', 'zone of combat' and so on, are to be characterized rather as *factual terms* describing the actual theatre of war in which the fighting is taking place, but that the 'geographical scope of application' is the correct legal standard when we want to decide whether IHL is applicable.

However, especially with regard to the armed conflicts in Afghanistan and Iraq, but also in the context of the use of drones in Pakistan, Yemen and Somalia, especially in US-American, and more and more in the

---

[40] ICRC, Commentary Article 14, Fourth Geneva Convention; Commentary Article 23, First Geneva Convention; Commentary Article 19, Third Geneva Convention.

[41] Protocol Additional to the Geneva Conventions of 12 August 1949, and relating to the Protection of Victims of International Armed Conflicts (Protocol I), Article 5(2)(c); see also N Lubell and N Derejko, 'A Global Battlefield? Drones and the Geographical Scope of Armed Conflict' (2013) 11 J Intl Crim Just 73.

[42] Additional Protocol to the Geneva Convention of 12 August 1949, and Relating to the Protection of Victims of Non-International Armed Conflicts (Protocol I), Article 33(4); see also Lubell and Derejko (n 41) 73.

[43] Schöberl (n 34) 69.

[44] Blank (n 3), referring to A N Guiora, 'Military Commissions and National Security Courts after Guantanamo' (2008) 103 Northwestern Univ L Rev Colloquy 199–200.

[45] Blank (n 3) 15.

international, academic literature the question discussed has been what the geographical scope of application of IHL is, in which drones are to be used in order to legally target objects and persons.[46] While '[t]here is little doubt that Afghanistan and Iraq form part of the zone of combat and a corresponding recognition that the entire territory of each country forms part of that zone of combat'[47] as long as we have an ongoing armed conflict in either of the two states, a more problematic issue is whether the zone of combat can be extended to countries and areas (like Western Pakistan) which are neither party to an IAC, nor is a traditional NIAC taking place on the respective territory. But it is clear why political and military leaders want to extend the scope of the zone of combat for the use of unmanned aerial vehicles. In essence, the appeal of invoking armed conflict is obvious: the law applicable in armed conflict arguably has more permissive rules for killing than does human rights law or a state's domestic law, and generally provides immunity to state armed forces.[48]

2.3 Decisive Criterion for the Geographical Scope: the Concept of 'Armed Conflict'

Therefore, the crucial factor for the applicability of IHL is and remains the existence of an 'armed conflict'. The factor which decisively links together concepts such as the 'combat zone' and the term 'armed conflict' is the *geographical scope* of the latter. That means that in the end the question concerning the scope of the combat zone needs to be transferred to the geographical scope of the armed conflict. However, if we look at the respective IHL provisions that deal with the general scope of applicability of the law of armed conflict, we have to realize that there are rather few or no indications concerning the specific content of the geographical scope of either an IAC or NIAC.[49]

There are only a few hints at the geographical scope of application which can be found, for example, in Common Article 2(2) of the Geneva Conventions (GC) which talks about 'occupation *of the territory* of a

---

[46] See O'Connell (n 37); Blank (n 3); Lewis (n 3); ICRC, 2016 Commentary Article 3, para 454, fn 169, First Geneva Convention.

[47] Blank (n 3) 15, with reference to N Balendra, 'Defining Armed Conflict' (2008) 29 Cardozo L Rev 2467, 2502.

[48] P Alston, Report of the Special Rapporteur on extrajudicial, summary or arbitrary executions, Study on targeted killings, UN Doc A/HRC/14/24/Add.6, 28 May 2010, para 47.

[49] Lubell and Derejko (n 41) 74; Pejic (n 33) 93.

High Contracting Party'[50] mentioning a first important aspect, namely that the application of IHL in an IAC is *linked to the territory of a state party*. With regard to NIACs, Common Article 3 GC states that it has to be an 'armed conflict not of an international character *occurring in the territory* of one of the High Contracting Parties', and again the indication goes into the direction that a NIAC has to be connected to the territory of a state party, and one could assume that this should be the state party which is involved in the conflict.[51] Furthermore, for IACs Article 1(3) Additional Protocol (AP) I refers to 'situations referred to in Article 2 common' to the Geneva Conventions, while Article 1(1) APII for NIACs again makes a reference to the *territory of a High Contracting Party* when stating that the Protocol:

> shall apply to all armed conflicts which are not covered by Article 1 [API] [...] *and which take place in the territory of a High Contracting Party* between its armed forces and dissident armed forces or other organized armed groups which, under responsible command, exercise such *control over a part of its territory* as to enable them to carry out sustained and concerted military operations and to implement this protocol.[52]

What becomes obvious from these provisions is that the respective treaty law in the area of IHL seems to require a certain link to the territory of a High Contracting Party to the conflict, while not giving any more indications on what the exact geographical scope is beyond the mere naming of the term 'territory of a High Contracting Party'. With regard to common Article 3 of the Geneva Conventions it is, however, not clear whether this is supposed to be on the territory of only exactly 'one' High Contracting Party, or whether this is rather meant as a 'jurisdictional' limitation in the sense that Common Article 3 is applicable on 'any' of the territories of a High Contracting Party.[53]

2.4 Definition of the Geographical Scope of Application in International Jurisprudence

While looking at the conventional rules does not give a lot of indication about the exact definition of the geographical scope of application, it might be helpful to look at the jurisprudence of the international criminal

---

50 Emphasis added.

51 Emphasis added.

52 Emphasis added.

53 See, for further references on this discussion, 2016 ICRC Commentary on the First Geneva Convention, Article 3, paras 465 et seq.

tribunals and especially to see how the International Criminal Tribunal for the former Yugoslavia (ICTY) has defined the geographical scope of 'armed conflicts'.[54] The starting point for this examination is usually the ICTY's statement in the 1995 *Tadić* Decision on Jurisdiction with regard to the definition of armed conflict as such, which says that: 'an armed conflict exists whenever there is resort to armed force between States or protracted armed violence between governmental authorities and organized armed groups or between such groups within a State'.[55]

This definition has been confirmed by Article 8(2)(f) of the International Criminal Court (ICC) Statute (with regard to NIACs), and by the International Committee of the Red Cross (ICRC) opinion paper on the definition of the armed conflict.[56] However, this general definition of the term 'armed conflict' does not include any reference to the geographical scope as such. But the ICTY elaborated on this specific aspect as well when stating that: 'the temporal and geographical scope of both internal and international armed conflicts *extends beyond the exact time and place of hostilities*'.[57] This helps insofar as it indicates that it is not necessary that in the area affected constant fighting needs to take place. The Chamber furthermore noted that: 'Although the Geneva Conventions are silent as to the geographical scope of international "armed conflicts", the provisions suggest that at least some of the provisions of the Conventions apply *to the entire territory of the Parties to the conflict, not*

---

[54] On the jurisprudence of the ICTY in the area of IHL see R Heinsch, *Die Weiterentwicklung des humanitären Völkerrechts durch die Strafgerichtshöfe für das ehemalige Jugoslawien und Ruanda* (Berlin 2007); see also R Heinsch, 'Judicial "Law-Making" in the Jurisprudence of the ICTY and ICTR in Relation to Protecting Civilians from Mass Violence: How Can Judge-Made Law be Brought into Coherence with the Doctrine of the Formal Sources of International Law?' in P Ambach, F Bostedt, G Dawson and S Kostas (eds), *The Protection of Non-Combatants During Armed Conflict and Safeguarding the Rights of Victims in Post-Conflict Society, Essays in Honour of the Life and Work of Joakim Dungel* (Brill/Nijhoff 2015) 297–331.

[55] *Prosecutor v Dusko Tadić*, IT-94-1/AR72, Decision on the Defence Motion for Interlocutory Appeal on Jurisdiction, 2 October 1995, para 70.

[56] International Committee of the Red Cross, 'How is the Term "Armed Conflict" Defined in International Humanitarian Law?' (ICRC 2008) 2, 4.

[57] *Prosecutor v Dusko Tadić*, IT-94-1/AR72, Decision on the Defence Motion for Interlocutory Appeal on Jurisdiction, 2 October 1995, para 67 (emphasis added).

just to the vicinity of actual hostilities'.[58] This position has also been adopted by the Trial Chamber of the ICTR in *Akayesu*, where the Chamber acknowledged that Common Article 3 and Additional Protocol II could apply 'over the whole territory hence encompassing massacres which occurred far from the "war front"'.[59]

If we apply these statements directly to counter-terrorism actions, for example, to a drone attack against a terrorist suspect in Pakistan, this would mean under a very broad interpretation that it is theoretically possible to include an area outside of the existing NIAC taking place in Afghanistan, covering also certain stretches located in Pakistan. The Trial Chamber in the *Blaskić* case referred to this definition of armed conflict provided by the *Tadić* Jurisdiction Decision as a 'criterion' stating that it is 'applicable to all conflicts whether international or internal. It is not necessary to establish the existence of an armed conflict within each municipality concerned. It *suffices to establish the existence of the conflict within the whole region of which the municipalities are a part*'.[60] This was reiterated in the *Kordić and Cerkez* Trial Chamber Judgment: 'in order for norms of international humanitarian law to apply in relation to a particular location, there need not be actual combat activities in that location. All that is required is a showing that *a state of armed conflict existed in the larger territory of which a given location forms a part*'.[61]

If we consider that the conflict existed in the 'larger territory of which a given location forms part' in this context not as the territory of a state party but rather as the geographical term, one could conclude that this might also encompass a neighboring country. The ICTY Appeals Chamber in *Kunarać* went into the same direction when stating that 'the Prosecutor did not have to prove that there was an armed conflict in each and every square inch of the general area. The state of armed conflict is not limited to the areas of actual military combat *but exists across the entire territory under the control of the warring parties*'.[62] Finally, the

[58] *Prosecutor v Dusko Tadić*, IT-94-1/AR72, Decision on the Defence Motion for Interlocutory Appeal on Jurisdiction, 2 October 1995, para 68 (emphasis added).

[59] ICTR, *Akayesu* Trial Judgment, 1998, paras 635–636.

[60] *Prosecutor v Tihomir Blaskić*, IT-95-14-T, Trial Chamber, Judgment, 3 March 2000, para 64 (italics by author); see also *Prosectutor v Delalic*, IT-95-21-T, Trial Chamber, Judgement, 16 November 1998, para 185.

[61] *Prosecutor v Dario Kordić and Mario Cerkez*, IT-95-14/2-T, Trial Chamber, Judgment, 26 February 2001, para 319 (emphasis added).

[62] *Prosecutor v Kunarac et al*, Appeals Chamber Judgment, para 56 (emphasis by author).

Kunarać Appeals Chamber picked up on the original statement of the *Tadić* Appeals Chamber by stating that:

> There is no necessary correlation between the area where the actual fighting is taking place and the geographical reach of the laws of war. The laws of war apply in the whole territory of the warring states, or, in the case of internal armed conflicts, *the whole territory under the control of a party to the conflict*, whether or not actual combat takes place there, and continue to apply until a general conclusion of peace or, in the case of internal armed conflicts, until a peaceful settlement is achieved. A violation of the laws or customs of war may therefore occur at a time when and in a place where no fighting is actually taking place.[63]

This broad territorial interpretation of the geographical application of IHL is not only justified for jurisdictional reasons for courts and tribunals, but also serves a safeguard to prevent evasion of the application of IHL by relocating individuals or directing operations.[64] Nevertheless, such a broad territorial application of IHL has been subject to criticism.[65] Instead of a broad geographical application, a nexus-approach is brought forward,[66] a position based on the development made by the *Kunarać* Appeals Chamber:

> As indicated by the Trial Chamber, the requirement that the acts of the accused must be closely related to the armed conflict would not be negated if the crimes were temporally and geographically remote from the actual fighting. It would be sufficient, for instance, for the purpose of this requirement, that the *alleged crimes were closely related to hostilities occurring in other parts of the territories controlled by the parties to the conflict*.[67]

This nexus-approach allows certain acts to fall under the geographical scope of IHL, while allowing acts not related to the armed conflict to be

---

[63] *Prosecutor v Kunarac*, AC Judgment, para 57 (emphasis by author).

[64] Lubell and Derejko (n 41) 74; T Ferrero, 'The Geographic Reach of IHL' in Proceedings of the Bruges Colloquium, Scope of Application of International Humanitarian Law, 18–19 October 2012, Collegium No 43 (2013) 111; Pejic (n 33) 95.

[65] Lubell and Derejko (n 41) 76; S Sivakumaran, *The Law of Non-International Armed Conflict* (Oxford University Press 2012) 252; Schöberl (n 34) 75.

[66] For an extensive discussion of the nexus-approach with further reference, see 2016 ICRC Commentary on the First Geneva Convention, Article 2, para 460 et seq.

[67] *Prosecutor v Kunarać et al*, Appeals Chamber Judgment, para 57 (emphasis by author).

governed under the regime of domestic legal enforcement measures.[68] This position has been followed by the International Criminal Court in the *Katanga* and *Bemba* Trial Judgments. The Trial Chamber in *Katanga* noted that:

> In this connection, the Chamber is of the view that the perpetrator's conduct must have been *closely linked to the hostilities* taking place in any part of the territories controlled by the parties to the conflict. The armed conflict alone need not be considered to be the root of the conduct of the perpetrator and the conduct need not have taken place in the midst of battle. Nonetheless, the armed conflict must play a major part in the perpetrator's decision, in his or her ability to commit the crime or the manner in which the crime was ultimately committed.[69]

This first of all re-emphasizes that the scope of the combat zone does not necessarily need to be a limited area in which actual hostilities between the parties of the armed conflict are taking place. And especially the last two sentences of the statement in the *Kunarać* Appeals Judgment indicate that there can be some distance between the actual conduct of hostilities and the area that still is covered by the definition of an armed conflict. However, what seems to be crucial in NIACs is that the geographical scope of application only extends to an area that is under the 'control of a party to the conflict'. Accordingly, the decisive criterion seems to be that the respective territory is actually *under control of one of the warring parties*. Although the control over territory criterion in some ways sounds familiar to the scope of application of Additional Protocol II, according to the *Kunarać* Appeals Chamber it seems as if this condition also needs to be fulfilled in Common Article 3 situations if we want to extend the geographical scope of application. The ICTY stated this condition in a very general way, not making any differentiation between the different thresholds of armed conflicts of a non-international character. Coming back to the concrete case mentioned above: If a US drone is targeting a terrorist in Pakistan, then the question would be whether this area is under the control of one of the warring parties which in most cases probably would have to be denied.

---

68 ICRC, 2016 Commentary Article 3, para 460, First Geneva Convention.

69 *Prosecutor v Germain Katanga*, ICC-01/04-01/07, Trial Judgment, 7 March 2014, para 1176; *Prosecutor v Bemba*, ICC-01/05-01/08, Trial Judgment, 21 March 2016, paras 142–144; see also, 2016 ICRC Commentary on the First Geneva Convention, Article 3, para 460.

2.5 Geographical Scope Also Covering the Origin of the Attack?

Another question one needs to discuss is whether the location where a UAV has been launched (so for example from Nevada in the United States) is covered by the geographical scope of application. If there is an IAC between, for example, the United States and Afghanistan at the respective time this usually does not become problematic because the territories of both states actors would belong to the geographical scope of application of IHL. It seems to be more difficult when we are talking about NIACs. In those cases in which the control base is within the territory of the state in which the NIAC is taking place, then usually the whole territory is covered anyway, or at least we have to follow the criteria laid out by the ICTY in the above cited decisions in order to determine the geographical scope of application. This has also been confirmed in the literature, in which some academics stress that the 'theatre of war also encompasses the area where the UCAV has been fired from'.[70] However, one has to be careful here if the control base lies outside of the 'host' territory of the NIAC, because the traditional view assumed that a legitimate military target would be on the territory of the state in which the NIAC is taking place. If the base for firing the drone is outside of the territory where the NIAC is taking place, then the situation is more problematic and we would again face the issue of 'transnational armed conflicts'[71] or a 'global' NIAC.[72]

---

[70] Stroh (n 2) 76: 'Aus einer Gesamtschau der einschlägigen Vorschriften und Entscheidungen lässt sich also ableiten, dass das "Theatre of war" im nicht-internationalen bewaffneten Konflikt auch das Gebiet erfasst, von welchem aus UCAVs im Einsatz gesteuert werden'; Lubell and Derejko (n 41) 85; see also *Manual on International law Applicable to Air and Missile Warfare* (HPCR 2009) rule 29; Patrycja Grzebyk, 'Who Can Be Killed?: Legal Targets in Non-International Armed Conflicts', in Steven J Barela (ed), *Legitimacy and Drones: Investigating the Legality, Morality, and Efficacy of UCAVs* (Routledge 2015) 58; M Schmitt, 'Charting the Legal Geography of Non-International Armed Conflict' (2014) 90 Intl L Stud 16.

[71] See *supra* n 14.

[72] On this issue, see 2016 ICRC Commentary on the First Geneva Convention, paras 479 et seq.

3. THE CRITERIA FOR THE DETERMINATION OF AN ARMED CONFLICT IN PRACTICE

If we apply the above listed criteria to a situation like the conflict in Afghanistan, the question has to be raised whether an armed conflict exists in the moment an American drone enters Pakistani air space and kills a suspected terrorist. In most cases, it seems that the Pakistani government probably will have given their (tacit) consent to these actions—although there have been instances in which it has been reported that the Pakistani government rejected any such assumed consent. Any consent from the Pakistani side would prima facie exclude the existence of an IAC. But it also does not amount to a NIAC, since there is no 'protracted armed violence between government authorities and organized rebel groups'.[73] In these circumstances an application of the law of armed conflict has to be questionable.

3.1 Situations of International Armed Conflict

Nowadays we are looking at the use of drones in situations that are often *not* characterized as IACs per se, although even this cannot be assumed lightly. When, for example, an American drone enters Pakistani territory and hits a civilian object, one could first get the idea that the situation could be characterized as an IAC in the sense of Common Article 2, since we have two state parties involved (the United States and Pakistan). However, there are two reasons why the existence of an IAC is doubtful. First, there are reports which indicate that the Pakistani government is supporting the drone attacks at least tacitly, and therefore the consent[74] of the one state partly concerned already excludes a conflict between two states.[75] In addition, the Pakistani armed forces are clearly not involved

---

[73] *Prosecutor v Dusko Tadić*, IT-94-1/AR72, Decision on the Defence Motion for Interlocutory Appeal on Jurisdiction, 2 October 1995, para 70.

[74] For an overview of the meaning and consequence of 'consent' and its application to in Yemen, Somalia, and Pakistan see M Byrne, 'Consent and the Use of Force: An Examination of "Intervention By Invitation" as a Basis for US Drone Strikes in Pakistan, Somalia and Yemen' (2016) 3 J on the Use of Force & Intl L 97.

[75] The views on whether Pakistan has given consent vary from (tacit) approval of drone strikes, see Greg Miller and Bob Woodward, 'Secret memos reveal explicit nature of U.S., Pakistan agreement on drones', *Washington Post*, 24 October 2013; B Emmerson, 'Promotion and protection of human rights and fundamental freedoms while countering terrorism', UN Doc A/68/389, 18

in the situation. According to the 1952 official ICRC commentary on the Geneva Conventions, it is a requirement that we are faced with '[a]ny difference arising between two States and *leading to the intervention of members of the armed forces*'.[76] This requirement has been confirmed by the International Criminal Tribunal for the Former Yugoslavia in its famous *Tadić* decision. However, in the 2016 ICRC Commentary on Geneva Convention I, it is now stated that 'an international armed conflict arises between the territorial State and the intervening State when force is used on the former's territory *without its consent*',[77] thereby dropping the requirement that the official armed forces of two states need to be involved for the existence of an IAC. The ICRC Commentary justifies this new approach with reference to the International Court of Justice's (ICJ) decision in the *Armed Activities* case, 'in which the Court applied the law governing international armed conflict to the military actions undertaken by Uganda in the [DRC] outside the parts of the DRC it occupied'.[78] In this context, the ICJ concluded that the situation could nevertheless be qualified as an IAC, although 'Uganda claimed to have troops in the DRC primarily to fight non-state armed groups and not DRC armed forces'.[79] Therefore, in the example of drone attacks in Pakistan under the traditional approach one would exclude the possibility of a 'classical' IAC between states because it does not amount

---

September 2013. But there are also claims circulating regarding the rejection and/or withdrawing of consent by Pakistan to the drone strikes. See Bureau of Investigative Journalism, 'Pakistan "categorically rejects" claim that it tacitly allows US drone strikes'; Owen Bowcott, 'US drone strikes in Pakistan "carried out without government's consent"' *The Guardian*, 15 March 2013.

[76] Commentary on the Fourth Geneva Convention, para 1, 1958 (emphasis added). See also 2016 Commentary to the First Geneva Convention, Article 1 section 222, for lowering the threshold of the existence of an IAC, which could also be triggered by an unilateral attack of State directed to another State, even if the attacked State does not respond.

[77] 2016 Commentary on the First Geneva Convention, para 262 (emphasis added).

[78] Ibid para 261.

[79] Ibid, with reference to ICJ, Armed Activities on the Territory of the Congo case, Judgment, 2005, paras 108, 146 and 208ff. See also UN Commission of Inquiry on Lebanon, Report of the Commission of Enquiry on Lebanon pursuant to Human Rights Council resolution S-2/1, UN Doc A/HRC/3/2, 23 November 2006, paras 50–62, recognizing that an IAC took place in 2006 between Israel and Lebanon even if the hostilities only involved Hezbollah and Israeli armed forces.

to an 'armed violence *between* two States'.[80] Under the new approach, however, promulgated by the ICRC with reference to the ICJ jurisprudence in the *Armed Activities* case, one would come to the result that an IAC exists in the sense of Common Article 2 of the Geneva Conventions if Pakistan had not (tacitly) consented to the strikes.

This contemporary interpretation seems to put more emphasis, on the one hand, on the *cross-border aspect* of the armed activities as such. This at first glance seems to be in contrast with the very wording of Common Article 2 ('armed conflict *between two or more of the High Contracting Parties*'). With regard to the example of the conflict between Israel and Hezbollah in Lebanon, there was clearly no armed violence from the side of the Lebanese forces, and it also cannot be construed that the actions of Hezbollah can be attributed to Lebanon according to the Articles on State Responsibility (also not according to Article 8 of the Articles on the Responsibility of States for Internationally Wrongful Acts dealing with the attribution of private actors, since there were no clear indications that Hezbollah was 'in fact acting on the instructions of, or under the direction or control of, that State in carrying out the conduct'[81]). There are commentators who try to establish an attribution through the fact that the 'host' state—be it Lebanon or Pakistan—is not doing anything against the actions of the paramilitary activities of the respective rebel or terrorist group. However, this does not find a basis in the rules on state responsibility, and can also not be construed relying on the rules of neutrality.[82] In addition, one might argue that the rules governing an IAC were designed to be applied to interstate relations, and it will create problems if we apply every detailed rule to a situation where a state is facing a non-state actor just because there is a cross-border element involved in the current case.[83] On the other hand,

---

80 *Prosecutor v Dusko Tadic* (Decision on the Defence Motion for Interlocutory Appeal on Jurisdiction), para 70, IT-94-1, International Criminal Tribunal for the former Yugoslavia (ICTY), 2 October 1995.

81 Article 8 of the Articles on the Responsibility of States for Internationally Wrongful Acts.

82 There are however voices which want to use the principles from neutrality law in order to justify that the territorial scope of NIACs is extended beyond the borders of its 'host state'. Cf Lewis (n 2) 293, 313.

83 On the difference and consequence of the distinction between an IAC and NIAC, see D Akande, 'Classification of Armed Conflicts: Relevant Legal Concepts' in E Wilmshurst (ed), *International law and the Classification of Conflicts* (Oxford University Press 2012) 4–11; M Milanović, 'Transnational and Mixed Conflicts' in A Clapham, P Gaeta and M Sassoli (eds), *The 1949 Geneva Conventions: A Commentary* (Oxford University Press 2015) 52.

the new ICRC commentary makes a greater emphasis on the fact that an 'unconsented-to armed intrusion into the territorial sphere of sovereignty' can amount to an IAC,[84] therefore shifting the relevance of the criterion important for determining the existence of an IAC from the use of armed force between states to the intrusion on one state's sovereignty by armed activities from another state. The main argument the ICRC commentary raises in this context is that one should take into account that:

> the population and public property of the territorial State may also be present in areas where the armed group is present and some group members may also be residents or citizens of the territorial State, such that attacks against the armed group will concomitantly affect the local population and the State's infrastructure.[85]

While this argumentation has some benefits and indeed seems to reflect a tendency also represented in the *Armed Activities* case, there remain doubts whether this new interpretation is actually covered by current state practice. The wording of Common Article 2 and the respective ICTY jurisprudence rather indicates that the necessary armed violence has to be exercised by at least two High Contracting Parties of the Geneva Conventions.

3.2 Situations of Non-international Armed Conflict

The second option, of course, is that in these kinds of situations we are not dealing with an IAC, but a NIAC. As mentioned above, the general requirement for a Common Article 3 NIAC is that there is 'protracted armed violence between governmental authorities and organized armed groups or between such groups within a State'. These conditions have been elaborated upon by the ICTY and the ICTR respectively, especially with regard to the intensity of the armed violence and the organization of the armed group.[86] Since the requirement for both intensity and organization are to be seen as indicating a rather high threshold, it is doubtful whether the drone attacks in Pakistan fulfil this criterion. First, if the attacks only appear in selected cases and not in a constant manner, then the requirement of 'protracted armed violence' would not be fulfilled,

---

[84] 2016 ICRC Commentary on the First Geneva Convention, Article 2, para 261.

[85] 2016 ICRC Commentary on the First Geneva Convention, Article 2, para 261.

[86] For more details on this issue, refer to R Heinsch, 'Conflict Classification in Ukraine: The Return of the "Proxy War"?' (2015) 91 Intl L Stud 323, 334–40.

since the time element is missing, and it would probably also not reach the required intensity level.[87] One might also question whether the targeted suspected terrorists are under the given circumstances to be equalized with an organized armed group. In theory, terrorist organizations might very well fulfill this requirement, but if we are talking about singled-out members of these terrorist cells, it already becomes highly doubtful whether a sufficient degree of organization has been reached. At the moment, it does not look as if the situation in Pakistan is in any way comparable to the Hezbollah-Israel conflict, which obviously fulfilled the requirement of being a NIAC.

If we deny the existence of a NIAC as such on the territory of Pakistan, we still have the possibility that the existing NIAC in Afghanistan has 'spilled over' to Pakistani territory.[88] In this context, the very question we have to ask is how the geographical scope of IHL can be defined if there are not two states opposing each other, but when we are dealing with a conflict between a state agency on the one side, and non-state actors on the other side which are located in a third state. As mentioned earlier in section 2.4, the geographical scope of application of IHL is broad, with the result that IHL is also applicable outside an area of active hostilities, namely throughout the entire territory. Such a broad application of IHL has been accepted in an extra-territorial situation where there is a (sufficient) nexus between the acts or actors of the initial conflict acting in another territory. The consequence of this would be that in a situation of a NIAC, IHL would apply also across the border and into the territory into which the NIAC has spilled over. This position seems to have been accepted to a certain extent in the 2016 Commentary to the First Geneva Convention as well as in academia, with reference to the wordings used in the Geneva Conventions and the Additional Protocols, and the rationale of IHL.[89] Following the statements made above, it would be possible to extend the borders of the geographical scope of IHL

---

[87] See also Lubell and Derejko (n 41) 78; K Schöberl, 'Boundaries of the Battlefield: The Geographical Scope of the Laws of War' in Steven J Barela (ed), *Legitimacy and Drones: Investigating the Legality, Morality, and Efficacy of UCAV's* (Routledge 2015) 80.

[88] On the phenomenon of so-called 'spill-over' conflicts, see 2016 ICRC Commentary on the First Geneva Convention, para 476.

[89] 2016 ICRC Commentary on the First Geneva Convention, para 470; Schöberl (n 34) 75; Schmitt (n 70) 11; T Ferraro, 'The Applicability and Application of International Humanitarian Law to Multinational Forces' (2013) 95 Intl Rev of Red Cross 610; Akande (n 83) 57; Lubell and Derejko (n 41) 76–7; S Sivakumaran (n 65) 251. For opposing views supporting the geographical application of IHL only within the territory of a state, see Advisory

past the territory of the 'host state' of the NIAC, as this would seem to be compatible within the framework of the geographical scope of IHL. However, in such a situation of a NIAC which spilled over, it still remains unclear how far the geographical scope of IHL should reach in the state where the NIAC has spilled over.[90] What is necessary under these circumstances is a practical and workable solution. Based on the wording of the Geneva Conventions as well as the respective interpretations of the ICTY in the *Kunarać* Appeals Judgment, it is put forward that the spillover area will extend to the area of which the organized armed group has control.[91] If control over territory is not given, and for areas where there is no actual fighting taking place, the criteria is that the respective target or affected situation needs to have a 'nexus' with the original NIAC.[92]

This reasoning mentioned above would also sideline the discussion that possible fighters in third states might be attacked since the conflict 'follows the combatant'. The rationale behind this position is that otherwise possible fighters could just evade being suspected under the rules governing the law of armed conflict by switching territory.[93] However, this does not seem to be completely persuasive because there is no indication in either conventional or customary law that would support this.[94] Common Article 3 actually clearly states that it is applicable '[i]n the case of armed conflict not of an international character occurring *in the territory of one of the high Contracting Parties*', establishing a connection with the territory of the state where the conflict is taking place. This connection is equally established in the French version of

---

Committee on Issues of Public International law, Main Conclusions of Advice on: Armed Drones (2013) 2.

[90] 2016 ICRC Commentary on the First Geneva Convention, Article 3, para 476; Schöberl (n 34) 82.

[91] See above at n 68.

[92] See above at n 70.

[93] In this sense, see Lewis (n 3) 313; J D Brennan, 'Strengthening our Security by Adhering to our Values and Laws', Program on Law and Security, Harvard Law School, Cambridge (2011); United States Department of Justice, 'Lawfulness of a Lethal Operation Directed Against a U.S. Citizen who is a Senior Operational Leader of Al-Qa'ida, or an Associated Force' (2011) 2–3.

[94] 2016 ICRC Commentary on the First Geneva Convention, Article 3, paras 479–480, First Geneva Convention; Pejic (n 33) 103 fn 144; ICRC, 32nd International Conference of the Red Cross and Red Crescent: International Humanitarian Law and the Challenges of Contemporary Armed Conflicts (2015) 15; J Daskal, 'The Geography of the Battlefield: A Framework For Detention and Targeting Outside the "Hot" Conflict Zone' (2013) 161 U Penn L Rev 1194.

Common Article 3 where it is stated that '[e]n cas de conflit armé ne présentant pas un caractère international et *surgissant sur le territoire de l'une des Hautes Parties contractantes*'.

As one can see, we are reaching here one of the fault lines of current public international law, because we are applying a legal regime (IHL), which was designed in a time with sovereign states and clear borders, to a world where not only armed conflict no longer seems to know any borders, and in addition non-state actors are able to conduct actions on such a scale that it can seriously threaten the peace and security of states and the world community. In this regard, one might now be faced with the problem of adapting IHL accordingly, going past the expressive wording of the Geneva Conventions and allowing for the application of the law applicable in NIACs even outside the territory of the original conflict (in our case Afghanistan). This might especially be arguable if one of the non-state fighters immediately retreats after they have been in combat function on the territory of Afghanistan.[95] In this sense one could almost use the figure of 'hot pursuit'[96] of Article 111 of the United Nations Convention on the Law of the Sea. This would also be in line with the jurisprudence of the ICTY and ICTR stating that the combat activities do not need to be in the immediate vicinity of the actual battlefield.

However, what remains here is the conceptual hiccup that the Geneva Conventions actually do not know the concept of hot pursuit, and that it would mean stretching the notion of armed conflict beyond its original conceived scope. Since the Geneva Conventions and Additional Protocols still start from an approach of a sovereign state with a delimited territory, the current author believes in general one still should judge the situation in Pakistan separately from the situation in Afghanistan, coming to the conclusion that if on Pakistani territory there is no 'protracted armed violence between governmental authorities and organized armed groups', we in principle do not have a NIAC, and therefore IHL would not be applicable. The only exception to this would be the immediate border region to the country in which the original NIAC took place, if one can

---

[95] On this problem, see Blank (n 3) 1 et seq; Lewis (n 3) 293, 312–13.

[96] Mentioning this argument as well, see O'Connell (n 2) 20, with reference to J Northam, 'Airstrikes in Pakistan's Tribal Areas Legally Murky' NPR, 17 March 2009 (remarks of Harvey Rishikof, professor at the National War College). See also N Lubell, *Extraterritorial Use of Force against Non-State Actors* (Oxford University Press 2010) 72–3.

establish that because of a strong nexus to the armed conflict we witness a 'spillover'.[97]

3.3 Situations of Law Enforcement under the Human Rights Regime

If IHL, due to the lack of an armed conflict, is not applicable, then the question is raised which legal standards are applicable instead.[98] In general, the answer to this is that due to the lex specialis character of IHL, we have to look at the international human rights regime in those cases where IHL is no longer applicable. In this regard, the attack against an Al-Qaeda leader through the use of a combat drone, even if planned in a very military-like way,[99] would have to follow the stricter regulations of human rights law, taking into account especially the standards of necessity and proportionality. Although the application of the human rights regime triggers the issue of the extra-territorial application of human rights treaties,[100] in general one should follow the approach set up by the ICJ that the two systems are complementary. As the Court mentioned already in the 1996 Advisory Opinion on the Legality of Nuclear Weapons, '[t]he protection of the International Covenant on Civil and Political Rights does not cease in times of war, except by operation of Article 4 of the Covenant whereby certain provisions may be derogated from in a time of national emergency'.[101] And it further elaborated on this in the 2004 Advisory Opinion on the Construction of the Wall when saying that:

> [...] both branches of international law, namely international human rights law and international humanitarian law, would have to be taken into consideration. The Court further concluded that international human rights instruments

---

[97] See above at n 93.

[98] On the discussion of whether human rights standards also have to be applied in areas of a NIAC where IHL in theory is applicable, but the lack of actual combat activities rather indicates the application of human rights law, see 2016 ICRC Commentary on the First Geneva Convention, para 456.

[99] Along these lines is the argument of Corn and Jensen (n 23) 33.

[100] On the extraterritorial application of human rights see eg M Milanović, *Extraterritorial Application of Human Rights Treaties* (Oxford University Press 2011); M Gondek, *The Reach of Human Rights in a Globalising World* (Intersentia 2009).

[101] ICJ, Legality of the Threat of Use of Nuclear Weapons, Advisory Opinion, ICJ Reports 1996, para 25.

are applicable "in respect of acts done by a State in the exercise of its jurisdiction outside its own territory", particularly in occupied territories.[102]

The problem we have to face, however, in this kind of situation is the issue of extraterritorial application of human rights treaties.[103] For example, if the United States targets terrorist suspects who are not present on its own territory, but on Pakistani soil, the question arises whether the United States is bound by human rights law. As such, a link to a state's own territory is in general not absolutely necessary because most human rights treaties, including the International Covenant on Civil and Political Rights (ICCPR), require that '[e]ach State Party to the present Covenant undertakes to respect and ensure to all individuals within its territory and *subject to its jurisdiction* the rights recognized in the present Covenant'.[104] With regard to the requirement of 'subject to its jurisdiction' the recent jurisprudence of regional human rights courts like the European Court of Human Rights (ECtHR) have equalized this with 'effective control' over a certain territory. The ECtHR in particular has denied a mere bombing from the air to fulfill this condition in the *Banković* case[105] dealing with a situation of a bombing by the North Atlantic Treaty Organization (NATO) in the former Yugoslavia. However, this jurisprudence is questionable and also far from coherent at the moment,[106] because the killing of an individual must be seen as one of

---

[102] ICJ, Legality of the Threat of Use of Nuclear Weapons, Advisory Opinion, ICJ Reports 1996, para 2.

[103] On this issue, see eg M Szydło, 'Extra-Territorial Application of the European Convention on Human Rights After Al-Skeini and Al-Jedda' (2012) 12 Intl Crim L Rev 271; R Nigro, 'The Notion of "Jurisdiction" in Article 1: Future Scenarios for the Extra-Territorial Application of the European Convention on Human Rights' (2010) 20 Italian Ybook of Intl L 9; A Zimmermann, 'Extraterritorial Application of Human Rights Treaties: the Case of Israel and the Palestinian Territories revisited' in I Buffard (ed), *International Law between Universalism and Fragmentation: Festschrift in Honour of Gerhard Hafner* (Brill 2008).

[104] International Covenant on Civil and Political Rights, Article 2, 16 December 1966, 999 UNTS 171.

[105] *Banković, Stojanović, Stoimedovski, Joksimović and Suković v Belgium, the Czech Republic, Denmark, France, Germany, Greece, Hungary, Iceland, Italy, Luxembourg, the Netherlands, Norway, Poland, Portugal, Spain, Turkey, and United Kingdom*, Application No 52207/99, ECHR, reprinted in (2001) 123 ILR 94.

[106] See in general F Coomans and M Kamminga (eds), *Extraterritorial Application of Human Rights Treaties* (Intersentia 2004); M Dennis, 'Application of Human Rights Treaties in Times of Armed Conflict and Military Occupation'

the strongest cases of exercising jurisdiction. If a police officer were to kill a suspected bank robber while chasing him through the hills of the Grand Canyon, nobody would ever doubt that this is an exercise of jurisdiction merely because he is acting in his position as a police officer. And there is little reason to doubt that if this very state agent is controlling a drone attack against a suspected terrorist following the orders of his superiors that he is then not exercising jurisdictional power, no matter whether he has effective control of the territory or not. In this regard, the jurisprudence of ECtHR in the *Al-Skeini* case[107] seems to leave the door open for a broader interpretation, especially in the context of the so-called 'state agent authority exception to territorial jurisdiction',[108] in which the Court reaffirmed that, 'as an exception to the principle of territoriality, a Contracting State's jurisdiction under Article 1 may extend to acts of its authorities which produce effects outside its own territory'.[109] The Court in *Hassan v the United Kingdom*, with reference to *Al-Skeini*,[110] asserted the extra-territorial application of the ECHR in the situation of the applicant since he fell within the authority and control of the forces of the United Kingdom from the moment of his capture until his release.[111] In addition the Court rejected the claim by the United Kingdom that the application of international human rights law should not apply in the phase of active hostilities.[112] In *Jaloud v The Netherlands*, the Court held that the ECHR applied extra-territorially since the Netherlands exercised its jurisdiction 'for the

---

(2005) 99 Am J Intl L 119; R Lawson, 'Life After Bankovic: On the Extra-territorial Application of the European Convention on Human Rights' in Coomans and Kanninga, supra; M Happold, 'Bankovic, Belgium and the Territorial Scope of the European Convention on Human Rights' (2003) 3 Hum Rts L Rev 77–90.

107 *Al-Skeini and Others v The United Kingdom*, Application no 55721/07, ECHR, Judgment, 7 July 2011; *Serdar Mohammed v Ministry of Defence* [2014] EWHC 1369 (QB), paras 147–148; *Serdar Mohammed & Others v Secretary of State for Defence* [2015] EWCA Civ 843 (30 July 2015). The Court of Appeal cast its doubts on the notion of extra-territorial applicability of the ECHR as mentioned in *Al-Skeini v United Kingdom* (paras 91–106); at the time of writing the case is currently before the Supreme Court of the United Kingdom.

108 A good overview on the *Al-Skeini* case can be found in A Cowan, 'A New Watershed? Re-evaluating Bankovic in Light of Al-Skeini' (2012) 1 Cambridge J Intl & Comp L 213.

109 *Al-Skeini and Others v The United Kingdom*, para 133.

110 *Hassan v United Kingdom*, Application no 55721/07, ECHR, Judgment, 16 September 2014, para 75.

111 Ibid para 78.

112 Ibid paras 71, 77.

purpose of asserting authority and control over persons passing through the checkpoint'.[113]

4. A POSSIBLE APPROACH TO DEALING WITH THE CHALLENGES OF THE GEOGRAPHICAL SCOPE OF IHL

As we have seen above, we are at a crucial point in the development of the law of armed conflict. The type of conflict has changed from territorial wars between independent states, who wanted to enlarge their power and area of influence; through NIACs in which more or less organized groups fought against the government in order to gain independence, gain their right of self-determination or just in order to overthrow a government; up to today where we have now reached a situation in which the conflict is not necessarily connected to a certain territory at all. This fact poses certain challenges to the applicability of the law of armed conflict. The main reason for this is that IHL started as a regime of rules between states, in a classical horizontal way of balancing the 'laws of humanity' with the 'needs of military necessity' in armed conflict. After World War II the scope broadened towards NIAC, but at the same time setting up certain conditions for the organization of non-state actors (APII and *Tadić*), trying to bring it back in some ways to a horizontal relationship. And this is important to keep in mind, because the law of armed conflict is only applicable to an armed conflict that fulfills certain requirements, usually with regard to the organization of the non-state party and the connection to a certain territory. This is what makes the approach of IHL in some ways effective: to establish rules which regulate armed violence of certain intensity in which at least in some ways the parties enjoy a comparable status, albeit with regard to the requirement of organization.

As a consequence, this means that in situations of protracted armed violence such as between Israel and Hezbollah, we can confirm the existence of an armed conflict, because the armed violence has reached a certain level, and the non-state actors follow a certain organizational pattern. Furthermore, due to the fact that we have a non-state actor on the one side, it is necessary to apply the law for NIAC even if there is a cross-border aspect involved. Due to the fact that nowadays the regimes

---

113 *Jaloud v The Netherlands*, Application no 47708/08, ECHR, Judgment, 20 November 2014, paras 152–153.

for IACs and NIACs have assimilated to a great extent[114]—at least in theory—it does not make a great difference which regime we actually apply in practice. However, if we talk about targeted killings against possible terrorists, this is not easily comparable. Even if these actions are conducted by the military with heavy weaponry such as armed unmanned aerial vehicles, the general character of the situation is much more comparable with that of a vertical enforcement action. There is no protracted armed violence, because we are usually dealing with single incidents of selected strikes against criminals (that is, terrorists) who are hiding in specific locations. The argument that the armed conflict follows the combatant cannot be sustained because we do not have any indications for this in the wording of the respective treaty provisions or state practice. Therefore, when we are leaving the area of armed conflict and entering the sphere of law-enforcement action, we need to take into account the principles known from human rights law when a decision is made on the targeting of a possible terrorist. An exception has to be made if organized non-state actors were to launch attacks from the third state at a certain level of intensity towards the government forces; then it would transfer again into a situation of a NIAC, and IHL would be applicable.

Consequently, the overall conclusion is that we need to stick to a rather strict standard when judging the existence of an armed conflict and its geographical scope of application. By extending the combat zone lightly, one would make it easier to allow for a wider range of targeted killings, circumventing the stricter human rights regime. Even with the consent of the 'host state', there should not be a possibility to directly kill suspects outside of emergency situations.

> Thus, the principle of humanity, more rationally supports a narrow view of the zone of combat's parameters, one that seeks to protect the most people by keeping conflict, and the battlefield, away from their countries altogether [...]. Because the risk of mistake increases dramatically as we move farther away from the conventional battlefield, humanity and its accompanying limitations on the use of force are ever more critical.[115]

As one commentator stated '[a]rmed conflict inevitably ha[s] a limited and identifiable territorial or spatial dimension because human beings who participate in armed conflict require territory in which to carry out intense, protracted, armed exchanges'.[116]

---

114 See Heinsch (n 54) 129–85.
115 Blank (n 3) 30.
116 O'Connell (n 37).

5. CONCLUSION

With regard to modern warfare, there is a danger in extending the combat zone because of the possibility of reaching remotely located territories. However, just because there might be a certain link of objects and persons with an ongoing conflict somewhere else, there is *not* an automatic assumption that the combat zone also covers areas where, for example, terrorist attacks are being planned or rebel groups are trained and financed. Rather, the decisive criterion is still the question whether the respective objects and persons are covered by the geographical scope of either an IAC or NIAC. This has to be decided on a case-by-case basis according to the traditional criteria laid down in the Geneva Conventions and their Additional Protocols, which have been interpreted extensively by the jurisprudence of the *ad hoc* tribunals. If we come to the conclusion that there is not an armed conflict, this leaves us with the application of the respective human rights standards, which then brings us into the area of the law enforcement paradigm. Although the problem of extraterritorial application of human rights has not yet been solved in a satisfactory way, it must be clear that the targeting of suspected terrorists through drones cannot happen outside of a legal framework.

4. The characterization of remote warfare under international humanitarian law

Anthony Cullen

1. INTRODUCTION

This chapter examines the qualification of remote warfare as a form of armed conflict under international humanitarian law. It does so first by considering how armed conflict is defined and how the concept has evolved since the drafting of the Geneva Conventions of 1949. It then focuses on three modes of attack that are commonly associated with remote warfare: the use of remotely piloted vehicles, cyber operations, and autonomous weapon systems. Bearing in mind the challenges that each of these present to the applicability of the law, it will be argued that the concept of armed conflict needs to be interpreted in terms consistent with the object and purpose of international humanitarian law, in accordance with Article 31 of the Vienna Convention on the Law of Treaties.

2. DEFINING 'ARMED CONFLICT'

The most fundamental prerequisite for the applicability of international humanitarian law is the existence of armed conflict. Without armed conflict, this body of law is deprived of the material field for its application. Accordingly, the characterization of the situation as one of armed conflict is of pivotal importance for the protection provided by international humanitarian law. In this section, the concept of armed conflict will be analysed, providing the basis for the qualification of remote warfare under international humanitarian law.

The applicability of international humanitarian law is determined by the terms of Articles 2 and 3 common to the four Geneva Conventions of 1949. Common Article 2 states that the Conventions will apply to 'all cases of declared war or of any other armed conflict which may arise between two or more of the High Contracting Parties, even if the state of war is not recognized by one of them'. Common Article 3 sets out the applicability of a 'minimum' of provisions '[i]n the case of armed conflict not of an international character'. Together, common Articles 2 and 3 define the applicability of the Geneva Conventions to situations of international and non-international armed conflict. The use of the term 'armed conflict' in both provisions was significant. It was the first time that term had been used to define the applicability of a treaty. As noted by the ICRC Commentary on the first Geneva Convention:

> It fills the gap left in the earlier Conventions, and deprives the belligerents of the pretexts they might in theory invoke for evasion of their obligations. There is no longer any need for a formal declaration of war, or for recognition of the state of war, as preliminaries to the application of the Convention. The Convention becomes applicable as from the actual opening of hostilities. The existence of armed conflict between two or more Contracting Parties brings it automatically into operation.
>
> It remains to ascertain what is meant by 'armed conflict'. The substitution of this much more general expression for the word 'war' was deliberate. One may argue almost endlessly about the legal definition of 'war'. A State can always pretend, when it commits a hostile act against another State, that it is not making war, but merely engaging in a police action, or acting in legitimate self-defence. The expression 'armed conflict' makes such arguments less easy.[1]

In this way, the use of 'armed conflict' in common Articles 2 and 3 avoided issues surrounding the legal characterization of 'war'.[2] The applicability of the law of war was expanded. With subsequent developments in treaty law, and changes in the nature of armed conflict, the meaning associated with the term has continued to evolve. One of the most significant turning points in this context was the decision of the ICTY Appeals Chamber in the *Tadić* case. In its Decision on the Defence Motion for Interlocutory Appeal on Jurisdiction, the Appeals Chamber defined the concept of armed conflict as follows:

> [a]n armed conflict exists whenever there is a resort to armed force between States or protracted armed violence between governmental authorities and

---

[1] Jean S Pictet (ed), *Commentary I Geneva Convention for the Amelioration of the Condition of the Wounded and Sick in Armed Forces in the Field* (ICRC 1952) 32.

[2] Even so, it still 'remains the case that some States deny the existence of armed conflicts, rendering dialogue difficult on the humanitarian consequences of the conflict and the protection of those affected by it.' International Committee of the Red Cross, *International humanitarian law and the challenges of contemporary armed conflicts*, 32nd International Conference of the Red Cross and Red Crescent, 32IC/15/11, October 2015, 7.

> organized armed groups or between such groups within a State. International humanitarian law applies from the initiation of such armed conflicts and extends beyond the cessation of hostilities until a general conclusion of peace is reached; or, in the case of internal conflicts, a peaceful settlement is achieved.[3]

This definition of armed conflict filled a lacuna that had previously existed in the law. The Geneva Conventions of 1949 did not include a definition of armed conflict. Although definitions were included in the additional protocols of 1977,[4] these definitions referred to specific categories of armed conflict; they did not address the conditions required for the application of international humanitarian law more generally in situations of international or non-international armed conflict. The concept of armed conflict propounded by the ICTY thus embodied a very significant development of the law. Sonja Boelaert-Suominen commented on its significance as follows:

> The seemingly innocuous description by the Appeals Chamber of what constitutes an armed conflict was innovative in various respects. First, it covers a variety of hypotheses and caters explicitly for conflicts between non-state entities. Second, whilst it sets a low threshold for the application of humanitarian law in general, it is particularly important for its consequences in relation to internal armed conflicts. The definition of armed conflict suggested by the Appeals Chamber covers not only the classic examples of (a) an armed conflict between two or more states and (b) a civil war between a state on the one hand, and a non-state entity on the other. It clearly encompasses a third situation, (c) an armed conflict in which no government party is involved, because two or more non-state entities are fighting each other.[5]

---

[3] *Prosecutor v Tadić*, Decision on the Defence Motion for Interlocutory Appeal on Jurisdiction, 2 October 1995, ICTY Case No. IT-94-1-AR72, para 70. For discussion of the definition provided by the *Tadić* Appeals Chamber see: Anthony Cullen, *The Concept of Non-International Armed Conflict in International Humanitarian Law* (Cambridge University Press 2010) 115–58.

[4] Protocol Additional to the Geneva Conventions of 12 August 1949, and Relating to the Protection of Victims of International Armed Conflicts (Additional Protocol I), 1125 UNTS 3, 1977, Article 1(4); and Protocol Additional to the Geneva Conventions of 12 August 1949, and Relating to the Protection of Victims of Non-International Armed Conflicts (Protocol II), 1125 UNTS 609, 1977, Article 1.

[5] Sonja Boelaert-Suominen, 'The Yugoslav Tribunal and the Common Core of Humanitarian Law Applicable to All Armed Conflicts' (2000) 13 Leiden Journal of International Law 619 at 632–3. According to Christopher Greenwood,

The *Tadić* definition, included in obiter and credited to the presiding judge Antonio Cassese,[6] has become one of the most authoritative points of reference in the characterization of armed conflict under international humanitarian in law.[7] In doing so, it has broadened the applicability of the Geneva Conventions beyond the conditions considered by the drafters of these treaties in 1949. Nevertheless, it preserves the distinction introduced by the Geneva Conventions between international armed conflict (under common Article 2) and non-international armed conflict (under common Article 3).[8] The section that follows considers the relevance of the concept to the practice of remote warfare.

3. THE CHARACTERIZATION OF REMOTE WARFARE UNDER THE LAW OF ARMED CONFLICT

As illustrated in the introductory chapter to this volume, remote warfare has existed from time immemorial. In terms of legal regulation, the challenge has been one of responding to changes in the actual conduct of armed conflict. This section focuses on changes arising from three new

---

> The definitions of international and internal armed conflict are of considerable importance. Neither term is defined in the Geneva Conventions or other applicable agreements. Whereas there is an extensive literature on the definition of 'war' in international law, armed conflict has always been considered a purely factual notion and there have been few attempts to define or even describe it.

Christopher Greenwood, 'The Development of International Law by the International Criminal Tribunal for the Former Yugoslavia' (1998) 2 Max Planck Yearbook of United Nations Law 97 at 114.

6 Peter Rowe, 'The International Criminal Tribunal for Yugoslavia: The Decision of the Appeals Chamber on the Interlocutory Appeal on Jurisdiction in the *Tadic* Case' (1996) 45 International and Comparative Law Quarterly 691 at 697.

7 See generally: Anthony Cullen, *The Concept of Non-International Armed Conflict in International Humanitarian Law* (Cambridge University Press 2010) 117–58.

8 As noted by Lindsay Moir, '[t]he characterization of an armed conflict as being either international or non-international in nature remains a vital exercise for determining the applicability of different rules of IHL.' Lindsay Moir, 'The Concept of Non-International Armed Conflict' in Andrew Clapham, Paola Gaeta and Marco Sassòli (eds), *The Geneva Conventions: A Commentary* (Oxford University Press 2015) 391–414, 414.

categories of weapons: remotely piloted vehicles (drones), cyber weapons; and autonomous weapon systems. In doing so, it will consider the impact of each for the characterization of armed conflict under international humanitarian law.

Remotely Piloted Vehicles (Drones)

The use of drones for the targeted killing of suspected terrorists has been a subject considerable debate among scholars of international law, in particular since the killing of Qaed Salim Sinan al-Harethi in November 2002.[9] Feeding this debate has been discussions concerning the ethical, humanitarian and legal implications of US foreign policy.[10] For scholars, activists and professionals in the field of international humanitarian law, a significant part of the debate concerning the use of such weapons has centred on the context for their use and the characterization of this context as one of 'armed conflict'. In the absence of the conditions described in *Tadić*, questions have been raised concerning the lawfulness of attacks undertaken remotely using such weapons and the applicability

---

[9] Qaed Salim Sinan al-Harethi was killed by the CIA in Yemen using an unmanned Predator drone on 3 November 2002. See Chris Downes, '"Targeted Killings" in an Age of Terror: The Legality of the Yemen Strike' (2004) 9 Journal of Conflict and Security Law 277–94; A P V Rogers, *Law on the Battlefield* (Oxford University Press 2013) 50–51; Noam Lubell, 'The War (?) against Al-Qaeda' in Elizabeth Wilmshurst (ed), *International Law and the Classification of Conflicts* (Oxford University Press 2012) 450.

[10] See: Claire Finkelstein, Jens David Ohlin, and Andrew Altman, *Targeted Killings: Law and Morality in an Asymmetrical World* (Oxford University Press, 2012); Bradley Jay Strawser (ed), *Killing by Remote Control: The Ethics of an Unmanned Military* (Oxford University Press 2013); Dan Saxon, *International Humanitarian Law and the Changing Technology of War* (Brill, 2013); Sikander Ahmed Shah, *International Law and Drone Strikes in Pakistan: The Legal and Socio-political Aspects* (Routledge, 2014); James DeShaw Rae, *Analyzing the Drone Debates: Targeted Killings, Remote Warfare, and Military Technology* (Palgrave Macmillan, 2014); Sarah Knuckey (ed), *Drones and Targeted Killings: Ethics, Law, Politics* (International Debate Education Association 2014); Steven Barela (ed), *Legitimacy and Drones: Investigating the Legality, Morality and Efficacy of UCAVs* (Ashgate, 2015); Aleš Završnik, *Drones and Unmanned Aerial Systems: Legal and Social Implications for Security and Surveillance* (Springer International Publishing, 2015); Jammel Jaffer (ed), *The Drone Memos: Targeted Killing, Secrecy, and the Law* (New Press, 2016); and Bart Custers (ed), *The Future of Drone Use: Opportunities and Threats from Ethical and Legal Perspectives* (Information Technology and Law Series, Springer, 2016).

of international humanitarian law. The significance of characterizing the context as one of armed conflict was highlighted in the *Study on Targeted Killings* authored by the UN Special Rapporteur on extrajudicial, summary or arbitrary executions, Philip Alston:

> Outside the context of armed conflict, the use of drones for targeted killing is almost never likely to be legal. A targeted drone killing in a State's own territory, over which the State has control, would be very unlikely to meet human rights law limitations on the use of lethal force.[11]

The position adopted by the government of the United States has been a starting point for many discussions on the applicability of international humanitarian law to drone warfare. The position of the United States is that since 11 September 2001 it has been engaged in an armed conflict with 'al-Qaida and associated forces'. Although references to a 'war on terror' were avoided under the Obama administration, there was continuity with the Bush administration in the policy adopted regarding the characterization of the campaign as one of armed conflict. A statement of this position was provided in 2012 by John Brennan, then a Legal Advisor to President Obama:

> As the President has said many times, we are at war with al-Qa'ida … Our ongoing armed conflict with al-Qa'ida stems from our right—recognized under international law—to self defense.
>
> An area in which there is some disagreement is the geographic scope of the conflict. The United States does not view our authority to use military force against al-Qa'ida as being restricted solely to 'hot' battlefields like Afghanistan. Because we are engaged in an armed conflict with al-Qa'ida, the United States takes the legal position that—in accordance with international law—we have the authority to take action against al-Qa'ida and its associated forces without doing a separate self-defense analysis each time. And as President Obama has stated on numerous occasions, we reserve the right to take unilateral action if or when other governments are unwilling or unable to take the necessary actions themselves.

---

[11] UN Special Rapporteur on Extrajudicial, Summary or Arbitrary Executions, Study on Targeted Killings, UN Doc A/HRC/14/24/Add.6, 28 May 2010, para 85. For a discussion of the US position on this point, see: Max Brookman-Byrne, 'Drone Use 'Outside Areas of Active Hostilities': An Examination of the Legal Paradigms Governing US Covert Remote Strikes' (2017) 64 *Netherlands International Law Review* 3.

Brennan's statement, like others issued by representatives of the Obama administration,[12] conflates the law of armed conflict with the right of self-defense. In terms of scope, the campaign is open-ended. Although Obama distanced himself from the idea of perpetual war, the duration of the campaign against al-Qaida and associated forces is one that is not limited by the timeframe of a US government administration. When asked during a Senate hearing in 2013 about the anticipated duration of the campaign, the Assistant Secretary of Defense for Special Operations and Low-Intensity Conflict, Michael Sheehan answered 'at least 10 to 20 years'.[13] This appears to suggest no predetermined limit to the duration of the campaign.

With regard to the geographic scope of the campaign, it is generally accepted that international humanitarian law applies to the theatre of hostilities, to places where prisoners of war are detained and to areas under the control of a party to the conflict. However, as Brennan mentioned in his statement, the United States does not view its campaign as being confined to 'hot' battlefields like the ones in Afghanistan. The campaign crosses many national boundaries: Besides Afghanistan, attacks have been reported in Yemen, Somalia, Pakistan, Iraq, Mali and Libya.

When attacks using drones are undertaken in the context of a pre-existing local armed conflict—whether international or non-international in nature—it is undoubtable that international humanitarian law would apply to these operations. The situation is less clear where there is no pre-existing armed conflict and no consent for the use of UAVs against suspects terrorists from the authorities of the state in which the attack takes place. According to the US position, the 'armed conflict' is one that is global; it is one that follows wherever the use of lethal force is authorized by the US government, the exercise of which is justified on a continuing basis of self-defense.

In addition to targeting 'al-Qa'ida and its associated forces', the approach adopted by the United States has also been extended to the

---

[12] For collection of relevant US legal and policy documents (including John Brennan's statement), see: Jammel Jaffer (ed), *The Drone Memos: Targeted Killing, Secrecy, and the Law* (New Press, 2016).

[13] See: Glenn Greenwald, 'Washington gets explicit: its "war on terror" is permanent', *The Guardian*, 17 May 2013, accessed 4 June 2017 at https://www.theguardian.com/commentisfree/2013/may/17/endless-war-on-terror-obama. See also: Spencer Ackerman, 'Pentagon Spec Ops Chief Sees '10 to 20' More Years of War Against al-Qaida', *Wired*, 16 May 2013, accessed 4 June 2017 at https://www.wired.com/2013/05/decades-of-war/

Islamic State (ISIS). In remarks made at the US-ASEAN Press Conference on 16 February 2016, President Obama stated: 'I have been clear from the outset that we will go after ISIS wherever it appears, the same way that we went after al Qaeda wherever they appeared.'[14] This position on the extraterritorial use of lethal force has attracted expressions of concern from various quarters.[15] When questioned by the chair of the UK Parliament's Joint Committee on Human Rights, the UK Secretary for Defence, Michael Fallon, acknowledged differences between the position of the United States and that of the United Kingdom. With regard to the characterization of the campaign as a non-international armed conflict, Fallon stated:

> It is for the Americans to defend or describe their own definition. We would consider on a case-by-case basis, where there is an armed conflict between government authorities and various organised armed groups, and we would look at various factors case-by-case … such as the duration or intensity of the fighting.[16]

Recognizing differences in the legal positions of the United States and United Kingdom on the extraterritorial use of lethal force, the Joint Parliamentary Committee on Human Rights emphasized the urgent need for greater clarity from the government of the United Kingdom:

> The UK's support for this use of lethal force abroad by the US demonstrates the urgent need for the Government to clarify its understanding of the legal basis for the UK's policy. The US policy, in short, is that it is in a global armed conflict with ISIL/Da'esh, as it has been since 9/11 with al-Qaida,

---

14 Office of the White House Press Secretary, Remarks by President Obama at U.S.-ASEAN Press Conference, February 16, 2016, accessed 4 June 2017 at https://www.whitehouse.gov/the-press-office/2016/02/16/remarks-president-obama-us-asean-press-conference.

15 For example, see: Amnesty International, 'Doctrine of pervasive 'war' continues to undermine human rights: A reflection on the ninth anniversary of the AUMF', AI Index: AMR 51/085/2010, 15 September 2010, 2; Human Rights Watch, Letter to President Obama: Targeted Killings by the US Government, 16 December 2011, accessed 4 June 2017 at https://www.hrw.org/news/2011/12/16/letter-president-obama-targeted-killings-us-government; UN News Centre, 'UN independent expert voices concerns over the practice of targeted killings', 2 June 2010, accessed 4 June 2017 at goo.gl/yysKob; or UN News Centre, 'UN human rights expert questions targeted killings and use of lethal force', 20 October 2011, accessed 4 June 2017 at goo.gl/a5K3wC.

16 UK Parliament Joint Committee on Human Rights, Oral evidence: The UK Government's policy on the use of drones for targeted killing, HC 574, Wednesday, 16 December 2015, 3.

> which entitles it to use lethal force against it 'wherever they appear.' On this view, the Law of War applies to any such use of force against ISIL/Da'esh, wherever they may be. This is not, however, the position of the UK Government. As the Defence Secretary made clear in his evidence to us, the Government considers itself to be in armed conflict with ISIL/Da'esh only in Iraq and Syria.[17]

It is noteworthy in this context that the position adopted by the United States on characterization of its campaign departs from prevailing views of what armed conflict consists of. The International Committee of the Red Cross, an organization regarded as the 'guardian of international humanitarian law', has stated that it 'does not share the view that a conflict of global dimensions is or has been taking place'.[18] Consistent with the approach developed in the jurisprudence of the ICTY, the position of the ICRC is that the applicability of international humanitarian law is triggered by 'violence [reaching] the threshold of armed conflict, whether international or non-international'.[19] Accordingly, the characterization of a situation as one of armed conflict is to be determined on a case-by-case basis:

> [E]ach situation of organized armed violence must be examined in the specific context in which it takes place and must be legally qualified as armed conflict, or not, based on the factual circumstances. The law of war was tailored for situations of armed conflict, both from a practical and a legal standpoint. One should always remember that IHL rules on what constitutes

---

[17] UK Parliament Joint Committee on Human Rights, The Government's policy on the use of drones for targeted killing: Second Report of Session 2015–16, HC 574, HL Paper 141, 10 May 2016, 58.

[18] ICRC, *International Humanitarian Law and the challenges of contemporary armed conflicts*, 31st International Conference of the Red Cross and Red Crescent, 31IC/11/5.1.2, October 2011, 10. See also: Amnesty International, 'Doctrine of pervasive 'war' continues to undermine human rights: A reflection on the ninth anniversary of the AUMF', AI Index: AMR 51/085/2010, 15 September 2010, 2. ('[T]here is no place in international humanitarian and human rights law for a legal category of global and pervasive but non-international armed conflict, suspending the ordinary rule of law and human rights whenever and wherever an individual state deems necessary, as distinct from a series of specific geographic zones of international or non-international armed conflict.') For a contrary view, see: Noam Lubell, 'Fragmented Wars: Multi-Territorial Military Operations against Armed Groups' (2017) 93 *International Law Studies* 215 at 245.

[19] ICRC, *International Humanitarian Law and the challenges of contemporary armed conflicts*, 30th International Conference of the Red Cross and Red Crescent, 30IC/07/8.4, October 2007, 7.

> the lawful taking of life or on detention in international armed conflicts, for example, allow for more flexibility than the rules applicable in non-armed conflicts governed by other bodies of law, such as human rights law. In other words, it is both dangerous and unnecessary, in practical terms, to apply IHL to situations that do not amount to war.[20]

For Professor Christine Gray, '[i]t is the substantive law that is crucial, and it is here that the USA's position is weakest'.[21] Although armed conflict has evolved considerably since the drafting of the Geneva Conventions of 1949, the concept still possesses temporal and geographic scope. In failing to take this into account, the position adopted by the United States blurs the distinction between peace and war. As noted by Christof Heyns, '[t]he danger is one of a global war without borders'.[22] It also 'raises the question why other States should not engage in the same practices'.[23] Ultimately, as noted by the Netherlands Advisory Committee on Issues of Public International Law, to avoid:

> setting precedents that could be used by other states or entities in the fairly near future, it is vital that the existing international legal framework for the deployment of such a weapons system be consistently and strictly complied with. States need to be as clear as possible about the legal bases invoked when deploying armed drones.[24]

---

[20] ICRC, *International Humanitarian Law and the challenges of contemporary armed conflicts*, 30th International Conference of the Red Cross and Red Crescent, 30IC/07/8.4, October 2007, 8. On the applicability of international humanitarian law to drones, see the reports of UN Special Rapporteurs, Philip Alston, Ben Emmerson and Christof Heynes: Report of the Special Rapporteur on extrajudicial, summary or arbitrary executions, Philip Alston, Study on Targeted Killings, UN Doc A/HRC/14/24/Add.6, 28 May 2010; Report of the Special Rapporteur on the promotion and protection of human rights and fundamental freedoms while countering terrorism, Ben Emmerson, UN Doc A/68/389, 18 September 2013; Report of the Special Rapporteur on extrajudicial, summary or arbitrary executions, Christof Heyns, UN Doc A/68/382, 13 September 2013; Report of the Special Rapporteur on the promotion and protection of human rights and fundamental freedoms while countering terrorism, Ben Emmerson, UN Doc A/HRC/25/59, 11 March 2014.

[21] Christine Gray, 'Targeted Killings: Recent US Attempts to Create a Legal Framework' (2013) 66 Current Legal Problems 75–106, 106.

[22] UN News Centre, 'UN human rights expert questions targeted killings and use of lethal force', 20 October 2011, accessed 4 June 2017 at goo.gl/a5K3wC.

[23] Ibid.

[24] Advisory Committee on Issues of Public International Law (CAVV), *Advisory Report on Armed Drones*, Advisory Report No. 23, The Hague, July 2013, 27–8.

For the use of drones to be lawful as a form of remote warfare, the context must be one of armed conflict. The absence of clarity concerning the characterization of situations as such is detrimental not only to applicable legal regimes but also for the maintenance of international peace and security. The concern has also been raised that 'the use of armed drones for killings in remote places with little or no risk to one's own forces raises the issue of lowering the threshold to the point of trivialising such interventions and of accountability for the actual outcome of each strike'.[25] Considering the continued growth in the deployment of armed drones, and the frequently transnational nature of their use, it is arguable that more attention would be useful at an international level to strengthen compliance with the law. Christof Heyns, Dapo Akande, Lawrence Hill-Cawthorne and Thompson Chengeta contend that:

> There is an urgent need for the international community to gain greater consensus on the interpretation of the constraints that international law in all its manifestations places on the use of drones. This is important not only because of the implications for those who currently find themselves on the receiving end of drones, but in order to keep a viable and strong system of international security intact. A central component of such a security system is the rule of law. Drones should follow the law, not the other way around.[26]

As the context for the use of armed drones determines the applicability of international humanitarian law, so it is with other forms of remote warfare. The section that follows examines cyber operations and explores

---

[25] Arcadio Díaz Tejera, 'Drones and targeted killings: the need to uphold human rights and international law', a report issued to Committee on Legal Affairs and Human Rights of the Parliamentary Assembly of the Council of Europe, Doc. 13731, 16 March 2015, para 61.

[26] Written evidence from Christof Heyns, Dapo Akande, Lawrence Hill-Cawthorne and Thompson Chengeta (DRO0024), 'The Right to Life and the International Law Framework Regulating the Use of Armed Drones in Armed Conflict or Counter-Terrorism Operations', 10 December 2015, 46. See: Christof Heyns, Dapo Akande, Lawrence Hill-Cawthorne and Thompson Chengeta, 'The right to life and the international law framework regulating the use of armed drones' (2016) 65 *International and Comparative Law Quarterly* 791 at 826. See also: Summary of the Human Rights Council interactive panel discussion of experts on the use of remotely piloted aircraft or armed drones in compliance with international law: Report of the Office of the United Nations High Commissioner for Human Rights, UN Doc A/HRC/28/38, 15 December 2014, para 56.

issues surrounding the characterization of such as a form of armed conflict.

Cyber Operations

In its 2015 report *International humanitarian law and the challenges of contemporary armed conflicts*, the International Committee of the Red Cross defined 'cyber warfare' as 'operations against a computer or a computer system through a data stream, when used as means and methods of warfare in the context of an armed conflict, as defined under IHL'.[27] According to the *Tallinn Manual on the International Law Applicable to Cyber Warfare*: 'A cyber attack is a cyber operation, whether offensive or defensive, that is reasonably expected to cause injury or death to persons or damage or destruction to objects.'[28] Although consensus has yet to emerge on the characterization of cyber warfare, as a weapon employed in the context of armed conflict the applicability of international humanitarian law to cyber operations is beyond doubt. According to Gary Solis:

> If there is a circumstance in armed conflict that was unforeseen (and unforeseeable) by the 1949 Geneva Conventions, it is cyber warfare. Still, cyber warfare can be dealt with using traditional law of war tools, recognizing that today's jus ad bellum cyber war questions can instantly ripen into jus in bello issues. Cyber attacks are not per se LOAC violations. They are another strategy or tactic of warfare ... When considering their effect or use, they may be thought of as being similar to kinetic weapons.[29]

As noted by William Boothby, '[t]he law of armed conflict contains no ad hoc rules that ... permit, prohibit, or restrict the lawful circumstances

---

[27] ICRC, *International humanitarian law and the challenges of contemporary armed conflicts*, Doc. 32IC/15/11, Geneva, October 2015, 39.

[28] Michael N Schmitt (ed), *Tallinn Manual on the International Law Applicable to Cyber Warfare* (Cambridge University Press, 2013) 106. See also Michael N Schmitt, 'Classification in Future Conflict' in Elizabeth Wilmshurst (ed), *International Law and the Classification of Conflicts* (Oxford University Press, 2012) 455–77, 461 ('Intuitively, it seems that the determinative criterion must be consequence severity. Death, injury, damage or destruction clearly qualify an action as armed conflict, while inconvenience and irritation do not. But beyond that, the law is uncertain.').

[29] Gary Solis, *The Law of Armed Conflict: International Humanitarian Law in War* (Cambridge University Press 2016) 702.

of use of cyber weapons as such'.[30] However, it is clear that cyber weapons are to be governed by the same rules that regulate the use of weapons more generally under international humanitarian law. Principles of distinction, proportionality and military necessity would apply to attacks undertaken by way of cyber operations. As a form of remote warfare, cyber operations must comply with the relevant rules of international humanitarian law, including prohibitions on indiscriminate attacks or attacks likely to cause superfluous injury or unnecessary suffering. The issue of ensuring compliance with such rules is, however, thwarted by the secretive nature of cyber operations, the lack of transparency under which attacks are undertaken and the absence of a treaty specifically concerned with the regulation of cyberwarfare. Solis comments that:

> Defining many aspects of cyber warfare is problematic because there is no multinational treaty directly dealing with cyber warfare. That is because, so far, many aspects of cyber war are not agreed upon. The law of war, as well as customary international law, lacks cyber-specific norms, and state practice interpreting applicable norms is slow to evolve.[31]

As a form of remote warfare, there are many issues which impact on the characterization of cyber operations as armed conflict under international humanitarian law. Questions concerning the attribution of attacks, the nature of operations required for the threshold of cyber warfare, and the classification of armed conflicts initiated in this way, all pose challenges to ensuring compliance and prompt calls for the further development of the law. The section that follows explores another form of remote warfare which has similarly prompted calls for the development of international humanitarian law, to regulate a method of warfare not anticipated by the drafters of the Geneva Conventions of 1949.

Autonomous Weapon Systems

While the use of drones and cyber operations present their own distinct challenges to the conceptual basis for the characterization of armed conflict, the use of autonomous weapon systems has been described as a

---

30 William H Boothby, *Weapons and the Law of Armed Conflict* (Oxford University Press 2016) 241.

31 Solis (n 29) 673–4.

potential 'paradigm shift'.[32] Autonomous Weapon Systems (AWS), also referred to as Lethal Autonomous Weapon Systems (LAWS), have been defined by the ICRC as '[a]ny weapon system with autonomy in its critical functions. That is, a weapon system that can select (i.e. search for or detect, identify, track, select) and attack (i.e. use force against, neutralize, damage or destroy) targets without human intervention'.[33] However, as noted by Michael W Meier, the US government representative at the third CCW Meeting of Experts on LAWS in April 2016, 'views on what would constitute LAWS have varied greatly'.[34]

In terms of legal regulation, much of the debate has centred on the degree of 'autonomy' exercised in the use of lethal force.[35] In the absence of meaningful human control, questions have been raised as to whether compliance with international humanitarian law would actually be possible with Autonomous Weapon Systems. This was reflected in the report of the 2016 Convention on Conventional Weapons (CCW) meeting of experts that took place at the United Nations in Geneva from 11 to 15 April 2016. The report submitted by the chairperson, Ambassador Michael Biontino of Germany, states:

---

[32] Jakob Kellenberger, 'Keynote address' in Wolff Heintschel von Heinegg (ed), *International Humanitarian Law and New Weapon Technologies, 34th Round Table on Current Issues of International Humanitarian Law, 8–10 September 2011* (International Institute of Humanitarian Law 2012) 23–7, 27; and Robin Geiß, *The International-Law Dimension of Autonomous Weapons Systems* (Friedrich-Ebert-Stiftung, International Policy Analysis 2015) 3.

[33] ICRC, 'Views of the International Committee of the Red Cross (ICRC) on autonomous weapon system', Convention on Certain Conventional Weapons (CCW) Meeting of Experts on Lethal Autonomous Weapons Systems (LAWS), 11–15 April 2016, Geneva, 11 April 2016, 1.

[34] Michael W Meier, US Delegation Opening Statement, The Convention on Certain Conventional Weapons (CCW) Informal Meeting of Experts on Lethal Autonomous Weapons Systems, Geneva, 11 April 2016, 2. The following definition of 'autonomous weapon system' has been used by the US Department of Defense: 'A weapon system that, once activated, can select and engage targets without further intervention by a human operator. This includes human-supervised autonomous weapon systems that are designed to allow human operators to override operation of the weapon system, but can select and engage targets without further human input after activation.' Department of Defense Directive 3000.09, Autonomy in Weapon Systems, 21 November 2012, 13–14.

[35] For a discussion of the different aspects of autonomy, see Tim McFarland, 'Factors shaping the legal implications of increasingly autonomous military systems' (2015) *International Review of the Red Cross* 1. See also: Duncan Hollis, 'Setting the Stage: Autonomous Legal Reasoning in International Humanitarian Law' (2016) 30 *Temple International and Comparative Law Journal* 1.

> 44. It was of common understanding that, as with all weapon systems, the rules of IHL are fully applicable to LAWS. However, many delegations questioned whether weapons systems that select and attack targets autonomously would be able to comply with these rules.
>
> 45. A number of delegations argued that human judgment was necessary in order to assess the fundamental principles of proportionality, distinction and precautions in attack. For this reason, it was recognized that a human operator should always be involved in the application of force. Many delegations questioned if it would be possible to programme a legal assessment into a machine prior to its deployment. Given the rapidly changing circumstances in a conflict, it would be difficult to conceive of a LAWS distinguishing between lawful and unlawful targets. For example, it was unclear as to how LAWS could be programmed to recognize the surrender of a combatant or take feasible precautions in attack. Additionally, it was noted that a potential target may alter its behaviour in order to deliberately confuse assessments made by a machine.[36]

The report of the CCW expert meeting states that: 'Most delegations maintained that machines are simply incapable of executing legal judgements as required by IHL, especially in complex and cluttered environments typical in conflict scenarios.'[37] In addition to the absence of meaningful human control in the selection and attack of targets, significant issues of accountability are raised by the use of Autonomous Weapons Systems. Given the technology's state of development, it is currently not clear how the doctrine of command responsibility would apply to attacks undertaken using such weapons. This was also reflected in discussions at the CCW expert meeting:

> Accountability was highlighted as a central element of IHL. Doubts were raised over whether the required standards of accountability and responsibility for the use of force and its effects could be upheld with the deployment of LAWS. In the case of an incident involving LAWS, it was uncertain as to who

---

[36] CCW, Report of the 2016 Informal Meeting of Experts on Lethal Autonomous Weapons Systems (LAWS), Geneva, 11–15 April 2016, paras 44–45, accessed 4 June 2017 at goo.gl/Sy0y8W.

[37] CCW, Report of the 2016 Informal Meeting of Experts (n 36), para 46. See also: Statement of the International Committee of the Red Cross (ICRC), Convention on Certain Conventional Weapons (CCW) Meeting of Experts on Lethal Autonomous Weapons Systems (LAWS), 13–16 May 2014, Geneva, 13 April 2015, 3. ('Based on current and foreseeable technology, there are serious doubts about the ability of autonomous weapon systems to comply with IHL in all but the narrowest of scenarios and the simplest of environments. In this respect, it seems evident that overall human control over the selection of targets and use of force against them will continue to be required.')

> would be held accountable within the chain of command or responsibility, such as the commander, programmer, or operator. As a result, it was argued by some that legal grey zones could emerge, which in turn might be deliberately exploited and foster impunity. Others noted that this would not be the case, but that evidentiary issues may arise. It was proposed that there should be a requirement for LAWS to keep records of their operations. Other delegations responded that, if LAWS can be used in compliance with IHL, there would not be an accountability gap as any issues could be addressed under international criminal law and the law of State responsibility.[38]

In light of the above, it was recommended that further consideration be given to the question of 'legal and political responsibility and accountability'.[39]

With regard to the applicability of international humanitarian law, concerns have been expressed that autonomous weapon systems may lower the threshold required for the qualification of a situation as one of armed conflict.[40] In addition, the absence of human participation poses a challenge as to how the parties to armed conflicts are to be characterized. When two sides engage in hostilities through the use of autonomous weapons systems and there is no direct human participation in the conflict from either side, does the law of armed conflict apply? In other words, is it possible to qualify a situation as one of armed conflict if none of the parties directly engaged in hostilities are human beings? The answer to this question is arguably best addressed by considering rules relating to the interpretation of international humanitarian law under customary international law and the Vienna Convention on the Law of Treaties. The section that follows will explore how such rules could be potentially applied to autonomous weapons systems and to the other forms of remote warfare discussed above.

---

[38] CCW, Report of the 2016 Informal Meeting of Experts (n 36), para 52.

[39] CCW, Recommendations to the 2016 Review Conference, 11–15 April 2016, 2, accessed 4 June 2017 at goo.gl/qDdGgS.

[40] ICRC, Report of the Expert Meeting on Autonomous Weapon Systems: Technical, Military, Legal and Humanitarian Aspects, Geneva, 26–28 March 2014 (ICRC 2014), 17–18; Geiß (n 32) 23. It was recommended the 'effects on the threshold for armed conflicts' be given further consideration by the 2016 CCW meeting of experts: Recommendations to the 2016 Review Conference (n 39) 2.

4. RESPONDING TO THE CHANGING NATURE OF ARMED CONFLICT

Hersch Lauterpacht commented in the 1950s that 'if international law is, in some ways, at the vanishing point of law, the law of war is, perhaps even more conspicuously, at the vanishing point of international law'.[41] If the law of armed conflict has a vanishing point in the 21st century, it is arguably that of remote warfare. The challenges posed by drones, cyber operations and autonomous weapons systems to the applicability of international humanitarian law go well beyond the conditions of warfare contemplated by the drafters of the Geneva Conventions of 1949. On account of this, it is essential to consider rules that govern the interpretation of such treaties. As mentioned above, the concepts of international and non-international armed conflict are linked to Articles 2 and 3 common to the four Geneva Conventions. If international humanitarian law is to be deemed applicable to the different forms of remote warfare, it must be interpreted in terms consistent with the scope of these provisions. In this context, reference must be made to the terms of Article 31(1) of the Vienna Convention on the Law of Treaties which states the following general rule of interpretation:

> A treaty shall be interpreted in good faith in accordance with the ordinary meaning to be given to the terms of the treaty in their context and in the light of its object and purpose.[42]

The status of this rule as customary international law has been confirmed in a number of cases before the International Court of Justice, including the *La Grand* case (*Germany v the United States*) in 2001,[43] the *Wall* Advisory Opinion in 2004,[44] and the case concerning the *Application of the Genocide Convention (Bosnia and Herzegovina v Serbia and*

---

[41] Hersch Lauterpacht, 'The Problem of the Revision of the Law of War' (1952) 29 British Yearbook of International Law 360–82, 382.

[42] Vienna Convention on the Law of Treaties, 1155 UNTS 331, 8 ILM 679, entered into force 27 January 1980.

[43] *La Grand case (Germany v United States of America)*, Judgment of 27 June 2001, ICJ Reports 2001, 501, para 99.

[44] *Legal Consequences of the Construction of a Wall in the Occupied Palestinian Territory*, Advisory Opinion of 9 July 2004, ICJ Reports 2004, 174, para 94.

Montenegro) in 2007.[45] The significance of a treaty's 'object and purpose' is underscored by the fact that the term is used eight times in the Vienna Convention.[46] For applicability of international humanitarian law, the 'object and purpose' of the Geneva Conventions and the Additional Protocols thereto is of pivotal importance to the interpretation of what 'armed conflict' consists of. This leads to the question as to how the object and purpose of these treaties should be characterized. The ILC Guide to Practice on Reservations to Treaties describes the approach adopted by the International Court of Justice:

> [T]he International Court of Justice has deduced the object and purpose of a treaty from a number of highly disparate elements, taken individually or in combination:
>
> - From its title;
> - From its preamble;
> - From an article placed at the beginning of the treaty that 'must be regarded as fixing an objective, in the light of which the other treaty provisions are to be interpreted and applied';
> - From an article of the treaty that demonstrates 'the major concern of each contracting party' when it concluded the treaty;
> - From the preparatory works on the treaty; and
> - From its overall framework.[47]

Applying these elements to the Geneva Conventions of 1949 and their Additional Protocols, the object and purpose of international humanitarian law may be characterized as the protection of victims of armed conflicts. While the titles of the Geneva Conventions specify different categories of protected persons, collectively they have been referred to as 'International Conventions for the Protection of War Victims'.[48] The title

---

45 *Application of the Convention on the Prevention and Punishment of the Crime of Genocide (Bosnia and Herzegovina v Serbia and Montenegro)*, Judgment of 26 February 2007, ICJ Reports 2007, 110, para 160.

46 The 'object and purpose' of a treaty is of relevance not only for its interpretation but also with regard to obligations that exist prior to the entry into force of the treaty (Article 18), reservations (Article 19(c) and Article 20(2)), modifications (Article 41(1)(b)(ii)), and the possibility of suspending operation of the treaty (Article 58(1)(b)(ii)).

47 Report of the International Law Commission: Sixty-third session (26 April–3 June and 4 July–12 August 2011), UN Doc A/66/10/Add.1 (United Nations 2011) 360–61.

48 Final Record of the Diplomatic Conference of Geneva (Federal Political Department 1949), Vol I, 5.

of Additional Protocol I refers to 'the Protection of Victims of International Armed Conflicts', while the title of Additional Protocol II refers to 'the Protection of Victims of Non-International Armed Conflicts'.[49] The preamble of Additional Protocol I states the belief of High Contracting Parties that it is necessary to 'reaffirm and develop the provisions protecting the victims of armed conflicts and to supplement measures intended to reinforce their application', while the preamble of Additional Protocol II emphasizes 'the need to ensure a better protection for the victims of [non-international] armed conflicts'. The terms of these provisions are significant in that they set the *context* for the interpretation of operative provisions.

The article 'placed at the beginning' of each of the four Geneva Conventions (Article 1) states: 'The High Contracting Parties undertake to respect and to ensure respect for the present Convention in all circumstances.' Taken together with the titles (both individual and collective), the provision reinforces the overall framework of each treaty for protection of victims of armed conflict. The *travaux préparatoire* is also consistent with this, reflected in the positions expressed by States involved in drafting process. For example, the Mexican Ambassador at the 1949 Diplomatic Conference stated: 'This Conference was convened to examine the problem of protecting war victims. Each of our four working documents [that is the draft Conventions] has its own individual character; but they all have the same purpose – the protection of victims of war.'[50]

Proceeding from the premise that the object and purpose of international humanitarian law is to further the protection of the victims of armed conflict, how does this impact on the characterization of remote warfare? This question is arguably best addressed on a case-by-case basis by considering how the applicability of international humanitarian law is consistent with its object and purpose. For example, while the characterization of the campaign against 'al-Qa'ida and its associated forces' as a non-international armed conflict provides a context for the use of lethal force, it is not clear how it serves to realize the protection provided by the law. On the contrary, the potential exists for such a characterization—

---

[49] Protocol Additional to the Geneva Conventions of 12 August 1949, and Relating to the Protection of Victims of International Armed Conflicts (Additional Protocol I), 1125 UNTS 3, 1977; Protocol Additional to the Geneva Conventions of 12 August 1949, and Relating to the Protection of Victims of Non-International Armed Conflicts (Protocol II), 1125 UNTS 609, 1977.

[50] *Final Record of the Diplomatic Conference of Geneva of 1949* (Federal Political Department 1949), Vol 2-B, 332–3.

not limited by time or geography—to undermine rather than strengthen the protection provided, creating 'global war without borders, in which no one is safe'.[51]

If the concept of armed conflict is to be interpreted to accommodate new forms of warfare, this development must be consistent with the object and purpose of international humanitarian law. If not, then the integrity of the law and its utility in situations of armed conflict will be undermined. As noted in the conclusions of an Expert Panel convened in 2014 on the use of remotely piloted aircraft or armed drones:

> The starting point of any legal analysis on armed drones should be existing international law, in particular the prohibition against the arbitrary deprivation of life. Modifying well-established rules of international law to accommodate the use of drones might have the unintended long-term consequence of weakening those rules. The existing legal framework was sufficient and did not need to be adapted to the use of drones, rather, it was the use of armed drones that must comply with international law.[52]

With regard to cyber warfare, the issue of characterization is complicated by the lack of consensus on how it is to be defined, problems of attribution and the absence of an international agreement clarifying the applicability of international humanitarian law to cyber operations. Although it is clear that international humanitarian law would apply once the threshold of armed conflict is reached, different views exist on the characterization of cyber operations. According to Noam Lubell:

> Cyber operations are a classic example of an attempt to fit things into the laws of armed conflict where in fact they should not be addressed through these laws at all. The default classification of cyber operations, on one view, is that they amount to an armed conflict and so the laws of armed conflict apply. However, it is also argued that since such operations do not adhere to the definition of attack under international humanitarian law, the restrictions on attacks, imposed by the principle of distinction, do not apply … One of the

---

[51] UN News Centre, 'UN human rights expert questions targeted killings and use of lethal force' (n 22).

[52] Summary of the Human Rights Council interactive panel discussion of experts on the use of remotely piloted aircraft or armed drones in compliance with international law: Report of the Office of the United Nations High Commissioner for Human Rights, UN Doc A/HRC/28/38, 15 December 2014, para 56.

> main challenges is to identify … which type of operation should be addressed under the laws of armed conflict and which type should not.[53]

In deciding which operations should be addressed by international humanitarian law, the characterization of each situation should be guided by the object and purpose of this body of law: the protection of victims of armed conflict. Lowering the threshold for the use of lethal force would contravene this if it resulted in the applicable legal protections being rendered less effective, leaving those affected in a more vulnerable position. In the continuing development of international humanitarian law, it is likely, according to Michael Schmitt, that 'new norms will emerge to address phenomena that have so fundamentally changed that the existing classification architecture … reveal classificatory lacuna'.[54] He states that: '[s]ome aspects of conflict classification are likely to fall into desuetude … Other aspects will likely be reinterpreted to fit emerging contexts of armed conflict that were unanticipated.'[55] If cyber operations are to be accommodated, this must be undertaken in a manner that preserves the integrity of armed conflict as a concept of international humanitarian law, consistent with its object and purpose.

Likewise with regard to autonomous weapon systems, this is an area where consensus on the basis for characterization is urgently required. According to William Boothby:

> Future developments in weapons technologies are likely to enable attacks to be prosecuted remotely, automatically, potentially autonomously and, in either case, perhaps also anonymously. Some such developments cause one to wonder whether notions of remote attack will take us to a point at which there is a degree of dissociation between armed forces personnel and the hostilities for which they are responsible. Taken to an extreme, perhaps hostilities in which machines target one another autonomously and/or automatically would cease to be 'warfare' as that term has traditionally been understood.[56]

---

[53] Chatham House, International Law Meeting Summary: Classification of Conflicts: The Way Forward, Chatham House, London, 1 October 2012, 14–15, accessed 4 June 2017 at goo.gl/ffwHCZ.

[54] Michael N Schmitt, 'Classification in future conflict' in Elizabeth Wilmshurst (ed), *International law and the classification of conflicts* (Oxford University Press 2012), 455–77, 477.

[55] Ibid.

[56] William Boothby, 'The Legal Challenges of New Technologies: An Overview', in H Nasu and R McLaughlin (eds), *New Technologies and the Law of Armed Conflict* (T.M.C. Asser Press 2014) 21–8, 25.

Accordingly, it is conceivable that autonomous weapon systems could be deployed in hostilities against other autonomous weapon systems. In the absence of human participation, could such a situation be characterized as one of armed conflict? As with other forms of remote warfare, assessments would need to be undertaken on a case-by-case basis. Even if autonomous weapon systems were to be deployed in a context where human casualties did not arise directly from the conduct of hostilities, it should be recognized that victims also result from displacement and the destruction of property, including the damage to works and installations containing dangerous forces, such as nuclear power stations. The question of qualification for application of international humanitarian law would necessarily need to take into account the function that law serves not only in relation to the protection of the human person but also with regard to the protection of cultural property and the natural environment.

5. CONCLUSION

To respond to the challenges posed by remote warfare, it is necessary to be mindful of how the law has evolved and the importance of preserving the integrity of its interpretation. In order for this to be realized the concept of armed conflict must be interpreted in terms consistent with the object and purpose of international humanitarian law, that is, the protection of victims. As noted by Elizabeth Wilmshurst, '[t]he protection of victims of war depends upon the proper application of international humanitarian law and that depends upon the appropriate classification'.[57] Indeed:

> Legal complexities about the distinctions between categories of hostilities should not be allowed to get in the way of the objectives of international humanitarian law, either by making the application of the legal protections more difficult or by rendering the law so complex that none but the most sophisticated of armed forces can realistically apply it.[58]

In order to further the protection provided by the law, newer forms of warfare—including the use of armed drones, autonomous weapons systems and cyber operations—must be accommodated in the concept of armed conflict. The basis for doing so should be consistent with the existing framework that governs the conduct of hostilities, irrespective of

---

[57] Elizabeth Wilmshurst (ed), *International Law and the Classification of Conflicts* (Oxford University Press 2012) 500–501.

[58] Ibid.

how hostilities are characterized. As noted by the International Military Tribunal at Nuremburg, the laws that govern armed conflict 'are not static, but by continual adaptation follow the needs of a changing world'.[59]

---

[59] *Trial of Major War Criminals before the International Military Tribunal*, Nuremburg, 14 November 1945–1 October 1946, Vol. I, 221.

5. Remoteness and human rights law

Gloria Gaggioli

1. INTRODUCTION[1]

The legality of remote weapons systems, such as drones and autonomous weapons, is usually discussed through the lens of international humanitarian law (IHL), also called the law of armed conflicts. The ability of such remote weapons systems to respect the conduct of hostilities principles of distinction, proportionality and precautions as well as the risk that the use of remote warfare might lead to a global battlefield have been the subject of much legal debate. A less thoroughly discussed topic is whether human rights law (HRL) is relevant and may limit the use of force by drones or autonomous weapons. In that context, two broad sets of questions arise.

The first set of questions relates to HRL *as a complement to* IHL in regulating remote warfare. In this context, the key questions are: Does HRL bring something to the table when discussing remote warfare? Even assuming IHL constitutes the 'lex specialis' regarding the conduct of hostilities, are there human rights obligations that persist in armed conflicts and that may be considered as complementary? For instance, to what extent should the human rights concepts of accountability, transparency, or even dignity be taken into account when considering the use of force through remote weapons systems? Do these concepts have a proper legal role to play in such situations? Are these notions relevant as mere policy considerations in warfare? May the increasing reference to those human rights concepts ultimately influence warfare even beyond the use of drones or autonomous weapons?

The second set of questions addresses the relevance of HRL *instead of* IHL in law enforcement operations. Even in armed conflicts, some

[1] I would like to thank Professor Robert Kolb for having introduced me to Professor Jens D Ohlin and to his important work on remote warfare, as well as for having provided me with his precise and thoughtful comments on this chapter. I also thank the Stockton Center for the Study of International Law, which provided a very conducive and stimulating environment for the writing of this chapter.

situations are indeed governed by the human rights law enforcement paradigm rather than the conduct of hostilities paradigm under international humanitarian law. This is especially so in non-international armed conflicts and in occupations. In such situations, could autonomous weapons, for instance, be used to do the job without violating human rights law enforcement requirements, for instance to secure checkpoints/ detention centers, or to prevent entry into specific areas? There is a risk that remote weapons/methods will be increasingly used not only in but also outside armed conflicts to perform law enforcement functions. In such a case, is there not an inherent contradiction between, on one hand, 'law enforcement' and the idea that law enforcement officials are there 'to serve and to protect' and, on the other hand, 'remoteness'?

Before discussing these issues, the first section will be dedicated to the concept of remoteness and to an overview of related practice. It will discuss the various dimensions of remoteness and attempt to identify the key commonality and distinctive criterion between various remote weapons systems. It will also provide an overview of the actual and potential use of remote weapons systems—including as law enforcement tools—and briefly present states' main policies and positions in relation to autonomous weapons. The practice of human rights bodies addressing remote weapons systems will also be summarized as well as their main positions in this respect. The human rights that have been identified as potentially affected by remote weapons systems will be elicited, although the focus of the chapter will be on the use of force against persons (rather than objects).

The chapter will conclude with a critical assessment of the relevance of HRL for addressing the legality and suitability of remote weapons in warfare and law enforcement.

2. REMOTENESS IN PRACTICE

(a) Definitional Issues and the Concept of 'Remoteness'

During the last 20 years, one of the most tremendous changes in warfare has been the development of remote weapons systems, through notably the development of drones, that is, remotely controlled unmanned aerial vehicles, that are able to target and kill surgically individuals while their operator is located far away from the battlefield. Semi-autonomous weapons are already in use and it is anticipated that in the next 10 to 20 years, some states will be able to replace soldiers with so-called 'killer-robots' who would be able to make life and death decisions on the battlefield without human involvement.

An obvious commonality between these 'remote weapons system' is that they introduce an increasingly important physical/geographical distance between human operators and the delivery of lethal or potentially lethal force.[2] Beyond this basic common feature, which is not necessarily unique to drones and autonomous weapons, differences between various remote weapons systems are numerous and difficult to categorize. Internationally agreed-upon definitions and typologies of various remote weapons systems are lacking. Even mere autonomous weapons have been defined and categorized differently by various stakeholders.[3]

The US Department of Defense makes a distinction between (1) autonomous weapon systems, (2) human supervised autonomous weapon systems and (3) semi-autonomous weapon systems.[4] While autonomous weapon systems can select and engage targets without further intervention by a human operator,[5] semi-autonomous weapon systems are intended to only engage individual targets or specific target groups that have been selected by a human operator. Human supervised autonomous weapon systems are a form of autonomous weapon system whose operation can be overridden by a human being, including in the event of a weapon system failure. In this typology, remote-controlled drones are not included, as the selection and targeting is human-controlled. In the future it is, however, not excluded that drones will be given the capability

---

[2] See Chapter 1 in this book by Jens D Ohlin, 'Remoteness and Reciprocal Risk' 1: 'Each of these categories of weapons allows the attacking force to inflict military damage while the operators of the weapon remain safely shielded from the theater of operations'.

[3] International Committee of the Red Cross (ICRC), *Expert Meeting Report: Autonomous Weapons Systems: Technical, Military, Legal and Humanitarian Aspects* (ICRC 2014) 7.

[4] United States Department of Defense (US DoD), *Autonomy in Weapon Systems, Directive 3000.09* (2012) 14, accessed 3 May 2017 at http://www.dtic.mil/whs/directives/corres/pdf/300009p.pdf.

[5] The UN Special Rapporteur's report to the Human Rights Council on autonomous weapon systems—or 'Lethal Autonomous Robots'—provides a similar definition of autonomous weapons under the label of 'Lethal Autonomous Robotics', and does not address semi-autonomous weapons directly. See *Report of the UN Special Rapporteur on extrajudicial, summary or arbitrary executions, Christof Heyns, to the Human Rights Council* (A/HRC/23/47, 2013) §38.

to independently select and/or attack targets.[6] In such a case, they would become semi-autonomous or even autonomous weapons.[7]

Human Rights Watch divides 'unmanned robotic weapons' into three categories based on the amount of human involvement in their actions: (1) 'human-in-the-loop weapons', that is, 'robots that can select targets and deliver force only with a human command'; (2) 'human-on-the-loop weapons', that is, robots that can select targets and deliver force under the oversight of a human operator who can override the robots' actions' and finally (3) 'human-out-of-the-loop weapons', that is, 'robots that are capable of selecting targets and delivering force without any human input or interaction'.[8] The last category is referred to as a 'fully autonomous weapon'—or 'killer robots'—while the penultimate type of weapons are said to fall also under that category if the human supervision is very limited. The first category includes remote-controlled drones.

The International Committee of the Red Cross (ICRC) distinguished in 2011 between an 'autonomous' weapon system and an 'automated' weapon system.[9] In this terminology, an automated weapon system can 'independently verify or detect a particular type of target object and then fire or detonate'.[10] A truly autonomous weapon would additionally be able to 'learn and adapt its functioning in response to changing circumstances in the environment in which it is deployed' and would require artificial intelligence to do so.[11] Remote-controlled drones are excluded and would constitute a separate category. In its 2014 Expert Meeting Report on Autonomous Weapon Systems, the ICRC recognized nevertheless that 'the distinction between autonomous and automated weapon systems is not always clear since both have the capacity to independently select and attack targets within the bounds of their human-determined programming'.[12] For the purpose of this expert meeting, it therefore switched to a single definition of autonomous weapon systems as 'weapon systems for which critical functions (ie acquiring, tracking, selecting and attacking targets) are autonomous'.[13] In its 2016 (second)

---

[6] ICRC Autonomous Weapons Report (n 3) 64.

[7] Ibid.

[8] Human Rights Watch and International Human Rights Clinic, *Losing Humanity: The Case Against Killer Robots* (November 2012) 2.

[9] ICRC, *International Humanitarian Law and the Challenges of Contemporary Armed Conflicts* (31IC/11/5.1.2, October 2011) 39.

[10] Ibid.

[11] Ibid.

[12] ICRC Autonomous Weapons Report (n 3) 64.

[13] Ibid.

Expert Meeting on autonomous weapons, it used a similar working definition.[14]

All of these definitions have been criticized as being overly simplistic.[15] It is indeed difficult to fit the range of possible unmanned weapons into clearly defined boxes (for example, remote-controlled, automated, semi-autonomous, autonomous and so on). The degree of human involvement might differ depending on various functions of the weapon system. For instance, the system may be able to 'observe', that is, gather data, autonomously, but it may not be able to make any 'decision' or 'act' without human involvement.[16] On the other hand, even a remote-controlled weapon for the function of targeting might, in reality, involve a very low level of human control if the human operator is relying heavily or exclusively on intelligence collected by the weapon system. Moreover, depending on the objective of classification (for example, for technical, operational or legal purposes), the distinctive criteria that should be chosen to distinguish between various remote weapon systems might vary greatly.

For the purpose of this chapter, it will not be necessary to attempt to provide a new or over-complicated typology of remote weapon systems. It will suffice to say that the key distinctive criterion between various remote weapons systems is the existence or non-existence and amount of human control/intervention and supervision into the identification, selection and engagement of targets. At the lower end of remote weapon systems are current drones, that is, armed unmanned air systems, whose target-selection and engagement functions are remotely controlled by a human operator, who is often located far away from the battlefield. At the high end are fully autonomous weapons that can select and engage targets without further intervention by a human operator; some of which, with artificial intelligence, could adapt to changing circumstances and

---

[14] ICRC, *Expert Meeting Report: Autonomous Weapon Systems: Implications of Increasing Autonomy in the Critical Functions of Weapons* (ICRC 2016) 8 and 71: 'An autonomous weapon system is: Any weapon system with autonomy in its critical functions. That is, a weapon system that can select (i.e. search for or detect, identify, track, select) and attack (i.e. use force against, neutralize, damage or destroy) targets without human intervention.'

[15] Christopher M Ford, 'Autonomous Weapons and International Law' (forthcoming) South Carolina Law Review. See also Alan L Schuller, 'At the Crossroads of Control: The Intersection of Artificial Intelligence in Autonomous Weapon Systems with International Humanitarian Law' (2017) 8 Harvard National Security Journal 392.

[16] Ibid.

'choose' their targets with great 'freedom'. In-between, multiple distinctive criteria can help to further refine the scale. For instance, can a human supervise the autonomous weapons and interrupt its operation if need be? To what extent is the human being involved in the selection of the target? Can the weapon adapt to a changing environment or can it operate only in limited areas?

The concept of 'remoteness' in the context of weapon systems can thus be seen as gravitating around the concept of human control. On one hand, the degree of human control is the key distinctive criterion between various remote weapon systems. On the other hand, it is a commonality between these various systems that human control is 'remote', either physically (especially for drones and other remote-controlled weapons systems) or both physically and substantially in the case of autonomous weapons, in the sense that the person programming the weapon system is far away from the battlefield and that human control of the weapon system is decreasing.

There are also additional, maybe less immediately visible dimensions of remoteness, which in turn influence human control. There is a temporal dimension to remoteness. Certain remote weapon systems are programmed to perform certain functions (for example, engaging certain identified targets) at a certain period in time (t1) and will perform these functions at a later period (t2).[17] The period of time that can elapse between t1 and t2 might be substantial and therefore the release of lethal or potentially lethal force will be temporally remote from the moment of programming, which may be an issue from both a conduct of hostilities and law enforcement perspective.

Finally, arguably, remote weapons systems create a sort of 'psychological' or 'mental' remoteness. This can be understood in two ways. A first one, that is common to all remote weapons, is that they create a 'psychological distance' between the individuals on behalf of whom force is used and the potential targets or victims of the use of force. This is true not only for individuals involved in the programming or controlling of the weapon system but also for the society at large and states deciding to resort to force. In the case of robots with artificial intelligence, this 'psychological' or 'mental' remoteness potentially reaches a much higher level. The robot being able to learn from its own experience would be able to make choices that were not anticipated by its human creators. The 'logic' of the robot, or the manner in which it will adapt and make its choices over time, will not only differ among the 'robots' population', but

---

[17] Schuller (n 15) 389.

it might also drastically differ from a human 'logic'. This form of psychological or mental remoteness would, in turn, impact the ability of the human creator to re-establish its control over the machine.

In brief, remoteness is much more than just physical remoteness. It is also, and mainly, remoteness of the human control over the machine and situation leading to the resort to lethal or potentially lethal force. This level of control is influenced in turn by temporal and psychological or mental remoteness. As we shall see, all these dimensions of remoteness must be taken into account to understand the human rights impacts of remote weapons systems.

(b) Actual and Potential Use of Remote Weapons Systems

Remote or unmanned technologies have mostly been used in armed conflict situations. They can perform various functions ranging from surveillance, reconnaissance, checkpoint security, securing areas or military objectives, neutralization of an improvised explosive device, biological or chemical weapon sensing, removal of debris, search and rescue, street patrols to the direct use of lethal force against targets or in self-defense.[18] It is now a cliché to say that, when weaponized, these remote technologies raise particularly intricate legal, ethical and societal issues.

Since 2000, unmanned combat aerial vehicles have been increasingly used in the global fight against terrorism. The United States of America has been at the forefront of this evolution with drone strikes conducted in Afghanistan, Pakistan, Yemen, Somalia, Iraq, the Philippines, Libya and Syria.[19] Media report that the US military already has more than 10,000 drones in its inventory and another 12,000 on the ground.[20] If only three states are held to have deployed 'combat drones', that is, the United States, the United Kingdom and Israel, many more are known to possess and manufacture such drones (including Italy, China, India, Pakistan and Turkey) and some 80 countries have some kind of drone capability.[21]

---

[18] *Interim report of the Special Rapporteur on extrajudicial, summary or arbitrary executions, Philip Alston, to the General Assembly* (A/65/321 2010) 10.

[19] Ed Pilkington, 'Former US military personnel urge drone pilots to walk away from controls' *The Guardian* (17 June 2015).

[20] Heather M Roff and P W Singer, 'The Next President Will Decide the Fate of Killer Robots—and the Future of War' *Wired* (9 June 2016).

[21] Peter Bergen and Emily Schneider, 'Hezbollah armed drone? Militants' new weapon' *CNN* (22 September 2014), accessed 3 May 2017 at http://edition.cnn.com/2014/09/22/opinion/bergen-schneider-armed-drone-hezbollah/.

While early drones, like the Predator, were almost entirely remote controlled, later versions have increasingly gained intelligence and autonomy.[22] Functions such as takeoff and landing, navigation and even target acquisition and tracking can now be automated.[23]

Other semi-autonomous (or even autonomous, depending on how this term is defined) weapons systems, that is, which feature a level of autonomy in target selection and attack, already exist and have been used in armed conflict situations.[24] They may be fixed weapon systems in stationary roles or mobile unmanned systems.[25] They may be ground-based, air-based or maritime based.[26] They include notably sentry guns, ground weapon systems for the purpose of bomb disposal, missiles and various other 'fire and forget' munitions, unmanned maritime (surface or underwater) vehicles for anti-submarine or surface warfare.[27] Although these weapons systems are usually used for defensive rather than offensive operations, to target objects rather than individuals and in simple, static rather than dynamic, so-called 'cluttered' environments, these features could easily change together with evolutions in technologies and state policies.[28]

For the time being, fully autonomous weapon systems with artificial intelligence that can make independent decisions, learn from their experience and adapt to their environment do not yet exist.[29] The likelihood that such weapon systems or soldier-robots will be developed in the not too distant future is, however, allegedly high. In an open letter signed by Elon Musk, Stephen Hawking and Steve Wozniak, artificial intelligence and robotics researchers wrote that artificial intelligence has advanced far enough that deployment of 'weapons [that] select and engage targets without human intervention … [such as] armed quadcopters that can search for and eliminate people meeting certain predefined criteria … are feasible within years, not decades, and the stakes

---

22 Ibid.

23 Ibid. See also ICRC Autonomous Weapons Report (n 3) 67.

24 ICRC Autonomous Weapons Report (n 3) 7.

25 Ibid 65–69.

26 Ibid.

27 Ibid. See also Dustin A Lewis, Gabriella Blum, and Naz K Modirzadeh, *War-Algorithm Accountability* (August 2016) 32–50, accessed 3 May 2017 at https://pilac.law.harvard.edu/waa/.

28 ICRC Autonomous Weapons Report (n 3) 7.

29 ICRC Autonomous Weapons Report (n 3) 7.

are high: autonomous weapons have been described as the third revolution in warfare, after gunpowder and nuclear arms'.[30]

In any case, the race for the development and acquisition of remote weapons systems is already real and concrete. Media report that the US military is working on more than 20 projects to increase the autonomy of weapons systems.[31] In 2014, the US Under-Secretary of Defense, Frank Kendall, requested the Defense Science Board to produce a Study on Autonomy 'to permit greater operational use of autonomy across all war-fighting domains' but also to explore 'the bounds—both technological and social—that limit the use of autonomy across a wide range of military operations'.[32] This study, which was released in June 2016, concluded that 'autonomy will deliver substantial operational value—in multiple dimensions—across an increasingly broad spectrum of DoD [Department of Defense] missions, but the DoD must move more rapidly to realize this value'. It remains to be seen if and how the new US Government will move forward in that direction. The United Kingdom is developing the 'Taranis', a combat drone capable of autonomously flying, identifying targets, and hitting them after having been authorized to do so by a human operator.[33] China is developing new generations of cruise missiles which would integrate a high level of artificial intelligence and automation and enable military commanders to 'tailor-make missiles' depending on combat situations.[34] Russia's Foundation for Advanced Studies—an advanced military research agency—is developing a remote-controlled humanoid military robot ('Iron Man', also called 'Ivan the Terminator') with the aim to 'replace the person in the battle or in emergency areas where there is a risk of explosion, fire, high background radiation, or other conditions that are harmful to humans'.[35] Even Iraq has developed a remotely-controlled ground mini-tank that is armed with an automatic machine gun and a rocket launcher, under the

---

30 Ibid. The open letter is available at http://futureoflife.org/ai-open-letter/ (accessed 3 May 2017).

31 Roff and Singer (n 20).

32 Defense Science Board, *Summer Study on Autonomy* (June 2016), terms of references 102, accessed 3 May 2017 at https://www.hsdl.org/?view&did=794641.

33 Lewis et al (n 27) 39.

34 Zhao Lei, 'Nation's next generation of missiles to be highly flexible' *China Daily* (19 August 2016).

35 Nick Enoch, 'Rise of the Russian robo-soldier: Iron Man military hardware is one step closer to reality as Putin's scientists reveal Ivan the Terminator' *Daily Mail* (27 May 2016).

name of 'Alrobot' (Arabic for robot), which is already in use in the ongoing fight against ISIS.[36]

Non-state actors may also acquire—and, even if this is less likely, develop in the future—unmanned and increasingly autonomous systems. For instance, in September 2014, CNN reported that Hezbollah used a 'combat drone'.[37] Hamas proudly claims it can now produce armed drones.[38] ISIS uploaded a video to YouTube that showed aerial views of a Syrian military base allegedly destroyed by a drone.[39] The risk of proliferation of armed drones, given their relative inexpensiveness and the rapid evolution in technology, has already been highlighted elsewhere.[40] The same risks apply to autonomous weapons.[41]

To date, only two states have officially published policies on autonomous weapons, which may somehow limit or constrain their development or use.[42] According to a US Department of Defense's Directive, 'autonomous and semi-autonomous weapon systems shall be designed to allow commanders and operators to exercise appropriate levels of human judgment over the use of force'.[43] The Directive does not define the 'appropriate level of human judgment' but further limits autonomy in existing and future weapon systems in a careful manner and provides that autonomous weapons without human supervision may be used to apply only 'non-lethal, non-kinetic force'.[44] However, this does not mean that the US policy is to never develop or use fully autonomous weapons to apply lethal force. A specific additional review and approval process is merely required to develop or use autonomous or semi-autonomous weapons in a manner that falls outside the policy.[45] It is also worth noting

---

[36] Mark Prigg, 'The remote controlled robot tank fighting ISIS: Iraqi military confirms Alrobot has been deployed in Mosul' *Daily Mail* (8 November 2016). See also Roff and Singer (n 20).

[37] Peter Bergen and Emily Schneider, 'Hezbollah armed drone? Militants' new weapon' *CNN* (22 September 2014), accessed 3 May 2017 at http://edition.cnn.com/2014/09/22/opinion/bergen-schneider-armed-drone-hezbollah/.

[38] Ibid.

[39] Ibid.

[40] Gloria Gaggioli, 'Lethal Force and Drones: The Human Rights Question' in Steven James Barela (ed), *Legitimacy and Drones* (Ashgate 2015) 113.

[41] Sasha Radin and Jason Coats, 'Autonomous Weapon Systems and the Threshold of Non-International Armed Conflict' (forthcoming) 30 Temple Intl & Comp L J 136–137.

[42] ICRC Autonomous Weapons Report (n 3) 8.

[43] US Department of Defense (n 4) §4.a.

[44] Ibid §§ 4.c.(1)–4.c.(3).

[45] Ibid § 4.d.

that this US directive which was adopted in 2012 has a five-year limit,[46] which means that the newly elected President, Donald Trump, who has so far remained silent on the issue, will need to decide in 2017 if this policy should be renewed, modified or abrogated.[47] The UK policy seems less elaborated but, at first sight, more restrictive and is that its '... operation of weapon systems will always ... be under human control'.[48] In other words, 'every release of weapons is [and will be] authorized by a human'.[49] The notion of 'human control' has not been further elaborated.[50] The United Kingdom clearly stated that it had no intention of developing (fully) autonomous weapons systems.[51] However, 'autonomous weapons systems' are understood narrowly as being systems which are 'capable of understanding higher level intent and direction ... [and that are] capable of deciding a course of action, from a number of alternatives, without depending on human oversight and control ... [and whose] individual actions may not be [predictable]'.[52] This definition has been criticized as limiting autonomous weapons systems to 'higher-level and futuristic' weapons system, which are not yet technically achievable, thus rendering the policy of limited practical usefulness.[53]

The so-called CCW informal expert meetings[54] held annually since 2014 at the United Nations Office in Geneva on 'lethal autonomous

---

[46] Ibid § 7.b.

[47] Roff and Singer (n 20).

[48] See *Official Report, House of Lords* (26 March 2013), Vol. 744, c. 960. See also House of Lords November 2014 written questions, Unmanned Air Vehicles: Written question – HL2710, asked by Lord West of Spithead.

[49] House of Lords November 2014 written questions, ibid.

[50] Article 36, *Background Paper: The United Kingdom and Lethal Autonomous Weapons Systems* (April 2016) 1, accessed 3 May 2017 at http://www.article36.org/wp-content/uploads/2016/04/UK-and-LAWS.pdf.

[51] United Kingdom Ministry of Defence, *Joint Doctrine Note 2/11: The UK Approach to Unmanned Aircraft Systems* (3 March 2011) §508; Foreign & Commonwealth Office, *United Kingdom of Great Britain and Northern Ireland Statement to the Informal Meeting of Experts on Lethal Autonomous Weapon Systems* (11 April 2016). See also *Official Report, House of Lords* (26 March 2013) Vol. 744, c. 960. See also House of Lords November 2014 written questions (n 48).

[52] United Kingdom Ministry of Defence, *Joint Doctrine Note 2/11* (n 51).

[53] Article 36, *Background Paper* (n 40).

[54] These meetings take place within the ambit of the 1980 Convention on Prohibitions or Restrictions on the Use of Certain Conventional Weapons, which may be deemed to be Excessively Injurious or to have Indiscriminate Effects (CCW).

weapons systems' offer a good overview of the variety of states' positions on the subject and of their continuing evolution.[55] They range from: autonomous weapons are unlawful to autonomous weapons should be banned, regulated, always operated under human supervision, subjected to careful legal review under Article 36 of Additional Protocol I or a combination thereof.[56] It will not be necessary to summarize these positions here: this has been done elsewhere.[57] It will suffice to say that the trend seems to be towards the need to keep 'meaningful', 'appropriate' or 'effective' human control/involvement over the release of deadly force.[58] Whether this should materialize through a treaty banning or regulating lethal autonomous weapons remains to be seen. The Fifth Review Conference of the High Contracting Parties to the CCW, which was held from 12 to 16 December 2016, decided—based on the recommendation of the 2016 Informal Meeting of Experts[59]—to establish an open-ended Group of Governmental Experts (GGE) on Lethal Autonomous Weapons Systems (LAWS), which will meet in 2017 (21–25 August and 13–17 November 2017) to 'explore and agree on possible recommendations on options related to emerging technologies in the area of LAWS'.[60] The United States and United Kingdom supported the creation of this GGE, while Russia considered this decision as premature.[61] China for the first time said that it sees a need for a new

[55] States' positions are available at http://www.unog.ch/80256EE600585943/(httpPages)/37D51189AC4FB6E1C1257F4D004CAFB2, accessed 3 May 2017.

[56] Lewis et al (n 27) 151–223, appendix II.

[57] Ibid.

[58] ICRC, 'Decisions to Kill and Destroy are a Human Responsibility', *Statement read at the Meeting of Experts on Lethal Autonomous Weapons Systems, held in Geneva from 11–16 April* (11 April 2016), accessed 3 May 2017 at https://www.icrc.org/en/document/statement-icrc-lethal-autonomous-weapons-systems; Chris Ford and Chris Jenks, 'The International Discussion Continues: 2016 CCW Experts Meeting on Lethal Autonomous Weapons', *Blog Post on Just Security* (20 April 2016), accessed 3 May 2017 at https://www.justsecurity.org/30682/2016-ccw-experts-meeting-laws/.

[59] *Report of the 2016 Informal Meeting of Experts on Lethal Autonomous Weapons Systems (LAWS) Submitted by the Chairperson of the Informal Meeting of Experts* (CCW/CONF.V/2, 10 June 2016). See also Ford and Jenks (n 58).

[60] *Report of the 2016 Informal Meeting of Experts on Lethal Autonomous Weapons Systems (LAWS), Submitted by the Chairperson of the Informal Meeting of Experts* (CCW/CONF.V/2, 10 June 2016) 14.

[61] 'Une Prix Nobel dénonce l'usage d'engins incendiaires en Syrie', *Swissinfo* (13 December 2016), accessed 3 May 2017 at http://www.swissinfo.ch/fre/une-prix-nobel-dénonce-l-usage-d-engins-incendiaires-en-syrie/42759828.

international instrument on lethal autonomous weapons systems and 19 states endorsed the call to ban lethal autonomous weapons.[62]

Although most unmanned weapons systems are currently being developed for warfare situations and most discussions revolve around the use of these weapons systems in such contexts, there is a growing industry making unmanned systems to conduct law enforcement operations in peacetime, and potentially in wartime as well.[63] Compared to the actual and potential use of remote weapons systems in warfare, this development remains under-scrutinized.

In certain developed countries, drones and other robots have already largely been used in peacetime law enforcement for a number of purposes.[64] They can perform highly advanced surveillance and can help to detect fires, collect data on suspected offenders or for relief personnel working in areas affected by natural disasters.[65] They can also do border control and security operations. For instance, drones are used along the United States-Mexico border to detect and track drug smugglers and human traffickers.[66]

The likelihood that armed robots will be commonly used in order to maintain or restore public security, law and order in the near future should not be underestimated. As highlighted by Christof Heyns, former

---

[62] Campaign to Stop Killer Robots, *Formal Talks Should Lead to Killer Robots Ban*, 16 December 2016, accessed 3 May 2017 at https://www.stopkillerrobots.org/2016/12/formal-talks/.

[63] Gaggioli (n 40) 92–94; Christof Heyns, 'Human Rights and the Use of Autonomous Weapons Systems (AWS) During Domestic Law Enforcement' (2016) 38 Human Rights Quarterly 358.

[64] 'Law Enforcement Agencies Using Drones List Map' *Governing the States and Localities* (16 January 2014), accessed 3 May 2017 at http://www.governing.com/gov-data/safety-justice/drones-state-local-law-enforcement-agencies-license-list.html.

[65] Peter Maurer (Web Interview), *The Use of Armed Drones Must Comply with Laws* (10 May 2013), accessed 3 May 2017 at https://www.icrc.org/eng/resources/documents/interview/2013/05-10-drone-weapons-ihl.htm. Dan Roberts, 'FBI Admits Using Surveillance Drones over US Soil' *The Guardian* (Washington, 19 June 2013), accessed 3 May 2017 at http://www.theguardian.com/world/2013/jun/19/fbi-drones-domestic-surveillance.

[66] 'Groups Concerned Over Arming of Domestic Drones' *CBSDC* (Washington, 23 May 2012), accessed 3 May 2017 at http://washington.cbslocal.com/2012/05/23/groups-concerned-over-arming-of-domestic-drones/; Aliya Sternstein, 'Obama Requests Drone Surge for U.S.-Mexico Border' *Defence One* (9 July 2014), accessed 3 May 2017 at http://www.defenseone.com/threats/2014/07/obama-requests-drone-surge-us-mexico-border/88303/.

UN Special Rapporteur on extrajudicial, summary or arbitrary executions: 'The use of armed drones happen[s] in conflict and counter-terrorism situations, but also increasingly in ordinary policing and law enforcement'.[67] Unmanned systems and potentially autonomous weapons may be portrayed as useful in a variety of law enforcement contexts: crowd-control, hostage situations, avoiding escapes of dangerous prisoners, areas/building-protections, border-controls, checkpoints, fighting against drug-lords, 'terrorists' and organized crime in general.[68]

Actually, a number of armed robots for law enforcement purposes already exist. For instance, in 2009 Technorobot, a company that is based both in Spain and the United States, created the 'Riotbot', a kind of miniature tank that is remote-controlled and that can incorporate video-equipment and be equipped with rubber bullets to assist law enforcement officials to control riots, maintain law and order in prisons or to conduct 'urban warfare'.[69] An Israeli firm, General Robotics Ltd, has developed a similar light robotic 'watch dog', called 'the Dogo', which provides live video reconnaissance, and which can 'neutralize' threats remotely by means of pepper spray, dazzling light module causing temporary blindness or even of a 9 mm Glock pistol.[70] Law enforcement robots can also be airborne. In 2010, a United States company, Vanguard Defense Industries, had already developed a remotely operated helicopter, named Shadowhawk, which can maintain aerial surveillance of an area (that is, a house, vehicle, person, and so on) at 700 feet without being heard or seen and which can be equipped with a Taser with the ability to fire four barbed electrodes that can be shot to a distance of 100 feet or armed with grenade launchers or shotguns with laser designations.[71] It has been used

---

[67] *Human Rights Council Holds Panel on Remotely Piloted Aircraft or Armed Drones in Counterterrorism and Military Operations.* (22 September 2014) accessed 3 May 2017 at http://www.ohchr.org/EN/NewsEvents/Pages/DisplayNews.aspx?NewsID=15080.

[68] Heyns (n 63) 359. See also the websites of the firms developing law enforcement drones below. They often provide law enforcement scenarios where their law enforcement robots may be used. See e.g. http://www.technorobot.eu/en/riotbot.htm (accessed 3 May 2017).

[69] Ibid. See also Heyns (n 63) fn 43.

[70] See the website of General Robotics Ltd, accessed 3 May 2017 at http://www.glrobotics.com/slideshow. See also April Glaser, '11 Police Robots Patrolling Around the World' *Wired* (24 July 2016), accessed 3 May 2017 at https://www.wired.com/2016/07/11-police-robots-patrolling-around-world/.

[71] See the website of Vanguard Defense Industries, accessed 3 May 2017 at http://unmanned.wixsite.com/vanguarddefense/applications. See also Heyns (n 63) 360.

in Afghanistan and East Africa against suspected terrorists and can be purchased for law enforcement purposes,[72] although the grenade launchers and shotguns are available for military use only.[73] In May 2014, a South African company created a 'crowd-control' drone that is able to shoot pepper spray and non-lethal paintballs to mark offenders.[74] It can also employ strobe lights and on-board speakers to send verbal warnings.[75] It was reported that Turkey and India purchased this 'riot drone'.[76]

These are just a few examples among many others. Numerous companies all around the world—in the United States, Europe, Asia, Africa and the Middle East—are developing such law-enforcement robots. For the time being, they remain remote-controlled and are usually portrayed as 'less-than-lethal' weapons, although they may well be lethal in practice.[77] It is well known, for instance, that Tasers, with which these law enforcement robots can be equipped, may, under certain circumstances, be lethal. Non-lethal robots can also be easily fitted out to kill.[78] For instance, in July 2016, a Texan SWAT team used a ground robot and loaded it with a pound of C-4 explosive to neutralize a military trained sniper who was shooting at the police from a building in downtown Dallas.[79] In some cases (for example, in Israel), law enforcement robots are even designed to kill for law-enforcement purposes.[80]

The trend towards law enforcement robots with more autonomy and lethality is clear. This might not be overly surprising given the amount of force that states are ready to use in internal disturbances and tensions, for

---

72 Paul Joseph Watson, 'Big Sis Gives Green Light For Drone That Tazes Suspects From Above' *Prisonplanet* (24 August 2011), accessed 3 May 2017 at http://www.prisonplanet.com/big-sis-gives-green-light-for-drone-that-tazes-suspects-from-above.html.

73 http://www.uavglobal.com/shadowhawk (accessed 3 May 2017).

74 'Riot Control Drone Armed with Paintballs and Pepper Spray Hits Market' *RT* (19 June 2014), accessed 3 May 2017 at http://rt.com/news/167168-riot-control-pepper-spray-drone/.

75 Ibid.

76 'Turkey Cracks Down: Skunk Riot Drone Will Fire Paint and Pepper Balls' *WorldTribune.com* (Ankara, 27 June 2014), accessed 3 May 2017 at http://www.worldtribune.com/archives/turkey-buys-skunk-riot-drone-payload-including-paint-pepper-balls/. See also Glaser (n 70).

77 Heyns (n 63) 358.

78 Glaser (n 70).

79 Sara Sidner and Mallory Simon, 'How Robot, Explosives Took Out Dallas Sniper in Unprecedented Way' *CNN* (12 July 2016), accessed 3 May 2017 at http://www.cnn.com/2016/07/12/us/dallas-police-robot-c4-explosives/.

80 See note 70.

instance, in the fight against criminality to combat drug cartels.[81] The increasing use of combat weapons such as M-16 rifles, armored trucks and grenade launchers in the context of peacetime law enforcement has been a subject of concern for some years now.[82] One case in point is the widely reported event in Ferguson where demonstrators protesting the fatal police shooting of a teenager were confronted with heavily armed, militarized, police officers.[83] The acquisition by police units, such as SWAT units (Special Weapons and Tactics), of armed robots would seem to be just one further step in this worrying evolution where policemen are armed, equipped and operating like soldiers.

In brief, unmanned weapon systems, be they remotely controlled or autonomous, are evolving at a fast pace and being used and expected to be used both in wartime and in peacetime, in conduct of hostilities and law enforcement operations.

(c) Human Rights Bodies' Practice and Positions on Remote Weapons Systems

Specific issues related to remote weapon systems—especially fully autonomous weapon systems rather than drones or semi-autonomous weapons—have been discussed by human rights non-governmental organizations and by United Nations Charter-based bodies, especially the Special Rapporteur on extrajudicial, summary or arbitrary executions, since 2010.[84] Human rights courts or treaty/expert bodies did not address the issue in much detail as individual complaints on such matters did not

---

81 Jane Perlez, 'Chinese Plan to Kill Drug Lord With Drone Highlights Military Advances' *The New York Times* (20 February 2013), accessed 3 May 2017 at http://www.nytimes.com/2013/02/21/world/asia/chinese-plan-to-use-drone-highlights-military-advances.html?_r=0.

82 Clyde Haberman, 'The Rise of the SWAT Team in American Policing' *The New York Times* (7 September 2014).

83 Ibid. Alcindor Yamiche and Bello Marisol, 'Police in Ferguson Ignite Debate about Military Tactics' *USA Today* (19 August 2014).

84 See, eg, Alston, Interim Report to the General Assembly 2010 (n 18). The Special Rapporteur recommended that urgent consideration be given to the legal, ethical and moral implications of the development and use of robotic technologies, especially but not limited to uses for warfare. He also recommended the creation of an expert group to ensure that robotic technologies are optimized in terms of their capacity to promote more effective compliance with HRL and IHL. These recommendations were not upheld at the time by the General Assembly.

arise yet and as—by nature—these bodies tend to deal with issues in retrospect, once violations have occurred.[85]

In general, human rights bodies focused on remote weapon systems as means and methods of warfare resorted to in armed conflict situations and in conduct of hostilities operations. Although it has been argued that remote weapon systems can affect many human rights, including the right to privacy (because of surveillance), the right to an effective remedy for both the direct victims and their next-of-kin (because of the lack of transparency and accountability), to a fair trial (presumption of innocence in particular), the freedom of assembly and expression (because of the fear they may stir up in the public), the focus of human rights bodies has been understandably put on the right to life (or the prohibition of arbitrary killings).[86]

Regarding the rules and principles on the use of force regulating remote weapons systems in armed conflicts, even human rights organs focused on the IHL principles of distinction, proportionality and precautions, thus indicating that IHL is the lex specialis in that respect.[87] Some human rights bodies concluded that drones or autonomous weapons were

---

[85] Philip Alston, 'Lethal Robotic Technologies: The Implications for Human Rights and International Humanitarian Law' (2012) 21 J of L, Information and Science 35. The recent General Comment of the African Commission on Human and Peoples' Rights on the right to life as well as the draft General Comment of the Human Rights Committee on the right to life make however brief references to remote weapons systems, such as unmanned aircrafts and autonomous weapons. See below (n 108 and 109).

[86] See, eg, Stimson Report 2014 (n 143) 36; Human Rights Watch and International Human Rights Clinic, Human Rights Program at Harvard Law School, *Shaking the Foundations: The Human Rights Implications of Killer Robots* (2014), accessed 3 May 2017 at https://www.hrw.org/sites/default/files/reports/arms0514_ForUpload_0.pdf; *Joint Report of the Special Rapporteur on the Rights to Freedom of Peaceful Assembly and of Association, Maina Kiai, and the Special Rapporteur on Extrajudicial, Summary or Arbitrary Executions, Christof Heyns, on the Proper Management of Assemblies to the Human Rights Council* (A/HRC/31/66, 2016); Nils Melzer, *Targeted Killing in International Law* (Oxford University Press 2008) 426.

[87] See, eg, Human Rights Watch and International Human Rights Clinic (n 8) 30–36; Heyns (n 5) §§ 63–67; Report of the Special Rapporteur, Ben Emmerson, on the promotion and protection of human rights and fundamental freedoms while countering terrorism to the General Assembly (A/68/389, 2013); Report of the Special Rapporteur, Ben Emmerson, on the promotion and protection of human rights and fundamental freedoms while countering terrorism to the Human Rights Council (A/HRC/25/59, 2014) § 23.

at risk of violating IHL principles,[88] while others went as far as considering that autonomous weapons would always violate IHL.[89] In addition to the analysis of IHL conduct of hostilities rules, human rights bodies have also referred to human rights principles such as the principles of transparency, accountability and dignity to either restrict or outlaw remote weapon systems.[90]

Based on these analyses, several human rights non-governmental organizations, led by Human Rights Watch, launched the 'Campaign to Ban Killer Robots',[91] while Special Rapporteur Heyns recommended states to adopt moratoria on the development of lethal autonomous weapons.[92] During the debates before the Human Rights Council in Geneva on Heyns' report on lethal autonomous weapons in 2013, a number of states considered that the Human Rights Council is not the appropriate forum to address autonomous weapons.[93] In practice, discussions thus continued in the context of the CCW rather than in the Human Rights Council; thus removing the topic from proper human rights fora.

Although not a human rights body, it is useful to recall here the position of the ICRC on autonomous weapons. Unlike some human rights non-governmental organizations, the ICRC does not consider that drones or autonomous weapons are prohibited per se by IHL or that their use is in essence contrary to IHL.[94] Nevertheless, it raised 'serious doubts about the capability of developing and using autonomous weapon systems that would comply with IHL in all but the narrowest of scenarios and the simplest of environments, at least for the foreseeable future'.[95] The ICRC has called on states to 'set limits on autonomy in weapon systems to ensure that they can be used in accordance with international

---

[88] Heyns (n 5) §§ 63–74.

[89] Human Rights Watch and International Human Rights Clinic (n 8) 46.

[90] See below, next section.

[91] See website of the campaign to stop killer robots https://www.stopkillerrobots.org (accessed 3 May 2017).

[92] Heyns (n 5) § 111.

[93] 'Consensus Killer Robots Must be Addressed' *Campaign to Stop Killer Robots* (28 May 2013), accessed 3 May 2017 at http://www.stopkillerrobots.org/2013/05/nations-to-debate-killer-robots-at-un/.

[94] ICRC, *International Humanitarian Law and the Challenges of Contemporary Armed Conflicts* (32IC/15/11, December 2015) 44–47, accessed 3 May 2017 at https://www.icrc.org/en/document/international-humanitarian-law-and-challenges-contemporary-armed-conflicts; ICRC Challenges Report 2011 (n 9) 36–40; Maurer (Web Interview) (n 65).

[95] ICRC Challenges Report 2015 (n 94) 45. See also: ICRC Autonomous Weapons Report 2016 (n 14) 79–80.

humanitarian law (IHL) and within the bounds of what is acceptable under the principles of humanity and the dictates of public conscience'.[96] To that effect, it has stressed the importance of conducting legal reviews of new weapons in accordance with Article 36 of Additional Protocol I.[97] It also holds the position that for legal, ethical or military-operational reasons, human control over weapon systems and the use of force must be retained and it makes the case for the need to better determine the kind and degree of human control over weapon systems that is necessary from a legal, ethical and policy perspective.[98] The ICRC does not openly support a ban to stop killer robots or moratoria.[99] The willingness of the ICRC to appear as 'neutral' as possible and to differentiate itself from human rights non-governmental organizations can probably explain this positioning.

So far, not many human rights organs considered the risk that remote weapons be used in law enforcement operation in wartime or the clear trend towards the increasing development and use of remote weapons in peacetime law enforcement. In 2014, though, Human Rights Watch and Harvard Law School's International Human Rights Clinic wrote a report on the human rights implications of killer robots.[100] The report focuses only on fully autonomous weapons, also called killer robots or lethal autonomous robots and recommends their ban, based on an analysis that such weapons, if used in law enforcement operations, would face obstacles to meeting the human rights principles of necessity and

---

[96] ICRC, Working Paper: Views and Recommendations for the Fifth Review Conference of the Convention on Certain Conventional Weapons (UN Doc. CCW/CONF.V/WP.3, 26 September 2016) §21.

[97] Ibid § 23. A provision similar to Article 36 of Additional Protocol I to the Geneva Conventions does not exist in the Convention on Certain Conventional Weapons, but the ICRC considers that 'such reviews are a logical and necessary element of CCW implementation'. For discussions on challenges pertaining to the legal review of autonomous weapons, see: ICRC Autonomous Weapons Report 2016 (n 14) 17, 21–22, 42, 58–59.

[98] Ibid § 20. See also 2015 ICRC Challenges Report (n 94) 45. See also the discussions during the 2016 ICRC Expert Meeting on the notion of "meaningful human control": ICRC Autonomous Weapons Report 2016 (n 14) 18–20, 46–56 and 83–85.

[99] See, e.g., Presentation by Kathleen Lawand, head of the arms unit, ICRC, Seminar on fully autonomous weapon systems, Permanent mission of France, Geneva, Switzerland, 25 November 2013, available at https://www.icrc.org/eng/resources/documents/statement/2013/09-03-autonomous-weapons.htm, last accessed 7 June 2017.

[100] Human Rights Watch and International Human Rights Clinic (n 86).

proportionality.[101] The crux of the argument is that such weapons—lacking human judgment—would be unable to properly assess existing threats, to strategize best alternatives to the use of force or to effect the balancing exercise required by the principle of proportionality and to decide the amount of force required by the situation.[102] In addition, the report identified (others would say invented) the human rights 'principle of dignity' and contended that allowing fully autonomous weapons to make determinations to take life away would conflict with this principle.[103] This idea was further elaborated in a 2014 report of Special Rapporteur Heyns to the UN General Assembly, which considered notably the issue of less lethal and unmanned weapons in law enforcement.[104] The report questioned 'whether remote-controlled weapons systems should be as readily viewed as legal weapons in the law enforcement context as in armed conflict'.[105] It highlighted that autonomous weapons may challenge the right to life and the 'right to dignity', although it did not elaborate much on this.[106]

In a 2016 joint report on the proper management of assemblies to the Human Rights Council, the Special Rapporteurs on the rights to freedom of peaceful assembly and of association, Maina Kiai, and on extrajudicial, summary or arbitrary executions, Christoph Heyns, held that 'where advanced technology is employed, law enforcement officials must, at all times, remain personally in control of the actual delivery or release of force' and recommended that 'autonomous weapons systems that require no meaningful human control … be prohibited and remotely controlled force … only ever be used with the greatest caution'.[107] These recommendations echoed findings of the African Commission on human and peoples' rights in its 2015 General Comment on the right to life.[108] It

[101] Ibid 9–14.

[102] Ibid.

[103] Ibid 23–24.

[104] Report of the UN Special Rapporteur on Extrajudicial, Summary or Arbitrary Executions, Christof Heyns, to the General Assembly (A/69/265, 2014).

[105] Ibid § 83.

[106] Ibid §§ 77–89.

[107] Kiai and Heyns Joint Report 2016 (n 86) §§ 56 and 67 f.

[108] African Commission on Human and Peoples' Rights, General Comment No. 3: The Right to Life (Article 4) (2015) §31: 'Where advanced technology is employed, law enforcement officials must remain personally in control of the actual delivery or release of force, in a manner capable of ensuring respect for the rights of any particular individual, as well as the general public.' See also § 35: 'The use during hostilities of new weapons technologies such as remote

remains to be seen if and to what extent new technologies will be addressed in the new general comment on the right to life of the Human Rights Committee.[109] So far, the draft simply states that 'lethal autonomous robotics should not be put into operation before a normative framework has been established with a with a view to ensuring that their use conforms with article 6 and other relevant norms of international law.'[110]

Despite these very important statements, and as rightfully stated in Heyns' 2014 report to the General Assembly, 'serious consideration needs to be given to whether unmanned systems, in particular autonomous weapons systems used in the context of law enforcement, whether with lethal or less lethal force, can be considered lawful weapons per se'.[111] This remains as true in 2016 as it was in 2014. The general lack of expertise and interest of human rights bodies in weapons systems and related new technologies[112] and what is in our view an exaggerated focus on fully autonomous weapons has relegated unmanned and semi-autonomous weapon systems in law enforcement to a secondary issue.

3. HUMAN RIGHTS LAW AS A COMPLEMENT TO IHL IN REGULATING REMOTE WARFARE

(a) Human Rights in Armed Conflicts and its Interplay with IHL: Assumptions Underlying this Chapter

Remote weapon systems are usually used and mainly developed for warfare situations, that is, international or non-international armed conflicts, for which IHL constitutes the main body of applicable law.

---

controlled aircraft should only be envisaged if they strengthen the protection of the right to life of those affected. Any machine autonomy in the selection of human targets or the use of force should be subject to meaningful human control. The use of such new technologies should follow the established rules of international law'.

[109] Draft General Comment n°36: Article 6 – Right to Life (CCPR/C/GC/R.36/Rev.2., 2015).

[110] Ibid § 13.

[111] Heyns (n 104) § 86.

[112] Alston, Interim Report to the General Assembly 2010 (n 18) 11; Alston (n 85) 36.

Although HRL was initially conceived as being applicable in peacetime,[113] that position has been reversed since the 1960s.[114] It is today generally accepted that HRL applies at all times.[115] Even if some rights can be derogated from 'in time of public emergency which threatens the life of the nation',[116] this is not the case notably for the prohibition of arbitrary deprivation of life, which is non-derogable and particularly relevant for the use of force by remote weapon systems.[117] Moreover, even derogable rights, such as the right to privacy and to an effective remedy, remain normally fully applicable if not derogated from by the relevant state.

In cases of warfare involving action beyond national borders (international armed conflicts or extraterritorial non-international armed conflicts) which are generally the situations in which remote weapon systems are used, it could be held, however, that the ordinary jurisdiction of the state over its territory or any equivalent control, which forms the basis for application of human rights treaties, is generally lacking.[118] Although a minority of influential states, notably the United States of America and Israel, does not accept the extraterritorial application of human rights treaties such as the International Covenant on Civil and Political Rights,[119] human rights bodies and the International Court

---

[113] Robert Kolb, 'Aspects Historiques de la Relation Entre le Droit International Humanitaire et les Droits de l'Homme' (1999) 37 Canadian Yearbook Intl L 67–69.

[114] Ibid 75 et seq.

[115] There is, however, a minority view according to which human rights law (HRL) does not apply to armed conflicts. See *Legality of the Threat or Use of Nuclear Weapons*, Advisory Opinion, ICJ Reports 1996, § 24.

[116] Article 4 of the International Covenant on Civil and Political Rights (ICCPR). See also Article 15 of the European Convention on Human Rights (ECHR); Article 27 of the American Convention on Human Rights (ACHR). However, the African Charter on Human and Peoples' Rights (ACHPR) does not include such a provision. No derogation is possible therefore. See *Commission Nationale des Droits de l'Homme et des Libertés v Chad* (African Commission on Human and Peoples' Rights, 1995) § 21.

[117] Article 4(2) ICCPR; Article 27(2) ACHR. In the ECHR however the right to life is non-derogable 'except in respect of deaths resulting from lawful acts of war': Article 15(2) ECHR.

[118] In fact, most of the HRL treaties contain a jurisdictional limitation. See Article 2(1) ICCPR; Article 1 ECHR; Article 1(1) ACHR.

[119] For the US, see eg, Statement of State Department Legal Adviser, Conrad Harper, 53rd session, 1405th meeting of the HRC (CCPR/C/SR 1405, 1995) para 20; US Department of State, Second and Third Periodic Report of the United States of America to the UN Committee on Human Rights Concerning the

of Justice (ICJ) have clearly recognized its extraterritorial application at least in cases of occupation or detention.[120]

Outside such situations, in particular when it comes to extraterritorial targeting (without control over the territory or person), the extraterritorial reach of HRL remains, however, a matter of ongoing legal debate. In the *Banković* case,[121] concerning the NATO bombing of Radio Television of Serbia in Belgrade, the European Court of Human Rights considered that the bombing's victims did not enter into the jurisdiction of the European allied states because this notion is essentially territorial. The Court subsequently seemed to soften its position, notably in the *Al-Skeini* case[122] but it remains unclear what the Court's decision would be if

---

International Covenant on Civil and Political Rights, Annex 1 (2005). United States Responses to Selected Recommendations of the Human Rights Committee (2007) 1–2. See also Human Rights Committee, Concluding observations on the fourth periodic report of the United States of America (CCPR/C/USA/CO/4, 2014) § 4a. For legal writings on which the US Government rely, see, eg Michael J Dennis, 'Application of Human Rights Treaties Extraterritorially in Times of Armed Conflict and Military Occupation' (2005) 99 Am J Intl L 119–141. For the position of Israel, see e.g. Human Rights Committee, Sixty-third session. Summary Record of the 1675th meeting: Consideration of the Initial Report of Israel (CCPR/C/SR.1675, 1998) §§ 21 and 27; Human Rights Committee, Addendum to the Second Periodic Report, Israel (CCPR/C/ISR/2001/2, 2001) § 8.

120 *Legal Consequences of the Construction of a Wall in the Occupied Palestinian Territory*, Advisory Opinion, ICJ Reports 2004, §§ 107–113; General Comment no 31: Nature of the General Legal Obligation Imposed on States Parties to the Covenant (CCPR/C/21/Rev.1/Add.13, 2004) § 10; *Al Skeini and others v United Kingdom*, App no 55721/07 (ECHR, 2011) §§ 130–150; *Al-Jedda v United Kingdom*, App no 27021/08 (ECHR 2011) §§ 74–86.

121 *Banković v Belgium and 16 other States*, App no 52207/99 (ECHR 2001) §§ 54–82.

122 *Al-Skeini* (n 120) § 142 (use of force in occupied territory). See also *Jaloud v Netherlands*, App no 47708/08 (ECHR 2014) §§ 152–153 (use of force at a checkpoint in Iraq). It should be noted however that these cases where substantially different from the *Banković* case as the respondent States were Occupying Powers at the time. Moreover, although the Court referred to the notion of 'authority and control over individuals killed' (§ 149) in the context of the extraterritorial use of force, certain (and notably the United Kingdom's Court of Appeal) have read this ambiguous case much more conservatively by holding that the Court 'found jurisdiction to have been established only in the context of security operations amounting to the exercise of public powers, ie powers which are normally exercised by a government authority'. See David Goddard, 'The UK's Al-Saadoon Case: Stepping Back from the Extraterritorial Application of the ECHR for Physical Force' *Just Security* (30 September 2016), accessed

another '*Banković*-type' case were to arise. Other human rights bodies, such as the Human Rights Committee and the Inter-American Commission on Human Rights, tend to accept the extraterritorial application of human rights treaties for a wider spectrum of situations.[123] The ICRC (like many commentators) held the view that, in any case, customary law prohibits the arbitrary deprivation of life without territorial limitation.[124] It is further submitted that the jurisdiction should be established not only through the state using force extraterritorially but also through the territorial state if the latter consented to the intervention. The territorial state cannot evade its own human rights obligations by consenting to the intervention of a third state. Put differently, it cannot allow the intervening state to violate the human rights of the persons under its jurisdiction. On the other hand, if the intervening state violates the human rights of the persons under the jurisdiction of the territorial state, this presumably constitutes a violation of the territorial state's consent and thus of its sovereignty. The intervening state therefore must, as a minimum, respect 'by proxy' the human rights obligations undertaken by the territorial state.

---

3 May 2017 at https://www.justsecurity.org/33296/uks-al-saadoon-case-stepping-extraterritorial-application-echr-arising-physical-force/.

[123] General Comment no 31 (n 120) §10; *Ecuador v Colombia*, Report no 112/10 (IACHR, 2010) § 99 (use of force during a security operations abroad); *Disabled Peoples' International v USA*, Case no 9213 (IACHR, 1987) (extra-territorial use of force); *Armando Alejandre, Carlos Costa, Mario de la Peña et Pablo Morales v Cuba*, App no 11589 (IACHR, 1999) (extraterritorial use of force). It should be noted however that the American Declaration on the Rights and Duties of Man contains no jurisdictional limitation.

[124] ICRC Challenges Report 2011 (n 9) 22. See also Melzer (n 86) 212. It is unclear whether states such as the United States of America and Israel may agree with such a position. Their arguments against the extraterritorial application of human rights law are mostly (if not only) based on the jurisdictional clauses in human rights treaties. According to Professor Michael Schmitt the United States has never rejected the extraterritorial application of the customary right to life. A counter-argument could be that if the right to life is certainly customary, it is difficult to establish practice and *opinio juris* to the effect that it is not intrinsically accompanied by any jurisdictional limitation. This *caveat* is often raised by Professor Marco Sassòli notably in his teachings and conferences. The fact that a number of non-binding human rights instruments such as the American Declaration on the Rights and Duties of Man or the Universal Declaration of Human Rights recognize the right to life without establishing jurisdictional limitation might nevertheless be seen as elements of relevant practice and *opinio juris*.

In brief, even if absolutist interpretations of the extraterritorial application of HRL were rejected, it is submitted that there are at least some situations in which HRL would theoretically apply to remote warfare. These include cases where lethal force is used in contexts of internal armed conflicts, occupation and, possibly, extraterritorial non-international armed conflicts with the consent of the territorial state.

But even then, the relevance of HRL for remote warfare may well be questioned taking into account the lex specialis character of IHL in armed conflict situations. This lex specialis-approach to the simultaneous application of HRL and IHL, which was initially crafted by the ICJ,[125] remains, however, shrouded in some ambiguity.[126] I had argued elsewhere that the lex specialis maxim does not operate in an absolute manner and that it makes no sense to consider that IHL as a whole is the more special law with respect to HRL.[127] Indeed, the concrete relations between the branches are much more multifaceted and complex than such a simple formula can express. The ICJ never had in mind an operation of the lex specialis rule resulting in mutual exclusiveness, since it clearly applied simultaneously to HRL and IHL. At the level of branches, its objective was rather to co-ordinate the two levels of protection of HRL and of IHL. This supposes a particular lex specialis maxim, situated on the level of interpretation rather than on that of conflict of norms, which could be better defined, if at all, as *lex specialis 'compleat' legi generali.*[128] The traditional '*lex specialis derogat*' maxim nevertheless has a proper place in the interplay between IHL and HRL to

---

[125] See *Nuclear Weapons Advisory Opinion* (n 115) §25; *Advisory Opinion on Legal Consequences of the Construction of a Wall in the Occupied Palestinian Territory* [2004] ICJ Rep §106.

[126] *Wall Advisory Opinion* (n 120), separate Opinions of Judges Higgins (§§ 24–25) and Kooijmans (§ 29). See also Anja Lindroos, 'Addressing Norm Conflicts in a Fragmented Legal System: The Doctrine of *Lex Specialis*' (2005) 74 Nordic J Intl L 42; Roger O'Keefe, 'Legal Consequences of the Construction of a Wall in the Occupied Palestinian Territory: A Commentary' (2004) 37 Rev Belge de Droit Intl 135; *Report of the Expert Meeting on the Right to Life in Armed Conflicts and Situations of Occupation,* organised by the University Centre for International Humanitarian Law (UCIHL), Geneva, 1–2 September 2005, 10 and 19.

[127] Gloria Gaggioli and Robert Kolb, 'A Right to Life in Armed Conflicts? The Contribution of the European Court of Human Rights' (2007) 37 Israel Yearbook on Human Rights 118–124; Gloria Gaggioli, *L'influence mutuelle entre les droits de l'homme et le droit international humanitaire à la lumière du droit à la vie* (Pedone 2013) 42–60.

[128] Gaggioli and Kolb (n 127) 121.

the extent that two, or more than two, *norms* of IHL and/or of HRL bearing on the same subject-matter are in conflict in such a way that a simultaneous application of both is impossible under the principle of non-contradiction.[129] Finally, even when IHL norms prevail over human rights norms on a specific subject-matter in a specific situation, this does not mean that the latter norms disappear. As highlighted in the ILC Study on the fragmentation of IHL, 'the more general rule remains in the background providing interpretative direction to the special one'.[130] The key question which remains is thus whether and how HRL could bring something to the table in relation to remote warfare.

(b) Relevance of Human Rights for Targeting in Remote Warfare?

Despite the simultaneous application of IHL and HRL in armed conflicts, it is difficult to negate the fact that when it comes to norms pertaining to means and methods of warfare, as well as to rules on targeting or the use of force in the conduct of hostilities, IHL is indeed a lex specialis.

Regarding means and methods of warfare, IHL may restrict or prohibit them either directly (usually through a specific treaty) or generally through the application of the 'cardinal principles'[131] of distinction (which prohibits indiscriminate means and methods of warfare)[132] and of the prohibition of superfluous injury or unnecessary suffering.[133] Although remote weapons systems like drones or autonomous weapons are not prohibited as such by specific IHL treaties or norms, the aforementioned cardinal principles remain relevant to assess their intrinsic legality as well as the legality in their use. For instance, if specific remote weapons systems were conceived in a manner that would run counter to these principles—for instance, if a 'killer robot' would be unable to distinguish properly between combatants and civilians—they would be prohibited as such by IHL. Even if not intrinsically unlawful,

[129] Ibid.

[130] International Law Commission, Report of the Study Group on Fragmentation of International Law: Difficulties Arising from the Diversification and Expansion of International Law (A/CN.4/L.682, 2006) §§ 102 and 103.

[131] ICJ *Nuclear Weapons Advisory Opinion* (n 115) § 78.

[132] Articles 48 and 51(4)(b)c) of Additional Protocol I to the Geneva Conventions (API). Jean-Marie Henckaerts and Louise Doswald-Beck, *Customary International Humanitarian Law (Vol. 1: Rules)* (ICRC 2005) rules 1, 11 and 12.

[133] Article 23(e) Hague Regulations of 1907; Article 35(2) API; Henckaerts and Doswald-Beck, ICRC Customary IHL Study (n 132) rule 70. See also 1868 St. Petersburg Declaration.

any weapon system must be used in a manner that respects these IHL principles. HRL does not contain equivalent rules and principles on means and methods of warfare. It is therefore necessary to rely on IHL to assess the intrinsic legality and the legality in their use of remote weapons systems as the ICJ did in its 1996 *Advisory Opinion on nuclear weapons*.[134] On the intrinsic legality of means and methods of warfare, there is arguably not even a conflict of norms between IHL and HRL rules, since the latter (HR norms) simply do not exist on this matter. This is so because HRL was initially conceived for peacetime situations. IHL can thus be portrayed as complementary to HRL in this context.

Regarding targeting or the use of force in armed conflicts, IHL provides clear rules pertaining to the conduct of hostilities that stem from the principles of distinction, proportionality and precautions.[135] HRL also contains a detailed framework for the use of force against persons that derives from the right to life and, which can be portrayed as flowing from five key principles: legality, necessity, proportionality, precautions and accountability.[136] However, these rules pertain to the use of force in law enforcement, not in the conduct of hostilities. Although some principles are common to both IHL and HRL (for example, the principles of necessity, proportionality, precautions), they operate differently in the two bodies of law and their application in specific situations may well lead to contradictory results (although this might not always be the case).[137] For instance, the human rights principle of 'absolute necessity' implies that the use of force must be the last resort in response notably to an imminent threat to life or limb, while under IHL the military necessity to attack legitimate targets such as combatants or fighters is presumed.[138] This IHL 'status-based' approach is at odds with the human rights 'threat-based' approach. In another example, while the IHL principle of proportionality protects surrounding civilians and civilian objects from damage which would be *excessive* in relation to the concrete and direct military advantage anticipated of an attack, the human rights strict

---

[134] ICJ *Nuclear Weapons Advisory Opinion* (n 115).

[135] See Articles 48, 51 and 57 API and corresponding customary IHL rules. Henckaerts and Doswald-Beck, ICRC Customary IHL Study (n 132) rules 1–24.

[136] See below, section entitled 'The human rights law enforcement paradigm and the use of force against persons'. See also Gaggioli (n 40) 100–105.

[137] Gloria Gaggioli, *Report of the Expert Meeting on the Use of Force in Armed Conflicts: Interplay Between the Conduct of Hostilities and Law Enforcement Paradigms* (ICRC 2013) 8–9.

[138] Ibid 9.

proportionality test protects not only bystanders much more broadly[139] but also the very person posing a threat and allows only minimal casualties.[140] These differences (among others) cannot be reconciled. There should be no doubt that when belligerents direct the use of force against persons/objects considered as legitimate military targets under IHL, conduct of hostilities rules under IHL would generally prevail as lex specialis over applicable and potentially relevant HRL rules, such as the right to life or the right to property (regarding the use of force against objects) even if the latter are more protective.[141] Another way of presenting this would be to say that although these human rights remain applicable in these situations, their content must be implied into IHL rules that were specifically made to regulate this type of situation. Regarding the use of force in the conduct of hostilities, be it by remote weapon systems or other means and methods of warfare, therefore, the principles regulating the use of force must be found in IHL, not in HRL.

This analysis should not be controversial since, as noted earlier, even human rights bodies did refer to the IHL principles of distinction, proportionality and precautions notably to assess the legality of the use of force by remote weapon systems in armed conflicts.[142] In practice, human rights bodies have relied mainly on three human rights principles to constrain the resort to remote weapon systems, without explaining, however, their interplay with IHL. These are the principles of transparency, accountability and dignity.

(c) Relevance of Human Rights Concepts of Transparency, Accountability and Dignity for Remote Warfare?

Transparency and accountability are the buzzwords when human rights experts discuss the use of force by drones.[143] Indeed, one of the biggest

---

[139] See below n 242.

[140] Ibid 9.

[141] Nils Melzer and Gloria Gaggioli, 'Conceptual Distinction and Overlaps Between Law Enforcement and the Conduct of Hostilities', in Terry D Gill and Dieter Fleck (eds), *The Handbook of the International Law of Military Operations* (Oxford University Press 2015) 79–80.

[142] Report of the Special Rapporteur on Extrajudicial, Summary or Arbitrary Executions, Philip Alston, to the Human Rights Council, Study on Targeted Killings (A/HRC/14/24/Add.6, 2010) § 29.

[143] See, eg, Alston, Report to the Human Rights Council 2010 (n 142) §§ 87–92; Alston, Interim Report to the General Assembly 2010 (n 18) §§ 11, 14–16, 30, 36; Heyns (n 5) § 111; Emmerson, Report to the General Assembly 2013 (n 87) §§ 41–45; Gen John P Abizaid and Rosa Brooks (co-chairs of the

problems identified with targeting by drones in armed conflicts is the total lack of information regarding who is targeted, when, where, by whom and what is the outcome.[144] In addition, the fact that drone strikes are often conducted in not easily accessible areas because of remoteness or security concerns, implies that it has been difficult or even impossible for human rights monitors and civil society to collect data directly in the field and thus to assess the legality of those strikes.[145] This in turn implies a risk of total lack of accountability on the part of those resorting to these new technologies. Given that one of the greatest advantages of remote weapons systems is precisely to operate in difficult areas where sending ground troops would be impossible or too dangerous, it can easily be deduced that challenges regarding the collection of information in the field would be particularly serious in the context of remote warfare in general (and not only in the context of drone strikes).[146]

Although human rights experts often refer to the notion of transparency in the context of remote warfare, its legal basis is rarely specified.[147] This notion does not appear explicitly in any human rights treaty. It would therefore be difficult to portray it as a right on its own. It is submitted, however, that it is a fundamental principle that underlies many human rights, such as the right to life under its procedural limb,[148] the right to an effective remedy[149] or the contested right to the truth.[150]

Task Force), *Recommendations and Report of the Task Force on US Drones Policy* (Stimson Report 2014) 32–33; Resolution 25/ … of the Human Rights Council, Ensuring use of remotely piloted aircraft or armed drones in counter-terrorism and military operations in accordance with international law, including international human rights and humanitarian law (A/HRC/25/L.32, 2014) § 2. At a panel discussion at the Human Rights Council on drones on 20 October 2014, many speakers called for more transparency and accountability. See http://www.ohchr.org/EN/NewsEvents/Pages/ArmedDrones.aspx (accessed 3 May 2017).

[144] Alston, Report to the Human Rights Council 2010 (n 143) § 91; Stimson Report (n 143) 32–33.

[145] Alston, Report to the Human Rights Council 2010 (n 143) § 91.

[146] Heyns (n 5) § 111.

[147] Philip Alston held that transparency is required by both IHL and HRL, without explaining why. Alston, Report to the Human Rights Council 2010 (n 143) § 88.

[148] Article 6 ICCPR; Article 2 ECHR; Article 4 ACHR; Article 4 ACHPR.

[149] Article 2(3) ICCPR; Article 13 ECHR; Article 25 ACHR; Article 7(1) ACHPR. This right is considered as non-derogable. See General Comment 29: Derogations during a State of Emergency (CCPR/C/21/Rev.1/Add.11, 2001) § 14; *Advisory Opinion on Judicial Guarantees in States of Emergency* (Inter-American Court of Human Rights OC-9/87) §§ 24–26.

Indeed, as so often and systematically repeated by human rights bodies, alleged violations of the right to life must be effectively investigated, including in armed conflict situations.[151] This requires notably the involvement of the next-of-kin[152] and some form of publicity,[153] and thus involves transparency. Access to an effective remedy for victims of remote warfare or their relatives is also obviously impossible without at least some degree of transparency. More broadly, transparency is an essential underpinning of the rule of law, which is intrinsically linked to HRL.[154]

While transparency can easily be accepted as a human rights principle, its validity in the context of international humanitarian law is far from obvious.[155] Not only can it not be found in IHL treaties—or literature/legal writings—but this body of law accepts methods of warfare, such as

---

[150] Anja Seibert-Fohr, *Prosecuting Serious Human Rights Violations* (Oxford University Press 2009) 70.

[151] See, e.g. among many others, *Pedro Pablo Camargo v Colombia* ('Guerrero') (CCPR/C/15/D/45/1979, 31 March 1982); *Mapiripán Massacre v Colombia* Series C no 134 (IACHR, 15 September 2005) §§ 216–241; Report of the Special Rapporteur on the Extrajudicial, Summary and Arbitrary Executions, Philip Alston, to the Commission on Human Rights (E/CN.4/2006/53, 2006) §§ 35–36; *Al-Skeini and others v United Kingdom* (ECHR, 7 July 2011) §§ 161–177; ACommHPR, *Commission Nationale des Droits de l'Homme et des Libertés v Chad*, § 22. See also UN Code of Conduct for Law Enforcement Officials [1979] (adopted by Resolution 34/169 of the UN General Assembly), commentary (c) to Article 3; UN Basic Principles on the Use of Force and Firearms by Law Enforcement Officials [1990] (adopted by the Eight UN Congress on the Prevention of Crime and the Treatment of Offenders and welcomed by Resolution 45/166 of the UN General Assembly), Principle 22.

[152] See, eg, *Güleç v Turkey* App no 21593/93 (ECHR, 27 July 1998) § 82; *Ximenes-Lopes v Brasil* (IACHR, 4 July 2006) § 193; *Prison de Miguel Castro-Castro v Peru* (IACHR, 25 November 2006) §§ 255–256.

[153] See, e.g., *Güleç* (n 152) § 78 and 80; *Bámaca Velásquez v Guatemala* (IACHR, 22 February 2002) § 77. The degree of publicity required may vary depending on the case. See *Isayeva v Russia*, App no 57950/00 (ECHR, 24 February 2005) § 214.

[154] See, eg, The UN Rule of Law Indicators, Implementation Guide and Project Tools (2011) 3–4.

[155] The Public Commission to Examine the Maritime Incident of 31 May 2010 (Turkel Commission), Israel's Mechanisms for Examining and Investigating Complaints and Claims of Violations of the Laws of Armed Conflict According to International Law, Second Report (February 2013) § 106.

ruses of war,[156] that require some form of secrecy. More broadly, it would be hard to conceive how the balancing exercise between the principles of military necessity and humanity may oblige belligerent states to divulge their military strategies, anticipated targets, military operations' planning and so forth. This may forfeit the belligerent's ability to effectively wage wars. Traditionally, IHL did not include either an individual right to an effective remedy, although it may be held that there have been important evolutions in that respect.[157] Nevertheless, the notion of transparency is not totally absent from IHL. For instance, in the context of missing persons, Additional Protocol I recognizes the right of families to know the fate of relatives,[158] which notably requires belligerent parties to account for the dead.[159] It may also be argued that transparency is closely linked to and derives from the principle of accountability,[160] which underlies every international legal framework, including IHL.[161] IHL violations entail both states' responsibility, and the individual criminal responsibility for war crimes. How could third states and the international community hold a belligerent state responsible for IHL violations without having information concerning the factual situation, for instance regarding the details of strikes conducted by way of remote weapon systems (who was the target, when the attack was conducted and so on)? There is, however, quite a leap between this reasonable statement and a supposed duty on the part of the belligerents to provide this information when the war is ongoing. Moreover, even if it is true that IHL implicitly provides for an obligation for belligerent states notably to investigate war crimes,[162] including in the context of the

---

[156] Ruses of war may include the use of camouflage, decoys, mock operations and misinformation. See Article 37 API. See also Henckaerts and Doswald-Beck, ICRC Customary IHL Study (n 132) rule 57.

[157] Gaggioli (n 127) 498–502.

[158] Article 32 of API.

[159] Gaggioli, 'International Humanitarian Law: The Legal Framework for Humanitarian Forensic Action' (forthcoming) *Forensic Science International, Volume on Humanitarian Forensic Action* 18.

[160] Emmerson, Report to the General Assembly 2013 (n 87) § 45 ('this obligation [the principle of transparency] ought to be viewed as an inherent part of the State's legal obligations of accountability under international humanitarian law and international human rights law').

[161] *Chorzow Factory* Case (Permanent Court of International Justice, 13 September 1928). See also Article 31 of the International Law Commission, *Draft Articles on State Responsibility* (2001).

[162] Articles 49–50 of the First Geneva Convention (GCI); Articles 50–51 of the Second Geneva Convention (GCII); Articles 129–130 of the Third Geneva

conduct of hostilities, such an obligation is much more limited than the human rights corresponding duty to investigate each and every violent death[163] (or at least each alleged violation of the right to life).[164] Typically, under IHL, the killing of a legitimate target (that is, combatants, fighters and civilians directly participating in hostilities), and even causing incidental civilian loss of life unwillingly or without knowing that the collateral damages would be excessive is not a war crime and therefore may not entail a full-fledged obligation to investigate.[165] This is so because 'IHL takes into account the fact that some deaths are inherent to the conduct of hostilities in armed conflicts'.[166] Moreover, unlike HRL, IHL says nothing about the criteria that need to be fulfilled to consider an investigation as effective and state practice does not seem

Convention (GCIII); and Articles 146–147 of the Fourth Geneva Convention. See also Articles 11 and 85–86 API. One may argue that the duty to suppress IHL violations also entails some form of investigation. See Article 49(3) GCI; Article 50(3) GCII; Article 129(3) GCIII; and Article 146(3) GCIV. In the same vein, an IHL obligation to investigate can also be derived from the duty of commanders to prevent and, where necessary, to suppress and to report to competent authorities breaches of IHL, with a view to initiate disciplinary or penal action against violators. See Article 87 API. See also Gaggioli, ICRC Use of Force Report (n 137) 49 and 52.

[163] During the ICRC Expert Meeting on the Use of Force, an expert made clear that: 'in the context of the European Convention on Human Rights, a State would have to investigate any violent loss of life—including in cases of killing of enemy combatants or of apparently lawful incidental civilian casualties under IHL—unless the State derogated to the right to life in accordance with Article 15 of the European Convention on Human Rights'. Gaggioli, ICRC Use of Force Report (n 137) 53.

[164] Ibid 49–51. See also Gloria Gaggioli, 'A Legal Approach to Investigations of Arbitrary Deprivations of Life in Armed Conflicts: The Need for a Dynamic Understanding of the Interplay between International Humanitarian Law and Human Rights' (2017) Zoom-in 36 Questions of International Law 27–51. Available at http://www.qil-qdi.org/legal-approach-investigations-arbitrary-deprivations-life-armed-conflicts-need-dynamic-understanding-interplay-ihl-hrl/, accessed 7 June 2017.

[165] To read what is considered as a war crime in the context of the conduct of hostilities, see Article 85(3)–(4) API. See also Article 50 GCI; Article 51 GCII; Article 130 GCIII; Article 147 GCIV. See also the 1998 Statute of the International Criminal Court, Article 8(2)(b) and (e).

[166] Gaggioli, ICRC Use of Force Report (n 137) 49.

to indicate that the involvement of the next-of-kin or wider publicity, transparency, are considered a must or priority in a wartime situation.[167]

In brief, if the human rights concepts of transparency and accountability are not absent from IHL, they are clearly less predominant or far-reaching than under HRL. What should then be the interplay between IHL and HRL in this context? In practice, the recognition of IHL as a lex specialis for the use of force in the context of remote warfare has not prevented human rights bodies from referring to the principles of transparency and accountability as legal constraints to be respected in remote warfare.[168] Most states did not object to this integration of human rights concept into warfare situations and some relied on these same principles.[169] In this context, it is interesting to note that the US Obama Administration released in December 2016 a 'Report on the legal and policy framework guiding the United States' use of military force and related national security operations', precisely to respond to the criticisms of the international community about lack of transparency and accountability surrounding drone operations notably.[170] The report makes clear, however, that it is 'as a matter of policy' that 'the United States ... frequently applies certain heightened policy standards and procedures that underscore its commitment to reducing civilian casualties and to enhancing transparency and strengthening accountability for its actions'.[171]

Legally speaking, these evolutions can be approached in various ways. The most conservative approach would be to consider that even if transparency and accountability (as understood in HRL) are important concepts to be taken into account for policy reasons, considering them as legal constraints in remote warfare is legally unsound as it is based on a misunderstanding of IHL obligations, which constitute the lex specialis for conduct of hostilities matters.[172] Another approach would be to

---

[167] Alston, Report to the Human Rights Commission 2006 (n 151) § 33. See also Gaggioli, ICRC Use of Force Report (n 137) 56–57; Gaggioli (n 127) 482; Gaggioli (n 164).

[168] See above n 143.

[169] Chris Wood, 'Consensus grows among UN states for greater transparency on drone civilian deaths', *The Bureau of Investigative Journalism* (26 October 2013).

[170] The White House, *Report on the Legal and Policy Frameworks Guiding the United States' use of military force and related national security operations* (December 2016).

[171] Ibid 24.

[172] This was the majority view during the ICRC Expert Meeting on the use of force. Gaggioli, ICRC Use of Force Report (n 137) 53.

consider that actually this is an example where human rights principles, as lex generalis, shall be taken into account when interpreting IHL rules.[173] This approach would be somehow consonant with the trendy and flexible approach towards 'harmonization' between IHL and HRL,[174] which may lead in this case to an expansive interpretation of IHL rules. The most progressive interpretation would be to consider that given solid evolutions in the field of HRL regarding accountability for every violent death and increased transparency—not only in non-binding UN reports but also in binding jurisprudence (especially the European Court of Human Rights)—HRL has become actually the most precise, specific and up-to-date law on the matter; and thus, is today the lex specialis.[175]

Unlike the previous one, this latter approach would not affect the interpretation of IHL rules, but only their interplay with HRL. In practice, the results of these two last approaches might be very similar, although the penultimate one might be diplomatically more palatable. They both rely on a dynamic, rather than static, understanding of the interplay between IHL and HRL.[176] To be realistic, however, they need to take into account the specific and inherent difficulties in armed conflicts for ensuring transparency and accountability. It is a reality that 'instability and insecurity in armed conflict situations can pose serious obstacles to investigations into each death, such as difficulties in gathering evidence on site or hearing witnesses'.[177] Human rights bodies have recognized that the context of an armed conflict needs to be taken into account to assess the effectiveness of an investigation.[178] For instance, as

---

[173] See, in this sense, Turkel Commission Report (n 155) § 108.

[174] As an example of this 'harmonization' approach in the context of detention, see *Hassan v United Kingdom* App no 29750/09 (ECHR 2014) § 102.

[175] Gaggioli (n 127) 474–504.

[176] Ibid.

[177] Gaggioli, ICRC Use of Force Report (n 137) 49.

[178] See, e.g., *Al Skeini and others v United Kingdom* App no 55721/07 (ECHR, 7 July 2011) § 168: 'The Court takes as its starting point the practical problems caused to the investigatory authorities by the fact that the United Kingdom was an Occupying Power in a foreign and hostile region in the immediate aftermath of invasion and war. These practical problems included the breakdown in the civil infrastructure, leading *inter alia* to shortages of local pathologists and facilities for autopsies; the scope for linguistic and cultural misunderstandings between the occupiers and the local population; and the danger inherent in any activity in Iraq at that time. As stated above, the Court considers that in circumstances such as these the procedural duty under Article 2 must be applied realistically, to take account of specific problems faced by investigators'. See also Turkel Commission Report (n 155) § 107.

regards transparency, it would be fair to accept that belligerent parties are not obliged to give details of each military operation during the armed conflict. Transparency might, however, imply obligations to keep and disclose the records of military operations after a certain period of time has elapsed following an armed conflict, which in turn would facilitate ex-post monitoring.[179] Transparency also means that, at least, states must acknowledge when they have conducted drone strikes or otherwise resorted to remote weapons systems and provide the underlying legal rationale for such strikes. This would imply for states to be transparent and forthcoming about the legal classification of situations of violence and about the applicable legal framework in a given situation. As former Special Rapporteur on extrajudicial killing, Philip Alston, highlighted, states are also expected to disclose information about the 'procedural and other safeguards in place to ensure that killings are lawful and justified, and the accountability mechanisms that ensure wrongful killings are investigated, prosecuted and punished'.[180] Similarly, accountability does not entail criminal prosecution or lengthy investigations for each use of force. It means at the very least that sufficient information about each military operation (who was the target, which collateral damages were expected, which precautions were taken, what was the outcome of the operation) must be collected and recorded, that preliminary assessments are conducted to determine whether further investigation is required and that appropriate procedures are put in place to ensure lessons are learned.[181] Fully-fledged criminal investigations are merely requested in the context of war crimes allegations. Such investigations should involve an appropriate degree of transparency to be considered as effective.[182]

---

179 See, in this sense, Marco Sassòli and Lindsey Cameron, 'The Protection of Civilian Objects—Current State of the Law and Issues de lege ferenda', in Natalino Ronzitti and Gabriella Venturini (eds), *The Law of Air Warfare: Contemporary Issues* (Eleven International Publishing 2006) 63–64; Alston, Report to the Human Rights Commission 2006 (n 151) § 43.

180 Alston, Report to the Human Rights Council 2010 (n 142) §87.

181 See in this sense: Emmerson, Report to the General Assembly 2013 (n 87) § 42; Turkel Commission Report (n 155) 102. Further work is needed to differentiate possible types of investigations, their triggers and criteria notably. A process has been launched in January 2014 by the Geneva Academy under the lead of Professor Noam Lubell on 'the duty to investigate under international law' and is ongoing. See https://www.geneva-academy.ch/our-projects/our-projects/armed-conflict/detail/3 (accessed 3 May 2017).

182 See, in this sense, Emmerson, Report to the General Assembly 2013 (n 87) § 42; Turkel Commission Report (n 155) § 106. Depending on the

While issues of transparency and accountability are relevant for any type of military operation, it is submitted that their importance in the context of remote warfare is all the more evident. Imagine: if requirements of transparency and accountability were brushed off in the context of remote warfare, this means that states would be able to conduct lethal attacks potentially anywhere on earth by means of remote weapons systems without acknowledging such strikes and without the possibility for the international community to (easily) attribute such strikes.[183] Without initial attribution, a legal analysis is impossible and accountability an empty shell. With the proliferation of remote weapon systems, such issues would become even more worrying.[184] Therefore, it is submitted that the human rights principles of transparency and accountability have to be adapted to armed conflict situations and complement the IHL rules applicable to remote warfare (and warfare in general).

Another principle has been invoked by some human rights experts to restrain or rather prevent and prohibit specifically lethal autonomous weapons. It is the principle of dignity, which has sometimes even been presented as a 'right'.[185] The principle of dignity was first suggested in the context of lethal autonomous weapons by Human Rights Watch and the Harvard Law School in a report on the 'human rights implications of killer robots' and has since been embraced by former Special Rapporteur on extrajudicial, summary or arbitrary executions, Christoph Heyns.[186] The argument proposed is that allowing machines to make life and death decisions would run against the principle of dignity as 'they could truly comprehend neither the value of individual life nor the significance of its

---

situation, however, involvement of the next of kin in all phases of the investigation may, however, be difficult or impossible. See Human Rights Council, Report of the Committee of independent experts in international humanitarian and human rights laws to monitor and assess any domestic, legal or other proceedings undertaken by both the Government of Israel and the Palestinian side, in the light of General Assembly resolution 254/64 including the independence, effectiveness, genuineness of these investigations and their conformity with international standards (A/HRC/15/50, 2010) § 33.

183 Gaggioli (n 40) 113.

184 Ibid.

185 See Explanations Relating to the Charter of Fundamental Rights, 2007/c, 303/02, Official Journal of the European Union, C303/17, 12 December 2007: 'The dignity of the human person is not only a fundamental right in itself but constitutes the real basis of fundamental rights'; Heyns (n 104) § 85; Heyns (n 63) 368.

186 See above, section entitled 'Human rights bodies' practice and positions on remote weapons systems'.

loss'.[187] In a recent article, Professor Heyns held that '[a] machine, bloodless and without morality or mortality, cannot fathom the significance of using force against a human being and cannot do justice to the gravity of the decision' and that 'death by algorithm means that people are treated simply as targets and not as complete and unique human beings, who may, by virtue of that status, meet a different fate. They are placed in a position where an appeal to the humanity of the person on the other side is not possible'.[188] He also held that '[w]hen someone comes into the sights of a computer, that person is literally reduced to numbers: the zeros and the ones of bits'.[189] It can be deduced from these quotes that the principle of dignity would require, on one hand, that only human beings make life and death decisions and, on the other hand, that targets of an attack shall always be in a position to appeal to the humanity of the attacker to save his/her life.

From a legal (not philosophical or theological) perspective, arguments based on the principle of dignity to consider autonomous weapons as unlawful raise several questions. First the legal basis of this construct may be questioned. Depicting dignity as a proper international human 'right' is questionable. No human rights conventions really include such a 'right to dignity' and evidence of its customary nature is lacking. Although some states may domestically recognize such a right,[190] it would be far-fetched to consider it as a general principle of law. Nevertheless, dignity is definitely at the center of international HRL. It is sometimes explicitly mentioned in human rights instruments[191] and it underlies many (if not all) human rights, in particular the right to life, the right to liberty, the prohibition of torture, inhumane and degrading treatment or the prohibition of enforced disappearances.[192] It is submitted

---

[187] Human Rights Watch and International Human Rights Clinic (n 8) 23.

[188] Heyns (n 63) 370.

[189] Ibid.

[190] See, eg, German Federal Constitutional Court case on legislation authorizing the shooting down of an aeroplane, 1 BvR 357/05, 15 February 2006, § 119.

[191] See Preamble of the United Nations Charter § 2; Article 1 of the Universal Declaration of Human Rights; Preamble of the ICCPR §§ 2–3; Preamble of the International Covenant on Economic, Social and Cultural Rights (ICESCR) §§ 2–3; Article 1 of the EU Charter of Fundamental Rights; Code of Conduct for Law Enforcement Officials (n 151) Article 2.

[192] See preamble of the ICCPR: 'Recognizing that these [inalienable] rights derive from the inherent dignity of the human person'. See also preamble of the ICESCR § 3. Some rights in these treaties also refer to dignity. See Article 10 ICCPR; Article 13 ICESCR; Article 5 of the African Charter on Human and

therefore that dignity can safely be described as an underlying principle of international HRL, and possibly of IHL through the prism of the principle of humanity.[193] The main issue is the exact content and applications of this principle, which remain largely unexplored.[194] Although everyone would agree that being tortured is an attack on dignity, it remains debatable whether being killed by an autonomous weapon is more disruptive of dignity than being killed by a traditional missile. Targets of traditional airstrikes are virtually as much 'reduced to numbers' and their ability to 'appeal to the humanity' of the pilots is frankly minimal.[195] If we were really serious about considering the principle of human dignity as requiring to always be able to appeal to the humanity of the attacker, a number of contemporary military technologies should be considered as prohibited, not only fully autonomous lethal weapons systems. Moreover, it is submitted that, even in the case of autonomous weapons, the decision to kill is ultimately a human decision. Humans have created and programmed the machine and remain ultimately legally and morally responsible for the actions of their creatures, even if they have totally lost control over the machine. In practice, it seems also unlikely that states will find an interest in developing fully autonomous lethal weapon systems, over which they would have no control whatsoever and which could potentially turn against them. Limiting the argument of dignity to such weapons is thus of limited practical value. In the context of autonomous weapons, the legal value of the principle of dignity is questionable, although it certainly constitutes a powerful advocacy argument to convince states to ban autonomous lethal weapons.

Peoples' Rights; International Convention for the Protection of All Persons from Enforced Disappearance, Article 19(2) and 24(5).

[193] Heyns (n 63) 367.

[194] Ibid 367. For one of the rare interdisciplinary books on the issue, see Christopher McCrudden (ed), *Understanding Human Dignity* (Oxford University Press 2014).

[195] Quotations from Heyns (n 63) 370.

4. HUMAN RIGHTS LAW INSTEAD OF IHL IN REGULATING REMOTE WEAPON SYSTEMS IN LAW ENFORCEMENT

(a) The Human Rights Law Enforcement Paradigm in Peacetime, Counter-Terrorism and Armed Conflicts[196]

In peacetime and in situations of violence not reaching the threshold of an armed conflict, it is clear that any use of remote weapon systems is regulated exclusively by international HRL (and domestic law), and not by IHL, where the great majority of provisions apply only in armed conflict situations.[197] But the human rights law enforcement paradigm regulates the use of force by remote weapons systems also in extra-territorial counter-terrorism operations not amounting to an armed conflict and, in certain cases, within situations of armed conflict as well.

For instance, in the global fight against terrorism, extraterritorial drone strikes have been conducted by states like the United States and the United Kingdom against 'Al-Qaeda, the Taliban and associated forces', including the Islamic State of Iraq and Syria (ISIS), not only in the context of an acknowledged armed conflict (for example, in Afghanistan or Iraq), but also outside 'hot battlefields' or in contexts whose legal classification is disputed (for example, in Pakistan, the Philippines, Somalia, Syria and Yemen). Some of those drone strikes have allegedly been conducted outside the territory of the parties to an ongoing armed conflict, that is, in 'non-belligerent States'.[198]

There are different legal opinions as to whether the remote targeting of an individual in the territory of a non-belligerent state is governed by the conduct of hostilities under IHL or law enforcement paradigm under

---

196 This section relies heavily on a previous article: Gaggioli (n 40) 92–99.

197 As highlighted by the ICRC President, Peter Maurer, it is clear that: 'If and when drones are used in situations where there is no armed conflict, it is the relevant national law, and HRL with its standards on law enforcement, that apply, not international humanitarian law'. See Maurer (Web Interview) (n 65).

198 Stuart Casey-Maslen, 'Pandora's box? Drone strikes under jus ad bellum, jus in bello, and international human rights law' (2012) 94 Intl Rev Red Cross 616; Jennifer C Daskal, 'The Geography of the Battlefield: A Framework for Detention and Targeting Outside the "Hot" Conflict Zone' (2013) 161 U Penn L Rev 1165, 1188.

HRL.[199] The answer to this question depends mainly on interpretations pertaining to the geographical scope of IHL. Under one school of thought, 'humanitarian law is not territorially delimited but governs the relations between the belligerents irrespective of geographical location'.[200] A person considered as a legitimate target in relation to an ongoing armed conflict would thus be targetable under the conduct of hostilities paradigm wherever that person is located. Pursuant to other views,[201] shared notably by the ICRC,[202] IHL would not apply outside the territories of the parties to an armed conflict. The use of force in a non-belligerent state would therefore need to comply with the rules pertaining to the human rights law enforcement paradigm.[203] This latter viewpoint relies mainly on the understanding that the customary right to life applies extraterritorially.[204] If this view were accepted, counter-terrorism operations conducted in the territory of a non-belligerent state would be governed by HRL.

There may also be other situations where the extraterritorial use of force, notably by remote weapons systems, would not amount to an armed conflict, or at least not be part of the hostilities under IHL. Consider a situation where there is a 'lone wolf' terrorist, who is about to launch a massive terrorist attack from the territory of State X against another country, State Y. This latter state may decide to respond by launching a drone strike in self-defense. Independently of hotly debated jus ad bellum considerations,[205] such a use of force would not amount to

---

[199] ICRC Challenges Report 2011 (n 9) 22; Jelena Pejić, 'Extraterritorial Targeting By Means of Armed Drones: Some Legal Implications' (2014) Intl Rev Red Cross.

[200] Nils Melzer, *Study on the Human Rights Implications of the Usage of Drones and Unmanned Robots in Warfare* (European Parliament, 2013) 21. See also Michael N Schmitt, 'Extraterritorial Lethal Targeting: Deconstructing the Logic of International Law' (2013) 52 Columbia J Transnatl L 99.

[201] Pejić (n 199). See also European Parliament Resolution on the Use of Armed Drones 2014/2567(RSP) (25 February 2014) para F; Concluding observations on the fourth periodic report of the United States of America (CCPR/C/USA/CO/4, 23 April 2014) § 9.

[202] ICRC Challenges Report 2011 (n 9) 22; ICRC Statement before the Human Rights Council, 22 September 2014.

[203] Gaggioli (n 40) 94–98.

[204] On the extraterritorial scope of application of HRL, see above, section entitled 'Human rights in armed conflicts and its interplay with IHL: assumptions underlying this chapter'.

[205] See, eg, Michael N Schmitt, 'Drone Law: A Reply to UN Special Rapporteur Emmerson' (2014) 55 Virginia J Intl L 15–19.

an armed conflict. This is so because non-international armed conflicts require the fulfillment of two criteria: first the belligerent parties must be sufficiently organized and the intensity of violence needs to meet a certain threshold. In this case, the 'lone wolf' terrorist cannot be equated with a non-state organized armed group and a single drone strike does not reach the threshold of violence required for a non-international armed conflict to happen.[206] The use of force against the 'lone wolf' must therefore be analyzed under a human rights law enforcement paradigm.

Even within situations of armed conflicts, not every use of force by states falls within the conduct of hostilities paradigm. In many contemporary armed conflicts, particularly in occupied territories and in non-international armed conflicts (NIAC), armed forces are increasingly expected to conduct both combat operations against the adversary and law enforcement operations to maintain or restore public security, law and order.[207] It is generally accepted that this latter type of operations is governed by applicable HRL rules and standards for the use of force, that is, the human rights law enforcement paradigm.

The difficulty then lies in finding the dividing line between the conduct of hostilities and law enforcement paradigms. This is a matter of ongoing legal debate and has been addressed elsewhere.[208] What is clear is that even in armed conflicts, there are situations—such as civilian unrest and other forms of civilian violence or criminal acts not amounting to direct participation in hostilities—that must be addressed under a law enforcement paradigm. For instance, when civilians riot in occupied territories or

---

[206] This same strike could nevertheless give rise to an international armed conflict if State X did not consent to it, since the threshold of violence required for international armed conflicts to take place is very low. See ICRC Commentary to common Article 2. But even then, there would be no *nexus* between the international armed conflict opposing States X and Y and the terrorist activity of the 'lone wolf'.

[207] Gaggioli, ICRC Use of Force Report (n 137) 1.

[208] Ibid; ICRC Challenges Report 2015 (n 94) 33–37. See also Tristan Ferraro, Expert Meeting Report on Occupation and Other Forms of Administration of Foreign Territory, Third Meeting of Experts: The Use of Force in Occupied Territory (ICRC 2012); Melzer and Gaggioli (n 141); Dieter Fleck, 'Law Enforcement and the Conduct of Hostilities: Two Supplementing or Mutually Excluding Legal Paradigms?' in Andreas Fischer-Lescano et al (eds), *Frieden in Freiheit Festschrift für Michael Bothe* (Nomos DIKE 2008) 391–407; David Kretzmer, Aviad Ben-Yehuda and Meirav Furth, 'Thou Shall Not Kill: The Use of Lethal Force in Non-International Armed Conflicts' (2014) 47 Israel L Rev 191–224; Ken Watkin, 'Use of Force during Occupation: Law Enforcement and Conduct of Hostilities' (2012) 94 Intl Rev Red Cross 295–296.

in a NIAC, force cannot go beyond what is authorized under a human rights law enforcement paradigm. If fighters hide in the crowd, then they might be targetable under the conduct of hostilities paradigm and the two paradigms may apply in parallel.[209] Also, when belligerents launch operations against common criminals, such as members of drug gangs or other forms of organized crime that are not a party to the armed conflict, such use of force must comply with the HR law enforcement paradigm.[210] Alleged criminals are civilians and remain protected against attacks under the IHL principle of distinction, unless and for such time as they directly participate in hostilities. These are just two obvious examples where the law enforcement paradigm kicks in. They are by no means meant to be exhaustive.

Different legal constructs can explain why the human rights law enforcement paradigm must be applied to such use of force. A first explanation would be to consider that such use of force has no *nexus* with the armed conflict situation and therefore is not governed by IHL.[211] HRL, as a lex generalis, would simply remain the governing paradigm and not be displaced by IHL. This legal reasoning can nevertheless be criticized because even if rioting civilians and drug lords are indeed conducting acts of violence that have no nexus with the hostilities, the use of force by a belligerent state as a response allegedly continues to be governed by IHL, in the sense that an excessive use of force by the belligerent state against the rioting civilians, for instance, would not only amount to an arbitrary deprivation of life under HRL but also to a war crime under IHL. Another legal reasoning would be to consider that the human rights law enforcement paradigm is actually the lex specialis when force is used against civilians who are not directly participating in hostilities.[212] Indeed, since IHL does not allow directing force against such civilians, the only possible authority for such use of force would be the HR law enforcement paradigm. Account must also be taken of the overall objective of the operation, that is, to maintain law and order, rather than to engage the enemy. In such situations, HRL would thus become the prevailing regime although IHL continues to govern the situation.

In brief, and whatever the legal reasoning adopted, it is clear that even in armed conflicts, in some cases, the human rights law enforcement paradigm is the one that must be applied to some operations pertaining to

---

209 Gaggioli, ICRC Use of Force Report (n 137) 25.
210 Ibid 29–33.
211 Ibid 25.
212 Ibid.

the use of force. If remote weapons were to be used in such situations—for instance to control riots, to deal with criminality, to secure checkpoints/detention centers, or to prevent entry into specific areas—then they would have to comply with the law enforcement paradigm.

(b) The Human Rights Law Enforcement Paradigm and the Use of Force against Persons

The human rights law enforcement paradigm is generally much more restrictive than the conduct of hostilities paradigm under IHL.[213] It revolves around five main principles.[214] Three of them pertain to the actual use of force:

(1) The principle of *necessity* requires that force must be the last resort (ultima ratio)[215] to reach a legitimate objective, such as self-defense or defense of others.[216] State officials must, as far as possible, apply non-violent means and may use force and firearms only if other means remain ineffective or without any promise of achieving the intended result.[217] If the use of force is unavoidable, only the smallest amount of force necessary may be applied. This implies

---

[213] The HR law enforcement paradigm for regulating the use of force against persons is based on the right to life as guaranteed in Article 6 ICCPR, Article 4 IACHR and Article 4 AfCHPR. It has been elaborated in soft law documents, such as the *Code of Conduct for Law Enforcement Officials* (n 151) and the *Basic Principles on the Use of Force* (n 151) and in human rights case law.

[214] Gaggioli (n 40) 99–105.

[215] UN Code of Conduct (n 151), Article 3; UN Basic Principles (n 151) principle 4. See, eg, *McCann and Others v United Kingdom*, App no 18984/91 (ECHR, 27 September 1995) § 149.

[216] The *Basic Principles on the Use of Force and Firearms* (n 151) specify that firearms—and more generally, lethal or potentially lethal force—may be permitted only if they pursue the following legitimate objectives: (1) Self-defense or defense of others against the imminent threat of death or serious injury; (2) To prevent the perpetration of a particularly serious crime involving grave threat to life and (3) To arrest a person presenting a danger of perpetrating such crimes and resisting the authority, or to prevent his or her escape. See, in the same sense, Article 2(2) ECHR.

[217] Ibid; UN Code of Conduct (n 151), Article 3, commentary §c. See also eg, *Montero-Aranguren et al v Venezuela*, Series C no 150 (IACHR, 5 July 2006) §§ 67–68; *Zambrano Vélez et al v Ecuador*, Series C no 166 (IACtHR, 4 July 2007) §§ 83–84; *Mouvement Burkinabé des Droits de l'Homme et des Peuples v Burkina Faso*, Comm. 204/97 (African Commission on Human and Peoples' Rights, 2001) § 43.

that state officials must strive to arrest suspected criminals by using non-violent means as far as possible ('capture-rather-than kill' approach); and, if force is unavoidable, there must be, as far as possible, a differentiated use of force (for example, verbal warning, show of force, 'less-than-lethal' force, lethal force).[218]

(2) The principle of *proportionality* requires that the kind and degree of force used must be commensurate to the seriousness of the offense and the legitimate objective to be achieved.[219] State officials must also strive to minimize damage and injury to human life.[220] Concretely, the principle of proportionality requires a balancing between the risks posed by the individual versus the potential harm to this individual as well as to bystanders. In particular, if the individual is not posing a threat of death (or serious injury), the use of lethal (or potentially lethal) force would not be considered as proportional, even if the necessity requirement were to be fulfilled.[221] Intentional lethal use of force may only be made when strictly unavoidable in order to protect life.[222]

(3) The principle of *precaution* is an additional principle of the law enforcement paradigm, which has been developed in human rights

---

[218] Basic Principles on the Use of Force (n 151), principles 4, 5, 9, 10; UN Code of Conduct (n 151), Article 3. For the case law, see, for example, *Pedro Pablo Camargo* (n 151) § 13.2; *Alejandre* (n 123) § 42; *Montero-Aranguren* (n 217) § 75.

[219] Basic Principles on the Use of Force (n 151) principle 5(a); UN Code of Conduct (n 151) Article 3, commentary §b. For the case law on the proportionality principle, see *Neira Alegria v Peru*, Series C no 20 (IACtHR, 19 January 1995) § 76. The case law does not always distinguish clearly between necessity and proportionality. See *Pedro Pablo Camargo v Colombia* (n 151) § 13.3; *Alejandre* (n 123) § 42; Finca *'La Exacta' v Guatemala* Case 11.382 (IACHR, 21 October 2002) § 43; *Montero-Aranguren* (n 217) §§ 68 et 74; *Zambrano Vélez* (n 217) §§ 84–85; *Carandiru v Brazil* Case 11.291 (IACHR, 13 April 2000) § 91.

[220] Basic Principles on the Use of Force (n 151) principle 5(b); *Stewart v United Kingdom* § 19; *Wolfram v RFA* App no 11257/84 (ECommHR, 6 October 1986) 213. See also *Montero-Aranguren* (n 217) §§ 68.

[221] See Basic Principles on the Use of Force (n 151) principle 9. See also UN Code of Conduct (n 151) commentary to Article 3 (§a and §c). For the case law, see, e.g., *Makaratzis*; *Natchova and Others v Bulgaria*, App no 43577/98, 43579/98 (ECHR, 6 July 2005); *Kakoulli v Turkey*, App no 38595/97 (ECHR, 22 November 2005). Regarding Article 2(2)(c), *see Stewart v United Kindgom*; *Güleç* (n 152). See also *de Oliveira v Brazil*, Case no 10/00 (IACHR, 24 February 2000) § 33; *da Silva v Brazil* Case no 11.598 (IACHR, 24 February 2000) § 34; *Finca 'La Exacta'* (n 219) §§ 41–42.

[222] Basic Principles on the Use of Force (n 151).

case law.[223] It implies that when planning and controlling law enforcement operations, states must take all feasible precautions to minimize recourse to deadly force; the aim being always to limit damage and injury, and respect and preserve human life.[224]

The HR law-enforcement paradigm involves also obligations before and after the actual use of force, that can be framed as deriving from the principles of legality and accountability.[225]

(4) The principle of *legality* inherent in the right to life requires governments to set an appropriate domestic legal framework (including not only clear and accessible laws but also directives such as rules of engagement).[226] This legal framework must restrict the use of force to the maximum extent possible and stipulate the limited circumstances in which state officials can use force in accordance with international law.[227] In order to ensure that the law is translated into practice, governments must provide an adequate training to their state officials, including in alternatives to the use of force and firearms (such as non-violent methods of arrest and techniques).[228] Governments must also equip state officials with various types of weapons and ammunition, including alternative

---

[223] *McCann* (n 215) §§ 150 and 194, §§ 202–214; *Ergi v Turkey* App no 23818/94 (ECHR, 28 July 1998) § 79; *Isayeva, Yusupova and Bazayeva v Russia*, App no 57950/00 (ECHR, 24 February 2005) §§ 188–201; *Neira Alegria* (n 219) § 62; *Montero-Aranguren* (n 217) § 82; *Zambrano Vélez* (n 217) § 89. See also Alston, Report to the Human Rights Commission 2006 (n 151) §§ 53–54. Other human rights bodies refer more vaguely to an obligation to 'prevent' violations of the right to life. See General Comment no 6: Right to life (Article 6) (HRI/GEN/1/Rev.8, 1982) § 3; *Rickly Burrell v Jamaica* (CCPR/C/57/D/546/1993, 1996) § 9.5; *National Commission on Human Rights and Freedoms v Chad* (AfCommHPR, 1995) § 22; *Montero-Aranguren* (n 217) § 71.

[224] Ibid. Basic Principles on the Use of Force (n 151) principle 5(b).

[225] Gaggioli, ICRC Use of Force Report (n 137) 43.

[226] All treaties protecting the right to life state that this right must be 'protected by law'. Article 6(1) ICCPR; Article 2(1) ECHR; Article 4(1) ACHR; Article 4 combined with Article I of the AfCHPR. See also General Comment No 6 (n 223) § 3; *Makaratzis* §§ 56–72; *Pedro Pablo Camargo* (n 148) § 13.3; *Montero-Aranguren* (n 217) § 75; *Zambrano Vélez* (n 217) § 86.

[227] Basic Principles on the Use of Force (n 151) principles 1 and 11.

[228] Basic Principles on the Use of Force (n 151) principles 18–21. For the case law, see, for example, *Hamiyet Kaplan and Others v Turkey*, App no 36749/97 (ECHR, 13 September 2005) §§ 51–55; *Rickly Burrell* (n 223) § 9.5; *Montero-Aranguren* (n 217) §§ 77–78; *Zambrano Vélez* (n 217) § 87.

means to firearms (such as water cannons and other 'less-lethal weapons'[229]). This allows a differentiated use of force and restrains the use of means capable of causing death or injury.[230] Governments must also equip state officials with self-defensive equipment (such as shields, helmets, bullet-proof vests and so on), in order to decrease the need to use weapons of any kind.[231]

(5) The *accountability* principle, as discussed above, requires that after a particular operation where the use of force has resulted in death or injury, state officials must report the incident promptly to their superiors.[232] An effective investigation must be conducted each time a person has been killed or at least each time there is a credible allegation of a violation of the right to life.[233]

(c) Legality and Suitability of Remote Weapons Systems under the Human Rights Law Enforcement Paradigm?

In light of the very demanding requirements of the human rights law enforcement paradigm, can remote weapons systems ever comply with it? Do they comport inherent features, which render them unlawful under HRL? Since remote weapons systems can be so different, and for instance take the shape of large aerial vehicles or miniature drones, with various levels of autonomy and operating in various contexts, a uniform answer to these questions would constitute a meaningless oversimplification. A case-by-case analysis is needed. Let us consider various scenarios.

Take the case of unmanned combat aerial vehicles (UCAV), or 'combat drones', conducting premeditated strikes in 'self-defense' against alleged members of criminal groups, such as terrorist groups or drug cartels outside an armed conflict situation.[234] Could these types of targeted

---

[229] Basic Principles on the Use of Force (n 151) principle 3.

[230] Basic Principles on the Use of Force (n 151) principle 2. For the case law, see, for example, *Güleç* (n 152); *Hamiyet Kaplan* (n 228).

[231] Ibid.

[232] Basic Principles on the Use of Force (n 151), principles 6 and 22; UN Code of Conduct (n 151), commentary (c) to Article 3.

[233] See eg among many others, Alston, Report to the Human Rights Commission 2006 (n 151) §§ 35–36; *McCann* (n 215) § 161; *Mapiripán Massacre v Colombia* (n 148) §§ 216–241; *Montero-Aranguren* (n 217) § 75; *Commission Nationale des Droits de l'Homme et des Libertés v Chad* (n 116) § 22.

[234] The first time a large UCAV has allegedly been used to target alleged terrorists outside the scope of the battlefield was in 2002 when the CIA killed six

killings comply with the law enforcement paradigm?[235] The answer would be negative in the vast majority of cases.[236] Although intentional lethal force does not necessarily violate the right to life, it must be 'strictly unavoidable',[237] meaning that a very high necessity threshold must be reached. This would imply notably that the alleged criminals pose an imminent threat to life. Under HRL, imminence is 'a matter of seconds not hours'.[238] Moreover, although it is tempting to consider these persons as necessarily constituting a threat to life, under the law enforcement paradigm, the threat posed by a specific individual depends on conduct, not membership within a group.[239] Temporal remoteness,[240] that is, planning far in advance to kill an alleged criminal, is completely at odds with the law enforcement rationale whose objective is always to arrest criminals not to kill them.[241] The human right principle of precaution requires the planning of law enforcement operations to avoid the use of force. Finally, given the important firepower of 'combat drones', they might (depending on where and how they are used) cause important casualties among bystanders, which would most probably be considered as unacceptable under the strict human rights proportionality test.[242] In brief, it seems obvious that a drone strike cannot lawfully be envisaged as a law enforcement operation per se. At best, the use of such 'combat weapons' may be contemplated, in the worst-case scenarios, as the ultima ratio at the very end of a law enforcement operation that turns out badly. If autonomous weapons were developed to conduct similar

---

purported Al Qaeda members in Yemen. See Casey-Maslen (n 198) 616; David Kretzmer, 'Targeted Killing of Suspected Terrorists: Extrajudicial Executions of Legitimate Means of Defence' (2005) 16 Eur J Intl L 171–172.

235 For a more detailed human rights analysis of this scenario, see Gaggioli (n 40) 105–109.

236 Gaggioli (n 40) 108–109 and references therein.

237 UN Basic Principles (n 151) principle 9.

238 Report of the Special Rapporteur on Extrajudicial, Summary or Arbitrary Executions, Christof Heyns, to the Human Rights Council (A/HRC/26/36, 2014) § 59.

239 Melzer, *Targeted Killing* (n 86) 425.

240 See above, section entitled 'Definitional issues and the concept of "remoteness"'.

241 Alston, Report to the Human Rights Commission 2006 (n 151) § 33; Melzer (n 124) 425.

242 Human rights bodies have accepted very limited and unforeseen casualties among bystanders in rare cases. See *Andronicou and Constantinou v Cyprus*, App no 25052/94 (ECHR, 10 September 1997) § 194; *Kerimova and Others v Russia*, App no 17170/04 (ECHR, 3 May 2011) § 246.

targeted killings, the above analysis would remain exactly the same. Even if the principles of legality and accountability were respected by putting in place strong legal frameworks to limit the targeting by remote weapon systems and to review ex post such strikes, this could not alter the conclusion that targeted killings are not appropriate law enforcement methods and should be relied on only in proper hostilities under an IHL framework.

Let us take a different example. Consider an automatic sentry gun that is used to protect an area, which has an important military value (for example, it contains an important military objective) in an armed conflict situation. If the sentry gun were programmed to target and kill anyone attempting to enter the area, this would violate international law. Suppose that some civilian attempts to enter the area and that his/her intentions are unclear. In case of doubt as to whether a person is a legitimate target, the presumption is that this person is a civilian that is not directly participating in hostilities.[243] The use of force in such a scenario is possible only under HRL to enforce 'law and order', in this case a military order. The use of force must be the ultima ratio and an escalation of force procedure is required. Note that some would also (or alternatively) consider that directly targeting such a civilian would violate the IHL principle of precaution which requires the doing of everything feasible to check whether a person is a legitimate target before engaging it.[244] This may allegedly include the obligation to proceed to a human rights-like escalation of force procedure. In any case, this does not mean that automatic sentry guns are necessarily outlawed under international law, but that their use—and the overall set-up to protect the area—must be carefully thought through. For instance, an initially larger area could be manned by service members, who could apply an escalation of force procedure and use deadly force as a last resort to prevent entry into the area, and an automatic sentry gun could constitute a back-up, closer to the military objective to be protected, if the threatening person was not prevented to enter the zone despite the graduated response by the service members defending the area. On the other hand, if, for instance, the

---

[243] Article 50(1) API. Further, the ICRC *DPH Guidance* includes as a recommendation that 'all feasible precautions must be taken in determining whether a person is a civilian and, if so, whether that civilian is directly participating in hostilities. In case of doubt, the person must be presumed to be protected against direct attack'. See Nils Melzer, *Interpretive Guidance on the Notion of Direct Participation in Hostilities under International Humanitarian Law* (ICRC 2009) 74 (recommendation VIII).

[244] Gaggioli, ICRC Use of Force Report (n 137) 40–41.

sentry gun were not automatic, but semi-automatic, and required authorization to target and kill any person attempting to enter the area, this would still be problematic under the human rights principle of necessity (and arguably under the IHL principle of precaution) if there were no opportunity to apply a graduated use of force anyway. In this example, therefore, the problem does not lie so much on the unmanned or automatic use of force by the sentry gun, but rather on the overall set-up put in place to ensure that defending an area does not rely on deadly force exclusively.

Actually, autonomous weapons may even allow better targeting in law enforcement contexts than law enforcement officials. A famous example is a hostage situation or suicide bombers case, where autonomous weapons equipped with facial recognition would be able to target and kill the alleged criminal even if he or she is exposed for a split second, while a law enforcement official (and even a well-trained sniper) in the same situation would have been too slow to react.[245] It is clear that recourse to such a weapon system should be a last resort, if all realistic possible negotiations by law enforcement officials are ineffective or without any promise of achieving the intended result.

As a last example, consider a riot situation in a remote location in a context of internal disturbances and tensions. Law enforcement officials might be outnumbered or may want to avoid direct confrontation that may endanger both state agents and the surrounding population.[246] They decide to use a remote-control weapon system that can employ strobe lights and on-board speakers to send verbal warnings, shoot pepper spray, solid plastic balls and non-lethal paintballs to mark offenders and transport an 80,000-volt Taser dart to zap criminals.[247] The system may even be equipped to fire real bullets as a last resort. The precision of the system is supposed to exist thanks notably to video and audio surveillance systems allowing, for instance, the operator to see precisely what is

---

[245] Heyns (n 63) 358.

[246] Ibid: 'The main attraction behind the introduction of unmanned weapons systems (remote controlled or autonomous) in the context of armed conflict and potentially in law enforcement is that it keeps one's own forces out of harm's way'.

[247] 'CUPID drone to "shock the world" with 80,000 volt stun gun' *RT Question More* (8 March 2014), accessed 3 May 2017 at http://rt.com/usa/drone-taser-gun-security-650/.

happening on the ground, if a preferable, or complementary, on-site assessment is not possible.[248]

There are actually no inherent flaws in such remote weapons systems that would lead to the conclusion that they always violate the human rights principles of necessity, proportionality and precautions. They have the ability to apply a graduation of force in accordance with the principle of necessity. They are programmed to respond to imminent threats rather than to kill preselected 'targets' depending on their membership. They are not necessarily less precise than other weapons and are thus not necessarily contrary to the principle of proportionality. In terms of precautions, such weapons systems seem to be developed notably to minimize death and injury. Obviously, depending on the situation, the use of such means might either be lawful or unlawful. If, for instance, a Taser-drone is used to arrest an offender despite the possibility of a non-violent arrest, the principle of necessity would be violated. If rubber bullets are fired by a remote weapon system in a disproportionate manner at a rioting crowd, then the principle of proportionality would be violated. The principle of precaution would not be respected if law enforcement operations are planned mostly or exclusively with remote weapon systems even though they do not allow minimizing death and injury. The fact that such systems are cheaper or permit law enforcement officials to keep safe does not constitute a sufficient excuse.

Although such remote weapon systems seem prima facie capable of respecting principles of necessity, proportionality and precautions, this does not mean that they do not pose any legal and policy issues. Under the principle of legality, it seems wise to suggest that such remote weapon systems and the manner in which they will be used should be strictly regulated in advance at the domestic level.[249] Although there is no equivalent to Article 36 of Additional Protocol I for law enforcement means under HRL, states should nevertheless carefully assess if and under which conditions new law enforcement means can respect the HR law enforcement paradigm before allowing their use. Appropriate training must be provided to law enforcement officials who are equipped with such new technologies. As for the use of firearms, such remote weapons systems should be subjected to strong reporting and accountability mechanisms. Procedures must be put in place to easily and transparently

---

[248] Eric Brumfield, 'Armed Drones for Law Enforcement: Why it Might be Time to Re-Examine the Current Use of Force Standard' (2015) 46 McGeorge L Rev 543–572.

[249] See a proposal in this sense by Eric Brumfield, ibid.

attribute the use of force through the intermediary of remote weapons systems.

Most importantly, from a policy perspective, it is submitted that the introduction of remote weapon systems into law enforcement should follow an open and democratic debate and buy-in. Even without going as far as considering hypothetical and futuristic fully autonomous weapon systems, or killer robots, for law enforcement, the existing and increasing reliance on remote-controlled or semi-autonomous weapons for law enforcement is not just one among many other technological evolutions in law enforcement. Because of the physical and psychological distance created between the state and the individuals it is meant to protect,[250] this evolution may actually change the face of law enforcement. It is a universal motto that law enforcement officials have the mission 'to serve and to protect'.[251] Proximity, in terms of physical and actual presence in communities, but also in terms of psychological/mental ability to see and understand the need of the population, to negotiate and convince alleged criminals to come to their senses and to de-escalate situations of violence is what is expected from law enforcement officials. The more autonomy is given to remote law enforcement weapon systems, the more prominent and frightening becomes this psychological/mental remoteness in law enforcement. Over-reliance on new technologies and remote weapon systems for law enforcement may actually lead to a sense of insecurity, create an atmosphere of fear, transform the image of democratic states from being protective to repressive and polarize rather than bring communities together.[252] This war mindset should be avoided at all costs, especially when terrorist threats seem to proliferate even in peaceful countries.

---

[250] See above, section entitled 'Definitional issues and the concept of "remoteness"'.

[251] 'To serve and to protect' or 'to protect and to serve' is a widely used motto for the police. It was allegedly launched by the Los Angeles Police Department and then borrowed by many police agencies around the word. See http://www.lapdonline.org (accessed 3 May 2017). See also ICRC publications on guidance for police behaviors, accessed 3 May 2017: https://www.icrc.org/en/publication/0698-serve-and-protect-human-rights-and-humanitarian-law-police-and-security-forces; https://shop.icrc.org/servir-et-proteger-guide-du-comportement-de-la-police.html.

[252] Alston, Interim Report to the General Assembly 2010 (n 18) 20. 'An important political consideration is whether the widespread use of robots in civilian settings, such as for law enforcement in cities, or in counter-insurgency operations, would alienate the very populations they were meant to assist.'

5. CONCLUSION: ASSESSING REMOTENESS, A ROLE FOR HUMAN RIGHTS LAW?

New technologies have led to the development of various remote weapons systems for both warfare situations and law enforcement purposes. These weapons can be very different in terms of shape, firepower and autonomy. They have in common the fact that they introduce a physical, psychological and sometimes temporal distance between those resorting to force and their 'targets'. HRL has a role to play in assessing the legality of the use of such remote weapon systems. This role may however differ considerably depending on the situation (peacetime or wartime) and on the context (conduct of hostilities or law enforcement).

In the context of remote warfare, the role of HRL remains limited. It is IHL—and the conduct of hostilities paradigm—which governs the rules pertaining to the use of force in such situations. Nevertheless, human rights underlying principles such as the principles of transparency, accountability and possibly dignity, play an important complementary role. Although this should be true irrespective of the type of means and methods of warfare employed, remote warfare certainly constitutes a paradigmatic case where total lack of transparency and accountability would almost inevitably lead to arbitrary deprivations of life. Because remote weapons systems open up options for using deadly force in remote and practically or security-wise unreachable areas anywhere on earth, and because it may be difficult or even impossible to attribute unacknowledged strikes by remote weapons systems, the actual conformity of such use of force with IHL would become unverifiable without transparency and accountability mechanisms put in place by belligerent states resorting to such technologies. Regarding the principle of dignity, it is debatable whether it can, by itself, prohibit autonomous weapons. It does, however, bring useful food for thought when states and the international community consider possible restrictions or prohibitions of autonomous weapons, and arguably of any other weapon systems.

In law enforcement situations—be they in peacetime, counter-terrorism operations outside armed conflicts or wartime (for example, riot, checkpoint scenario)—HRL is the only or main applicable international legal framework. The use of force by remote weapon systems in such situations has to be assessed under the restrictive human rights law enforcement paradigm, which revolves around the principles of legality, necessity, proportionality, precautions and accountability. Some remote weapon systems, such as UCAV, are not appropriate law enforcement means and would almost always lead to multiple violations of the right to life under HRL. Other remote weapon systems, such as 'law enforcement drones', which are able to apply a graduated use of force do not have inherent features which would lead to the conclusion that they are unlawful under the human rights law enforcement paradigm. In some cases, remote weapon systems may even improve compliance with the human rights law enforcement paradigm. Nevertheless, the increasing use of remote weapon systems for law enforcement must be closely monitored as over-reliance on remote systems may pervert the very mission of law enforcement officials, which is to serve and protect communities, through their reassuring physical presence among communities and their mental ability to understand human behaviors and criminal minds with a view to maintaining law and order while avoiding to the maximum extent the use of force. Ultimately, the question is not so much whether remote weapon systems could be conceived to do law enforcement but whether societies where manifestations and riots are 'pacified' by 'law-enforcement drones', detainees are guarded by robots and dangerous criminals are targeted and killed through remotely controlled weapons is really what we want for the generations to come.

6. Exploiting legal thresholds, fault-lines and gaps in the context of remote warfare

Mark Klamberg

1. INTRODUCTION

Conflicts increasingly involve action at a distance as opposed to traditional battlefield engagements. Development of new weapons, modern communications and growing economic interdependence between states push national decision-makers to adopt asymmetrical strategies, overt as well as covert. States may adopt such strategies to minimize the exposure to risk of their own forces while their opponents can be easily attacked and also for the purpose of avoiding attribution and retribution. Since international law is used as a tool for legitimizing state policies—in the words of Sari—legal thresholds, fault-lines and gaps will be used by states to portray their own actions as legal or at least belonging to a grey area but never illegal.[1]

These issues have been brought to the fore not least by increased tensions between the West and Russia. Russia states in its 2014 Military doctrine that the nature and characteristics of modern warfare conflict includes, inter alia:

> a) [i]ntegrated use of military force, political, economic, informational and other non-military measures nature, implemented with the extensive use of the protest potential of the population, and special operations forces ... h) participation in hostilities irregular armed groups and private military companies; i) the use of indirect and asymmetric methods Action; j) the use of externally funded and run political forces and social movements.[2]

---

[1] See terminology and approach of Aurel Sari, 'Legal Aspects of Hybrid Warfare' (2015), accessed 3 May 2017 at https://www.lawfareblog.com/legal-aspects-hybrid-warfare.

[2] Military Doctrine of the Russian Federation, adopted 25 December 2014, English translation accessed 3 May 2017 at https://www.offiziere.ch/wp-content/uploads-001/2015/08/Russia-s-2014-Military-Doctrine.pdf, para 15.

Russia perceives as one of the main military dangers 'subversive activities of special services and organizations foreign states and their coalitions against the Russian Federation'.[3] At the same time the Armed forces of Russia have been tasked to develop means of information warfare.[4] Similarly, in the US 2013 Military doctrine, warfare includes operations in the information environment (which includes cyberspace).[5] The doctrine describes information as an important instrument of national power and a strategic resource critical to national security. The doctrine states that it:

> is essential to our ability to achieve unity of effort through unified action with our interagency partners and the broader interorganizational community. Fundamental to this effort is the premise that key audience beliefs, perceptions, and behavior are crucial to the success of any strategy, plan, and operation. Through commander's communication synchronization (CCS), public affairs (PA), information operations (IO), and defense support to public diplomacy are realized as communication supporting capabilities.[6]

This chapter will examine legal thresholds, fault-lines and gaps (in sections 2–5) giving states the opportunity to adopt asymmetrical strategies which give them leverage against their opponents and the ability to avoid attribution and retribution. For the purpose of this chapter, remote warfare is understood to include computer network attacks, psychological operations, use of irregular and/or non-state groups, use of 'peace keeping' and/or pre-stationed forces, and expulsion of populations. The analysis is conducted in relation to recent and ongoing situations and conflicts such as Syria and the Ukraine-Russia conflict. The next sections will briefly introduce these means of remote warfare. Subsequent sections will discuss in more detail how remote warfare may exploit legal thresholds, fault-lines and gaps.

2. MEANS OF REMOTE WARFARE

Information operations (IO) includes the use of information technology, such as computer network attacks (CNA) or psychological operations

---

[3] Ibid, para 12.

[4] Ibid, para 46.

[5] Doctrine for the Armed Forces of the United States, adopted 23 March 2013, accessed 3 May 2017 at http://www.dtic.mil/doctrine/new_pubs/jp1.pdf (Defense Technical Information Center) x.

[6] Ibid 12–13.

(psyops, sometimes also referred to as influence operations), to influence, disrupt, corrupt, usurp, or defend information systems and the infrastructure they support. IO methods such as psyops seek alternative ways to accomplish larger strategic goals without resorting to force at all by convincing the adversary (or those who support it) to change their policies or positions.[7] In other words, psyops may be part of an 'information-related conflict at a grand [strategic level] between nations or societies. It means trying to disrupt, damage or modify what a target population "knows" or thinks it knows about itself and the world around it'.[8] Hollis has identified what he calls three 'near-fatal conditions' associated with IOs: (1) uncertainty (that is, states lack a clear picture of how to translate existing rules into the information environment); (2) complexity (that is, overlapping legal regimes threaten to overwhelm state decision-makers seeking to apply IO); and (3) insufficiency (that is, the existing rules fail to address the basic challenges of modern conflicts with non-state actors and to facilitate IO in appropriate circumstances).[9] These conditions will be revisited later.

Modern conflicts are to a large extent associated with the presence of irregular and/or non-state groups. This raises questions as to whether their actions can constitute an armed attack, the characterization of the conflict, whether their actions can attribute responsibility for states and who can be attacked.

The Ukraine-Russia conflict and the 2008 Georgia war provide examples of how 'peace keeping' and/or pre-stationed forces may be used in a conflict. Under the pretext that the forces are in the territory of a foreign state the argument can be made that their presence neither constitutes use of force nor transforms the situation into an international armed conflict.

Movement of migrants is not normally perceived as warfare. A migrant crisis could be created for deliberate purposes by a state to destabilize other states. Such strategies may be an efficient, albeit cynical, means to exert pressure while still staying below legal thresholds and thus avoiding retribution.

---

7 Duncan B Hollis, 'Why States Need an International Law for Information Operations' (2007) 11[4] *Lewis & Clark Law Review* 1032.

8 Douglas S Anderson and Christopher R Dolley, 'Information Operations in the Space Law Arena' (2002) 76 *International Law Studies* 265, 269.

9 Hollis (n 7) 1029.

3. INTERVENTION, USE OF FORCE, ARMED ATTACK

The first type of threshold concerns the difference between intervention on the one hand and use of force and armed attack on the other. Even if an act does not amount to use of force it may still violate international law as an illegal intervention. Information operations will in themselves typically not amount to use of force, it is more appropriate to discuss whether they are illegal interventions. It is arguably rational for actors using information operations, if possible, to take measures that accomplish the desired objectives while as staying below the mentioned thresholds in order to avoid attribution and retribution. The same logics apply to the reliance on irregulars and non-state groups. Russia went to great lengths in its illegal annexation of Crimea in 2014 to argue that the 'green men' taking control over various buildings were 'local self-defense forces' and not Russian troops,[10] contradicted by various international observers and later by president Putin himself.[11] Thus the next section will discuss intervention followed by an analysis of the concepts of use of force and armed attack in order to establish the lower as well as the higher thresholds.

Intervention

The non-intervention principle derives primarily from the international law notions of sovereignty and territory.[12] Going further back, the origins of the non-intervention principle go back to the writings of Vattel, Wolff

[10] *Russia says cannot order Crimean 'self-defense' units back to base*, Reuters (5 March 2014), accessed 3 May 2017 at http://www.reuters.com/article/2014/03/05/us-ukraine-crisis-lavrov-spain-idUSBREA240NF20140305; *Crimea crisis: Russian President Putin's speech annotated*, BBC (19 March 2014), accessed 3 May 2018 at http://www.bbc.com/news/world-europe-26652058.

[11] *Putin admits Russian forces were deployed to Crimea*, Reuters (17 April 2014), accessed 3 May 2017 at http://uk.reuters.com/article/2014/04/17/russia-putin-crimea-idUKL6N0N921H20140417; *Putin acknowledges Russian military serviceman were in Crimea*, Russia Today (17 April 2014), accessed 3 May 2017 at https://www.rt.com/news/crimea-defense-russian-soldiers-108/. See also Mark Klamberg, *Power and Law in International Society* (Routledge 2015) 155.

[12] Sean Watts, 'Low-Intensity Cyber Operations and the Principle of Non-Intervention' in Jens David Ohlin, Kevin Govern and Claire Finkelstein (eds), *Cyber War: Law and Ethics for Virtual Conflicts* (Oxford University Press 2015) 251.

and later Kant.[13] Non-intervention is at a middle ground of international wrongs—it is not insignificant, but also not supreme among wrongful acts.[14]

The Friendly Relations Declaration (FRD) mentions the non-intervention principle, that is, the duty of states not to intervene in matters within the domestic jurisdiction of any other state. This includes a prohibition to 'organize, assist, foment, finance, incite or tolerate subversive, terrorist or armed activities directed towards the violent overthrow of the regime of another State, or interfere in civil strife in another State'. The principle applies to economic, social, cultural, technical and trade fields.[15] The concept of *domaine réservé* describes activities and matters belonging to the internal or domestic affairs of states. The precise scope of matters forming the *domaine réservé* of states has always and probably will be in flux since states may surrender matters previously regarded as within their exclusive jurisdiction to the international legal system. The matter most clearly within the *domaine réservé* is the choice of states' political systems and their means of political organization.[16]

---

[13] Emmerich de Vattel, 'Le droit des gens ou principes de la loi naturelle—The Law of Nations or the Principles of Natural Law (1758)' in *The Classics of International Law* (Hein 1995), book I, ch. III, sec. 37: 'Enfin toutes ces choses n'intéressant que la Nation, aucune Puiffance Etrangère n'est en droit de s'en mêler, ni ne doit y intervenir autrement que par ses bons office, à moins qu'elle n'en soit requise, ou que des raisons particulières ne l'y appellent. Si quelqu'une s'ingère dans les affaires domestiques d'une autre, si elle entreprend de la contraindre dans ses délibérations, elle lui fait injure'; Christian Wolff, 'Jus gentium methodo scientifica pertractatum (1764), translation by Joseph H Drake' in James Brown Scott (ed), *The Classics of International Law* (Hein 1995), ch. I, sec. 256: 'Since by nature no nation has a right to any act which pertains to the exercise of the sovereignty of another nation, if any nation dares to do anything which belongs to the exercise of the sovereign power of another nation, it does this without right and contrary to the right of the other nation, and consequently does a wrong to it.'; Immanuel Kant, *Philosophical Essay on Perpetual Peace* (1795); see also Ann van Wynen Thomas and A J Thomas, *Non-Intervention: The Law and Its Import in the Americas* (Southern Methodist University Press 1956) 5 and 7.

[14] Watts (n 12) 250.

[15] UN General Assembly resolution 25/2625, A/RES/25/2625 (1970). Declaration on Principles of International Law concerning Friendly Relations and Co-operation among States in accordance with the Charter of the United Nations, 24 October 1970.

[16] Watts (n 12) 264–6.

The ICJ has stated that this principle is binding as part of customary international law. Moreover, 'assistance to rebels in the form of the provision of weapons or logistical ... may be regarded as a threat or use of force, or amount to intervention in the internal or external affairs of other States'.[17] When addressing the content of the non-intervention principle the ICJ stated the following:

> A prohibited intervention ... [is] one bearing on matters in which each State is permitted, by the principle of State sovereignty, to decide freely. One of these is the choice of a political, economic, social and cultural system, and the formulation of foreign policy. Intervention is wrongful when it uses methods of coercion in regard to such choices, which must remain free ones. The element of coercion, which defines, and indeed forms the very essence of, prohibited intervention, is particularly obvious in the case of an intervention which uses force, either in the direct form of military action, or in the indirect form of support for subversive or terrorist armed activities within another State. ... [However] strictly humanitarian aid to persons or forces in another country, whatever their political affiliations or objectives, cannot be regarded as unlawful intervention, or as in any other way contrary to international law.[18]

The ICJ has explained that if one state, with a view to the coercion of another state, supports and assists armed bands in that state whose purpose is to overthrow the government of that state, that amounts to an intervention by the one state in the internal affairs of the other, whether or not the political objective of the state giving such support and assistance is equally far-reaching.[19] In other words, many acts that are below the threshold of armed attack may still constitute illegal intervention or interference.[20] O'Connell notes that '[i]nterference with a state's economic sphere, air space, maritime space, or territorial space, even if

---

[17] *Military and Paramilitary Activities in and against Nicaragua (Nicaragua v United States of America)*, Judgment of 26 November 1984 (ICJ), para 73; *Military and Paramilitary Activities in and against Nicaragua (Nicaragua v United States of America)*, Judgment of 27 June 1986 (ICJ), paras 123, 176, 185, 195, 202, 206–209, 228, 239, 245–246, 263, 266.

[18] *Nicaragua*, 27 June 1986 (n 17), paras 205 and 242.

[19] Ibid, para 241.

[20] Pål Wrange, 'Intervention in National and Private Cyberspace and International Law' in Jonas Ebbesson et al (eds), *International Law and Changing Perceptions of Security* (Brill Nijhoff, 2014) 307–26, 313.

not prohibited by Article 2(4) of the UN Charter is prohibited under the general principle of non-intervention'.[21]

Most cyber operations have low-level impact, relatively small scale effects and as such primarily implicate the principle of non-intervention.[22] The principle of non-intervention extends, as a norm of customary international law, to states' actions in cyberspace.[23] It should be noted that the principle of non-intervention operates exclusively with respect to states which means that the principle does not operate on private actors as such.[24]

Not all violations of territory or sovereignty immediately concern the principle of non-intervention. A mere intrusion into another state's networks to gather information would certainly amount to a violation of sovereignty but without evidence on additional coercive action taken, the intrusion would not amount to an intervention.[25] From the principle of non-intervention and the concept of *domaine réservé* follows that a state may not use cyber means to support domestic or foreign agents' efforts to alter another state's governmental or social structure.[26]

The fact that most inter-state cyber operations will violate the principle of non-intervention is often overlooked in the analysis of and debate over cyber operations. However, not all interstate cyber operations will violate the principle. For example cyber espionage and cyber exploitation lacking a coercive element do not per se violate the non-intervention principle. As indicated above, mere intrusion into the system of another state does not violate the non-intervention principle.[27] Although it is legitimate to steal or corrupt the opponent's information, supply them with disinformation and sabotage their equipment during an armed conflict, the same acts prior to the absence of commencement of an

---

[21] Mary Ellen O'Connell, 'Cyber Mania' in Mary Ellen O'Connell, Louise Arimatsu and Elizabeth Wilmshurst (eds), *International Law: Meeting Summary: Cyber Security and International Law* (Chatham House 2012) 7.

[22] Watts (n 12) 251.

[23] Ibid 257.

[24] Ibid 254.

[25] Ibid 257.

[26] Ibid 266.

[27] *Tallinn Manual on the International Law Applicable to Cyber Warfare* (Michael N Schmitt (ed)) (Cambridge University Press 2013) 44–5; James A Green, 'The regulation of cyber warfare under the *jus ad bellum*' in James A Green (ed), *Cyber Warfare: A Multidisciplinary Analysis* (Routledge 2015) 96–124, 107–8.

armed attack constitute unlawful interference and retain the characteristics of an internationally wrongful act.[28] The reason why cyber operations are less discussed as violations against the principle of non-intervention and more focus is placed on prohibition against use of force may be explained by the fact that the first principle is perceived as a 'weaker' rule of international law. The principle of non-intervention is a rule of customary international law, it is not treaty-based and is more often breached.[29] States may thus perceive it as less costly to violate the principle of non-intervention.

State-sponsored propaganda towards the population of another state may constitute an intervention. Propaganda may be defined as communication to shape attitudes and behavior for political purposes. Propaganda can be used in peace as well as during war and as such is a broader concept than psyops, the latter primarily concerning military operations. State-sponsored propaganda towards the population of another state is not necessarily an intervention. Propaganda that seeks to persuade, offers incentives and favorable treatment to induce a certain course of action will not alone violate the principle of non-intervention. The line between criticism and hostile propaganda amounting to intervention is arguably fine but should be upheld even if it is only a matter of degree. When it comes to news, one may distinguish between news broadcasts which transmit 'facts' on the one hand and news commentaries that may constitute propaganda on the other,[30] but then one would need to define what are 'facts'. It is arguably only when the propaganda supports coercion tantamount to the level of organization, financial and logistical support recognized by the ICJ in the *Nicaragua* case, could the propaganda amount to a violation of the principle of non-intervention.[31] Schmitt has a somewhat different approach arguing that 'psychological operations directed against the civilian population that cause no physical harm are entirely permissible, as long as they are not intended to terrorize'.[32]

There appears to be agreement that 'each state has a duty to refrain from spreading propaganda in a friendly country hostile to the latter's government; but, aside from special treaty provisions, it is under no

---

[28] Heather Harrison Dinniss, *Cyber Warfare and the Laws of War* (Cambridge University Press 2012) 73.

[29] Green (n 27) 109–10.

[30] Thomas and Thomas (n 13) 273 and 278; Watts (n 12) 290–91.

[31] Thomas and Thomas (n 13) 273 and 278; Watts (n 12) 262–3.

[32] Michael N Schmitt, 'Wired Warfare: Computer Network Attack and the *Jus in Bello*' (2002) 76 *International Law Studies* 187, 195.

responsibility with respect to private propaganda activities.'[33] Bring argues that the prohibition on intervention covers state propaganda that has a subversive purpose (a subjective element) and which, if the propaganda is adhered to, means that the targeted population de facto is encouraged to overthrow its own government (an objective element). In such a situation the principle of people's right to self-determination is set aside since the foreign influence involves support to subversive powers. However, it will not amount to a violation of international law if the propaganda is of a general nature and the principle of people's right to self-determination is allowed to operate in the sense that the population concerned has the possibility of taking their own stand on the propaganda, whether it is information or disinformation. The only form of propaganda that is prohibited under the non-intervention principle is any that is part of a comprehensive campaign with the purpose of overthrowing a current government.[34] Thomas and Thomas note that there is no state responsibility for hostile propaganda emanating from a private source; however, states are arguably responsible for private propaganda activities in totalitarian systems where the private organization is dependent upon and under the control of the government.[35] They also argue that hostile propaganda in radio broadcasts or television broadcasts, although originating locally and primarily for a local purpose, which penetrates another state is prohibited under general international law.[36] That argument was made in the 1950s and in the modern era where the internet is even less 'local' the same reasoning would appear even more relevant.

Does a state have the right to respond to intervention with intervention as a countermeasure? The ICJ mentions this possibility in the *Nicaragua* case but does not rule on the matter.[37] Even if a computer network attack does not amount to an armed attack, states may arguably still respond with proportionate countermeasures.[38] The same applies to countermeasures against hostile propaganda.[39]

---

[33] Thomas and Thomas (n 13) 273.

[34] Ove Bring, *FN -stadgan och välrdspolitiken* (Third Edition, Stockholm: Norstedts Juridik 2000) 148–149; see also Pontus Winther, *Report: Några rättsliga aspekter på strategiska psykologiska operationer* (Försvarshögskolan, Institutionen för säkerhet, strategi och ledarskap (ISSL), Folkrättscentrum, 2011) 6.

[35] Thomas and Thomas (n 13) 275 and 278.

[36] Ibid 280.

[37] *Nicaragua*, 27 June 1986 (n 17), para 210.

[38] Harrison Dinniss (n 28) 105.

[39] Thomas and Thomas (n 13) 287.

Use of Force

The UN Charter uses notions such as 'use or threat of force', 'threat to the peace', 'breach of the peace', 'act of aggression', 'armed attack' and 'aggressive policy'.[40] The scope of the notion of 'force' has been disputed; does it also cover other forms of force than armed force such as political and economic coercion? Other provisions such as Article 44 support the view that it means 'armed force'. Moreover, proposals during the San Francisco conference to extend the prohibition to economic coercion were explicitly rejected. Finally, when dealing with the prohibition on the threat or use of force the FRD deals solely with military force.[41] However, the Definition of Aggression adopted by the General Assembly does not offer guidance; Article 1 states that '[a]ggression is the use of armed force' with no mention of the notion of 'use of force'. Article 6 clarifies that '[n]othing in this Definition shall be construed as in any way enlarging or diminishing the scope of the Charter, including its provisions concerning cases in which the use of force is lawful'.

The ICJ indicated that:

> while the concept of an armed attack includes the despatch by one State of armed bands into the territory of another State, the supply of arms and other support to such bands cannot be equated with armed attack. Nevertheless, such activities may well constitute a breach of the principle of the non-use of force and an intervention in the internal affairs of a State, that is, a form of conduct which is certainly wrongful. but is of lesser gravity than an armed attack.[42]

The prohibition of the threat or use of force is laid down in Article 2(4) of the UN Charter and is a norm of customary international law. The ICJ stated in the 1986 *Nicaragua* decision 'that both the Charter and the customary international law flow from a common fundamental principle outlawing the use of force in international relations'.[43] It covers not only

---

40 Charter of the United Nations, 26 June 1945, 1 UNTS XVI, Articles 39, 51 and 53; Albrecht Randelzhofer and Oliver Dörr, 'Article 2(4)' in Bruno Simma, and others (eds), *The Charter of the United Nations – A Commentary* (3rd edn, Oxford University Press, 2012) vol. I, 200–34, 208.

41 Harrison Dinniss (n 28) 41–42; Randelzhofer and Dörr (n 40) 209.

42 *Nicaragua*, 27 June 1986 (n 17), para 247.

43 Ibid, para 181; see also paras 34 and 188; *Legal Consequences of the Construction of a Wall in the Occupied Palestinian Territory*, Advisory Opinion of 9 July 2004 (ICJ), para 87; Randelzhofer and Dörr (n 40) 203, 229; Klamberg (n 11) 130.

when war is declared but extends the scope to all kinds of military force.[44] The general norm on the prohibition of force is hardly ever questioned. Instead the controversy concerns the scope and content of certain exceptions to this prohibition.[45]

Some argue that Article 2(4) also covers physical force of a non-military nature such as cross-frontier expulsion of populations, the diversion of a river up-stream from another state or computer network attacks. Randelzhofer and Dörr argue that this view can only be accepted within narrow limits, when such use of non-military force produces the effect of an armed attack prompting the right of self-defense laid down in Article 51.[46] US General Philip Breedlove claims that Russia and Syria's leader, Bashar al-Assad, were 'deliberately weaponising migration in an attempt to overwhelm European structures and break European resolve'. With reference to the use of barrel bombs—unguided weapons—against civilians in Syria, he argues that the 'only purpose of these indiscriminate attacks was to terrorise Syrian citizens and get them on the road to create problems for other countries'.[47] Even if Breedlove's allegations are true, such action would arguably fall short of amounting to use of force vis-à-vis the European states concerned. It would appear more appropriate to describe it as interventions aimed at changing policies of European states.

The use of force prohibition encounters real difficulty when translated into the IO context. There is no clear or explicit rule when IO constitutes a use of force, let alone an armed attack for self-defense. Hollis suggests three main approaches to the matter:

(1) the 'instrumentality approach': IO does not qualify as armed force because it lacks the physical characteristics traditionally associated with military coercion, which has support in the text of the UN Charter;[48]
(2) the 'target-based approach': IO constitutes a use of force or an armed attack whenever it penetrates 'critical national infrastructure' systems, even absent significant destruction or casualties; or

[44] Randelzhofer and Dörr (n 40) 207.
[45] Ibid, 204.
[46] Ibid, 210.
[47] *Migrant crisis: Russia and Syria 'weaponising' migration*, BBC (2 March 2016).
[48] UN Charter, Article 41: 'measures not involving the use of armed force … include … telegraphic, radio, and other means of communication'.

(3) the 'consequentiality approach: whenever IO intends to cause effects equivalent to those produced by kinetic force (death or destruction of property), it constitutes a use of force and an armed attack.[49]

Dinstein makes the argument that there is 'no reason to differentiate between kinetic and electronic means of attack ... the crux of the matter is not the medium at hand ... but the violent consequences of the action taken'.[50] Most CNAs will fall below this threshold. Harrison Dinnis notes that even if the denial of service attacks against Estonia in 2007 were disruptive and the use of Stuxnet in 2010 may have amounted to use of force, neither of the incidents reached the threshold for an armed attack.[51] All of the approaches above suffer from either under- or over-inclusion.

Article 2(4) is thus restricted to the prohibition of armed force. However, the notion of armed force does not only cover direct use of force, that is, incursion of regular military forces into the territory of another state or cross-border shooting but also the use of indirect armed force. The notion of indirect force refers to situations where a state allows its territory to be used for violent attacks against a third state as well as a state's participation in the use of force by unofficial bands organized in a military manner.[52] However, the ICJ has pointed out in the *Nicaragua* case that not every form of assistance amounts to the use of force. The Court considers that the mere supply of funds to the Contras, while undoubtedly an act of intervention in the internal affairs of Nicaragua, did not in itself amount to a use of force.[53]

Threat of force has received less attention in legal disputes and scholarly writings. One explanation may be that threats that have occurred often precede actual use of force.[54]

Armed Attack

Armed attack is a term used in Article 51 of the UN Charter and is distinct from the term 'threat or use of force' in Article 2(4). 'Armed

---

[49] Hollis (n 7) 1041–2.

[50] Yoram Dinstein, 'Computer Network Attacks and Self-Defense' (2002) 76 International Law Studies 103; see also Harrison Dinniss (n 28) 60.

[51] Harrison Dinniss (n 28) 81–2.

[52] Randelzhofer and Dörr (n 40) 211.

[53] *Nicaragua*, 27 June 1986 (n 17), para 228.

[54] Randelzhofer and Dörr (n 40) 218.

attack' is a narrower notion than 'threat or use of force' which has the consequence that not every use of force contrary to Article 2(4) may be responded to with self-defense.[55] The perception that a gap exists between these two notions has been confirmed by the ICJ in the *Nicaragua* case when it distinguished 'the most grave forms of the use of force (those constituting an armed attack) from other less grave forms'.[56] This was reaffirmed in the *Oil Platforms* case.[57] This has raised objections; such a gap would mean that 'there is not always effective protection against States violating the prohibition on the use of force, as long as they do not resort to an armed attack'.[58] Thus, the gap arguably has to be narrow.[59]

A controversy concerns self-defense against non-state actors. There is no requirement in Article 51 on self-defense that the armed attack is from a state. This raises the question as to against whom a state may use self-defense. Before the events of 11 September 2001, the assumption was that for the purposes of Article 51 an armed attack meant the attack by a state against another state. The requirement that the attacker has to be a state was (and maybe still is) a part of international customary law which complements Article 51.[60] It could be argued that the widespread acceptance of other states of US actions after the September 11 attacks and in relation to the subsequent use of force against targets in Afghanistan reflected a change in customary international law, meaning that international law now accepts the use of force against a non-state actor as a response to large-scale terrorist acts. A less radical approach is that the de facto government of Afghanistan at the time of the September 11 attacks were complicit and responsible for the attacks. This would mean that the traditional interpretation of Article 51 to a large extent is intact. At the same time such an interpretation would arguably exclude the use

---

55 Albrecht Randelzhofer and Georg Nolte, 'Article 51' in Bruno Simma et al (eds), *The Charter of the United Nations – A Commentary* (3rd edn, Oxford University Press 2012), vol. II, 1397–428, 1401.

56 *Nicaragua*, 27 June 1986 (n 17), para 191.

57 *Oil Platforms* (*Iran v United States of America*), Judgment of 6 November 2003 (ICJ), para 51; see also *Armed Activities on the Territory of the Congo* (*Democratic Republic of the Congo v Uganda*), Judgment of 19 December 2005 (ICJ), para 147; Malcolm N Shaw, *International Law* (7th edn, Cambridge University Press 2014) 822.

58 Randelzhofer and Nolte (n 55) 1402.

59 Dinstein (n 50) 99, 100; Harrison Dinniss (n 28) 79.

60 Klamberg (n 11) 132–3.

of force outside the territory of Afghanistan under the parole of 'war against terrorism'.[61]

Aggression

As indicated above, it is appropriate to distinguish between use of force, and the more narrow concepts of armed attack and aggression. The occurrence of an armed attack gives rise to the right to self-defense while aggression is one of the three alternative conditions for a decision of the UN Security Council (see Article 39 of UN Charter) to decide on enforcement measures. Both armed attack and aggression are serious forms of use of force. Even if it is uncertain whether the two concepts are synonymous they are adjacent.[62]

A definition of aggression is contained in the annex of General Assembly Resolution 3314 (XXIX) of 14 December 1974.[63] The resolution has proven to be of continued value to indicate whether certain forms of uses of force constitute 'armed attacks' in the sense of Article 51.[64] This is appropriate as the resolution is a consensual and time-tested document adopted by the General Assembly.[65] Article 1 of the annex defines an act of aggression in general terms based almost word for word on Article 2(4) of the UN Charter: 'the use of force by a State against the sovereignty, territorial integrity or political independence of another State, or in any other manner inconsistent with the Charter of the United

---

61 Mark Klamberg, 'International Law in the Age of Asymmetrical Warfare, Virtual Cockpits and Autonomous Robots' in Jonas Ebbesson et al (eds), *International Law and Changing Perceptions of Security* (Brill Nijhoff 2014) 152–70, 156.

62 Pål Wrange, *Aggressionsbrottet och Internationella brottmålsdomstolen* (Totalförsvarets folkrättsråd, Försvarsdepartementet 2011) 16; Klamberg (n 61) 153–54. However, Dinstein argues that armed attack is more narrow, it is 'a particular type of aggression'; Dinstein (n 50) 100.

63 *Definition of Aggression, General Assembly Resolution 3314 (XXIX)* (General Assembly 1974).

64 Randelzhofer and Nolte (n 55) 1408.

65 Stefan Barriga, 'Against the Odds: the Results of the Special Working Group on the Crime of Aggression' in Roberto Bellelli (ed), *International Criminal Justice* (Ashgate, 2010) 621–43.

Nations'. Article 2 provides that the gravity of the act and its consequences should be taken into consideration. This would suggest that minor border incidents do not constitute aggression.[66]

Turning to the example of pre-stationed forces, Article 3(e) provides that aggression includes '[t]he use of armed forces of one State which are within the territory of another State with the agreement of the receiving State, in contravention of the conditions provided for in the agreement or any extension of their presence in such territory beyond the termination of the agreement'. This form of aggression, a concession by the Great Powers to the anxieties of smaller states, excludes minor violations. Only if the breach of terms of the agreement has the effect of an actual invasion or occupation can an 'armed attack' trigger the right to self-defense.[67] Russia's military presence in Crimea and use of military forces prior to and during its illegal annexation is a good example of this type of aggression.

Incidents such as the overwhelming of government servers in Estonia in 2007 have fueled the debate about whether cyber-attacks or the causation of the self-destruction of Iranian uranium centrifuges by the 'Stuxnet' worm in 2010 are armed attacks and can trigger the right to self-defense.[68] The ICJ has stated that Articles 2(4) and 51 do not refer to specific weapons. They apply to any use of force, regardless of the weapons employed. The Charter neither expressly prohibits, nor permits, the use of any specific weapon.[69] This means that an armed attack does not require the use of kinetic weapons, but may, in principle, also be conducted by electronic weapons. In order to reach the standard of an 'armed attack' it must, however, produce substantial and immediate destructive effects.[70]

---

[66] The ICJ distinguishes minor armed exchanges or 'frontier incidents' from attacks that give rise to the right of self-defense, *Nicaragua*, 27 June 1986 (n 17), paras 102–104; see also Wrange (n 62) 15; Klamberg (n 61) 154.

[67] Randelzhofer and Nolte (n 55) 1413.

[68] Hollis (n 7) 1024–6; Harrison Dinniss (n 28) 37–9; Randelzhofer and Nolte (n 55) 1419.

[69] *Legality of the Threat or Use of Nuclear Weapons*, ICJ, Advisory Opinion, 8 July 1996, para 39.

[70] Dinstein (n 50) 103; Randelzhofer and Nolte (n 55) 1419.

4. SITUATIONS OF INTERNAL DISTURBANCES AND TENSIONS, NON-INTERNATIONAL ARMED CONFLICTS OR INTERNATIONAL ARMED CONFLICTS

Situations involving violence may be defined in any of the following three categories: (1) internal disturbances and tensions; (2) non-international armed conflicts; or (3) international armed conflicts. Different thresholds define to which category a certain conflict belongs and the applicable legal framework: rules of peace (human rights) or rules of war (international humanitarian law—IHL). The actors involved in the situation may seek to exploit these thresholds. The use of computer network attacks, psychological operations, irregular and/or non-state groups, 'peace keeping' forces, and expulsion of populations may keep a situation at the lower end, that is, as internal disturbances and tensions or as a non-international armed conflict.

Internal disturbances and tensions such as riots, isolated and sporadic acts of violence and other acts of a similar nature do constitute an armed conflict.[71] Banditry, unorganized and short-lived insurrections, or terrorist activities will thus fall below the threshold of armed conflict;[72] instead such violence should be evaluated under the law enforcement paradigm.[73]

The existence of an armed conflict within the meaning of IHL depends on factual criteria and is not dependent on formal declarations.[74] There is some ambiguity in the term 'armed conflict' which needs to be clarified. The intensity of the conflict and the organization of the parties to the conflict are factors relevant for the determination of whether it is an armed conflict.[75] The *Tadić* Appeals Chamber has stated that 'an armed conflict exists whenever there is a resort to armed force between States or protracted armed violence between governmental authorities and organized armed groups or between such groups within a State'. In other

---

[71] Protocol Additional to the Geneva Conventions of 12 August 1949, and relating to the Protection of Victims of Non-International Armed Conflicts (Protocol II), 8 June 1977, Article 1(2).

[72] *Prosecutor v Dusko Tadić* (Case No IT-94-1), ICTY T Ch, Opinion and Judgment, 7 May 1997, para 562.

[73] Klamberg (n 61) 158.

[74] Nils Melzer, 'Human rights implications of the usage of drones and unmanned robots in warfare' *Study for the European Parliament's Subcommittee on Human Rights* 2013, 19.

[75] *Tadić* (n 72), para 562.

words, certain 'intensity requirements applicable to both international and internal armed conflicts' have to be exceeded for IHL to be applicable.[76]

In an asymmetrical conflict the involved parties may for various reasons seek to have it classified under different legal frameworks. A non-state actor may pursue legitimacy and the status of an aspiring and/or future state. For that purpose the non-state actor may wish to have a situation classified as an armed conflict, with a preference for it to be 'international' since that will grant it status as a state with greater political and legal leverage. In contrast, states that are pitted against a non-state actor are reluctant to give them such status; for their purpose it is more convenient to have the situation classified as internal disturbances and tensions. However, states may perceive a need to use violence that is only allowed under the armed conflict paradigm. These opposing interests tend to push states to advocate that the situation should be classified as a non-international armed conflict where there are fewer rules restraining states: human rights law is derogable and prisoner of war status is not available. Moreover, a seemingly internal conflict may be rendered international where it is found that local armed groups are in fact acting on behalf of an external state,[77] as discussed in section 3 above.

Hollis notes that the law of war includes no provisions specifically addressing IO. This raises the possibility that IO could escape existing international law through the application of the *Lotus* principle—that is, what international law does not prohibit, it permits. However, states have declined to extend the *Lotus* principle to the law of war.[78] The 'Martens Clause' adopted by the 1899 and 1907 Hague Peace Conferences provides that the High Contracting Parties deem it expedient that in cases not included in the Regulations adopted by them, the inhabitants and belligerents remain under the protection and the rules of principles of law of nations as they result from the usages established amongst civilized nations, from the laws of humanity and the dictates of public conscience.[79]

---

[76] *Prosecutor v Dusko Tadić* (Case No IT-94-1), ICTY A Ch, Decision on the Defence Motion for Interlocutory Appeal on Jurisdiction, 2 October 1995, para 70; see also Klamberg (n 61) 161–2.

[77] Robert Cryer et al, *An Introduction to International Criminal Law and Procedure* (3rd edn, Cambridge University Press, 2014) 278.

[78] Hollis (n 7) 1035.

[79] *Lotus (France v Turkey)*, Ser. A., No. 10, PCIJ, Judgement, 7 September 1927: 'This way of stating the question is also dictated by the very nature and existing conditions of international law. ... Restrictions upon the independence of States cannot therefore be presumed'.

The Martens Clause is enshrined in Additional Protocol I to the 1949 Geneva conventions which provides that even if a certain conduct is not prohibited in a treaty 'civilians and combatants remain under the protection and authority of the principles of international law derived from established custom, from the principles of humanity and from the dictates of public conscience'.[80] In other words, the absence of a treaty provision explicitly prohibiting conduct during armed conflict does not mean that international law permits it.

States and commentators appear to favor an analogy approach where the existing legal framework applies to IO. However, this may be premature and the approach of analogizing existing rules to IO—such as those prohibiting the use of force, requiring civilian distinction—may be challenged.[81] Hollis identifies several thresholds and fault-lines within international humanitarian law that may be subject to exploitation. There are serious 'translation' problems with extending the existing rules to IO.

First, IHL may possibly be applied to IO in the case of an international armed conflict involving two or more states. However, in asymmetrical situations involving non-state actors and/or when the level of violence is short of an armed conflict it will be more difficult to determine the applicable legal framework.[82] Two significant challenges with analyzing IO under IHL is confusion surrounding (i) what IO triggers the civilian distinction requirement in relation to an attack—what is an attack?; and (ii) the dual-use nature of most information infrastructure. Where infrastructures have a 'dual-use' serving both civilian and military purposes, they qualify as military objectives subject to attack, even if their primary purpose is not military, but civilian.[83]

Second, IHL prohibits killing, injuring or capturing an adversary by resort to perfidy. The prohibition of perfidy has two purposes. First, it seeks to protect persons who wish to surrender, have protected status or are injured: the prohibition aims to prevent such misuse and further erosion of the protection for these persons. Second, it seeks to impose a minimum level of fairness to dealings between combatants.[84] More

---

[80] Protocol Additional to the Geneva Conventions of 12 August 1949 and Relating to the Protection of Victims of International Armed Conflicts (Protocol I) of 8 June 1977, Article 1(2).

[81] Hilaire McCoubrey, *International Humanitarian Law: Modern Developments in the Limitation of Warfare* (2nd edn, Dartmouth 1998) 281; Hollis (n 7) 1029, 1035–6.

[82] Hollis (n 7) 1039–41.

[83] Ibid 1043–4.

[84] Harrison Dinniss (n 28) 263.

specifically, it is '[a]cts inviting the confidence of an adversary to lead him to believe that he is entitled to, or is obliged to accord, protection under the rules of international law applicable in armed conflict, with intent to betray that confidence'.[85] For example, manipulating information systems so that enemy forces wrongly believe that troops are surrendering rather than gathering for an attack would be perfidious.[86] However, perfidy only applies if it results in injury to, or the capture of, adversaries. Hollis notes that that IO otherwise feigning protected status (for example, conducting CNA as if originating from a civilian source) does not constitute perfidy if it only produces physical damage but no casualties.[87]

Noting the translation problems above IHL may under certain conditions still be applicable to psyops and CNOs. However, even if there are rules applicable for psyops in armed conflict, many such operations are conducted when IHL is not applicable. The applicable framework in peacetime is based on international human rights treaties.[88] This raises the question whether a state can be responsible for human rights violations outside its territory. Protection for individuals under human rights treaties is normally limited to the 'jurisdiction' of a state party.[89] The European Court of Human Rights (ECtHR) has ruled that the territorial jurisdiction of a state extends beyond its national borders when that state exercises 'effective control of the relevant territory and its inhabitants abroad as a consequence of military occupation or through the consent, invitation or acquiescence of the Government of that territory, exercises all or some of the public powers normally to be

---

[85] API, Article 37(1).

[86] Harrison Dinniss (n 28) 263.

[87] Hollis (n 7) 1044–5; Pontus Winther, *Report: Rättsliga aspekter på psykologiska operationer inom ramen för NATO International Security Assistance Force (ISAF) verksamhet i Afghanistan. Delrapport 1—internationell humanitär rätt och Rules of Engagement* (Försvarshögskolan, Institutionen för säkerhet, strategi och ledarskap (ISSL), Folkrättscentrum, 2009) 13.

[88] Pontus Winther, *Report: Rättslig reglering av psykologiska operationer i nationella insatser* (Försvarshögskolan, Institutionen för säkerhet, strategi och ledarskap (ISSL), Folkrättscentrum, 2010) 4–5.

[89] European Convention for the Protection of Human Rights and Fundamental Freedoms adopted 4 November 1950 as amended by Protocol No 11, 213 UNTS 221, Article 1; International Covenant on Civil and Political Rights adopted 16 December 1966, 999 UNTS 171, Article 2(1); American Convention on Human Rights adopted 22 November 1969, 1144 UNTS 123, Article 1(2). cf African Charter on Human and Peoples' Rights, adopted 27 June 1981, 1520 UNTS 217, Article 1.

exercised by that Government'.[90] Even in the absence of territorial control, states have obligations under IHRL to the extent that their agents do, in fact, exercise physical power, authority or control over individuals.[91]

5. DISTINCTION BETWEEN OVERALL CONTROL AND EFFECTIVE CONTROL

Attribution of responsibility to a state may become an issue in the context of remote warfare. State responsibility does not only cover unlawful acts or omissions directly committed by the state and its officials.

In addition, the conduct of a person or group of persons shall be considered an act of a state under international law if the person or group of persons is in fact acting on the instructions of, or under the direction or control of, that state in carrying out the conduct.[92] The first instance involves private persons acting on the instructions of the state in carrying out the wrongful conduct. This includes cases where state organs supplement their own action by recruiting or instigating private persons or groups who act as 'auxiliaries' while remaining outside the official structure of the state. These include, for example, individuals or groups of private individuals who, though not specifically commissioned by the state and not forming part of its police or armed forces, are employed as auxiliaries or are sent as 'volunteers' to neighboring countries, or who are instructed to carry out particular missions abroad. The second instance deals with a more general situation where private persons act under the state's direction or control. The degree of control which must be exercised by the State in order for the conduct to be attributable to it was a key issue in the *Nicaragua* case.[93] In this the ICJ introduced the effective control test when it stated the following:

---

[90] *Banković and others v Belgium and 16 Other Contracting States* App no 52207/99, ECtHR, Judgment, 12 December 2001, para 71; *Al-Skeini and others v The United Kingdom* App no 55721/07, ECtHR, Judgment, 7 July 2011, para 135.

[91] *Issa and Others v Turkey* App no 31821/96, ECtHR, Judgment 16 November 2004, para 71; *Delia Saldias de Lopez v Uruguay* (Communication No 52/1979), HRC, Views, 29 July 1981, paras 12.1–12.3.

[92] Draft Articles on "Responibility of States for internationally wrongful acts", International Law Commission, 2001, Article 8.

[93] Draft Articles on "Responsibility of States for internationally wrongful acts" with commentaries, International Law Commission, 2001, 8.

> All the forms of United States participation mentioned above, and even the general control by the respondent State over a force with a high degree of dependency on it, would not in themselves mean, without further evidence, that the United States directed or enforced the perpetration of the acts contrary to human rights and humanitarian law alleged by the applicant State. ... For this conduct to give rise to legal responsibility of the United States, it would in principle have to be proved that that State had effective control of the military or paramilitary operations in the course of which the alleged violations were committed.[94]

The ICJ has also stated that the 'effective control' test is a norm of customary international law.[95] The 'effective control' test should be distinguished from the broader and more flexible 'overall control' test introduced by the ICTY in the *Tadić* case and repeated by the ICC in *Situation in Georgia*.[96] While the first test concerns attribution of state responsibility the second is used to characterize whether an armed conflict is non-international or international. Under the 'overall control' test it is not necessary to produce evidence of specific orders or instructions relating to particular military actions. It is sufficient to establish 'overall control going beyond the mere financing and equipping of such forces and involving also participation in the planning and supervision of military operations'.[97] The 'overall control' test is controversial and should be used to determine the nature of the conflict but not when determining state responsibility.[98] The ICJ stated the following in the *Genocide* case:

> logic does not require the same test to be adopted in resolving the two issues, which are very different in nature: the degree and nature of a State's involvement in an armed conflict on another State's territory which is required for the conflict to be characterized as international, can very well, and without logical inconsistency, differ from the degree and nature of involvement

94 *Nicaragua*, 27 June 1986 (n 17), para 115.

95 *Application of The Convention on the Prevention and Punishment of the Crime of Genocide (Bosnia and Herzegovina v Serbia and Montenegro)*, Judgment of 26 February 2007 (ICJ), para 401.

96 The overall control test was introduced in *Prosecutor v Dusko Tadić* (Case No IT-94-1), ICTY A Ch, Judgment 15 July 1999, paras 88–91, 120, 122, 131 and repeated in *Situation in Georgia* (No. ICC-01/15-12), ICC PT. Ch. I, Decision on the Prosecutor's request for authorization of an investigation, 27 January 2016, para 27.

97 *Tadić* (n 96), para 145.

98 Antonio Cassese, 'The Nicaragua and Tadić Tests Revisited in Light of the ICJ Judgment on Genocide in Bosnia' (2007) 18(4) *European Journal of International Law* 649; Cryer et al (n 77) 278.

required to give rise to that State's responsibility for a specific act committed in the course of the conflict.[99]

As the Ukraine-Russia conflict illustrates, a state such as Ukraine may simultaneously want to deny separatists recognition and thus characterize the conflict as non-international but also attribute actions of its opponents to another state (Russia). If it is shown that the non-state actor is a mere pawn and under the effective control of foreign state, both objectives (non-recognition and attribution) may be achieved. However, as shown by the case law from the ICJ, ICTY and the ICC, the threshold for effective control is high and may be difficult to prove.

6. DISTINCTION BETWEEN COMBATANT AND NON-COMBATANT

Even if an armed conflict exists, persons may only be targeted with military force if they are combatants.[100] Do a terrorist, a separatist or members of other non-state groups qualify as combatants? The principle of distinction requires that the parties to a conflict at all times distinguish between civilians and combatants.[101] This is a principle of customary international law.[102] Given the importance of the principle and its requirement to distinguish civilians and combatants, the two categories need to be defined.

Even though the Fourth Geneva Convention (GC IV) concerns the protection of civilians, it still lacks a definition of what is a civilian. Article 50 of the Additional Protocol I to the 1949 Geneva Conventions (API) may serve as a starting point as it states that a 'civilian is any person who does not belong to one of the categories of persons referred to in' Articles 4A(1), (2), (3) and (6) of the Third Geneva Convention (GC III) and Article 43 of API. Civilians are thus not (1) members of the armed forces of a party to the conflict; or (2) members of militias or volunteer groups.[103] In order to enjoy specially protected status as a civilian they are strictly prohibited from participating in the hostilities,

---

[99] *Genocide Case* (n 95), para 405.

[100] Nils Melzer, *Targeted Killing in International Law* (Oxford University Press 2008) 56.

[101] API, Articles 48 and 51(2); Melzer (n 100) 300–27.

[102] *Legality of the Threat or Use of Nuclear Weapons* (n 69), paras 78–79.

[103] A McDonald, *The Challenges to International Humanitarian Law and the Principles of Distinction and Protection from the Increased Participation of Civilians in Hostilities* (The Asser Institute 2004), s 2.1.

except in the exceptional case where they are participating in a *levée en masse*, in which case they shall be regarded as belligerents provided that they carry their arms openly and respect the laws and customs of war.[104] If civilians participate in the hostilities, they may be directly attacked as if they were combatants. The interpretive guidance of the International Committee of the Red Cross (ICRC) provides that in order to qualify as direct participation in hostilities, a specific act must meet the following cumulative criteria:

1. the act must be likely to adversely affect the military operations or military capacity of a party to an armed conflict or, alternatively, to inflict death, injury, or destruction on persons or objects protected against direct attack (threshold of harm);
2. there must be a direct causal link between the act and the harm likely to result either from that act, or from a coordinated military operation of which that act constitutes an integral part (direct causation);
3. the act must be specifically designed to directly cause the required threshold of harm in support of a party to the conflict and to the detriment of another (belligerent nexus).[105]

However, unlike combatants, civilians regain protection against direct attack as soon as their individual conduct no longer amounts to direct participation in hostilities.[106] This poses a potential problem in asymmetrical conflicts if the non-state actor uses a hit-and-run operation which would oblige the armed forces of the state to act purely reactively.

To summarize, civilians who directly participate in hostilities, namely by committing acts that meet the threshold of harm, requirements on direct causation and belligerent nexus, may be targeted under IHL. An alleged terrorist who resides in an area where there is an armed conflict but who is not directly participating in the hostilities is thus not a legitimate target under IHL.[107]

The principle of distinction also applies to cyber operations occurring within the context of an armed conflict. However, cyber operations may

---

[104] Hague Convention IV – Respecting the Laws and Customs of War on Land and annexed regulations adopted 18 October 1907, Article 2; McDonald (n 103), s 2.1.

[105] Nils Melzer, 'Interpretive Guidance on the Notion of Direct Participation in Hostilities under International Humanitarian Law' (2009) 90(872) *International Review of the Red Cross* 991, 995–6. See also Kevin Jon Heller, '"One Hell of a Killing Machine": Signature Strikes and International Law' (2013) 11(1) *Journal of International Criminal Justice* 89, 93.

[106] Melzer (n 100) 329.

[107] Klamberg (n 61) 163–4.

cause significant damage to systems, its component computers and its surroundings without causing physical damage. Thus there is debate as to whether such operations constitute an attack under IHL. It is generally agreed by commentators that cyber operations that result in death or injury to people, or physical damage or destruction to property will constitute attacks under IHL.[108] The targetable group for cyber operations is the same as under the general IHL rules, that is, members of the armed forces; members of armed groups who perform a continuous combat function; participants in *levée en masse*; and civilians who directly participate in hostilities. However, the current ability to target persons directly with cyber operations is limited.[109] Turning to propaganda, undermining the will of the enemies (that is, of the people, government and the armed forces), while preserving one's own, is an accepted military objective.[110]

7. CONCLUSIONS

This chapter shows that there is ample opportunity for states to act and still avoid direct confrontation as well as attribution. For states at the receiving end these legal thresholds, fault-lines and gaps are more unfortunate. The international system as such may also suffer if states circumvent rules that are at the foundation of the system. One dilemma is that these are somewhat digital in the sense that a certain defined action is required in order to trigger a right to use countermeasures. The rules on self-defense against armed attacks is an obvious example. An alternative approach would be more analogue, acknowledge that states have obligations and may commit violations against international law across a grayscale. By the same reasoning, systems of countermeasures need to be adaptive and be able to scale.

---

108 Heather Harrison Dinniss, 'The regulation of cyber warfare under the *jus in bello*' in James A Green (ed), *Cyber Warfare: A Multidisciplinary Analysis* (Routledge 2015) 125–59, 126–9.

109 Ibid 130–31.

110 Charles J Dunlap, 'Meeting the Challenges of Cyberterrorism' (2002) 76 *International Law Studies* 353, 363.

PART II

REMOTELY PILOTED VEHICLES AND CYBER WEAPONS

7. Drone strikes: a remote form of self-defence?

Nigel D White and Lydia Davies-Bright

1. INTRODUCTION

A Motyxia Sequoiae millipede glows with bioluminescence in order to warn off predators. Should it be faced with an imminent threat, it secretes a toxic mix of cyanide and chemicals in order to stave off a potentially fatal attack.[1] This capacity of the organism to defend itself is arguably one of the most basic and fundamental natural instincts—the desire to live and the ability to survive an external attack will ensure the continuance of the life form. If an organism capitulates in the face of external danger, it will soon be subsumed by others and be but a mere speck in the history of life on this planet. Thus, it could be said that every organism has the right to defend itself from external attack as to deny it the capacity to do so, is to condemn it to death and annihilation.

The right of individual persons to defend themselves against an imminent attack is recognised in law.[2] Similarly, a state, as an international legal person, has an 'inherent' right to defend itself.[3] As with the Motyxia Sequoiae, a state may passively warn would-be attackers that it is dangerous, for example, by having a well-equipped military. However, in the event of an imminent threat, a state will also launch appropriate counter-measures. This right to self-defence against 'unjust' attack was described in 1758 by Emmerich de Vattel as 'not only the right that every

---

[1] C Arnold, 'New Glowing Millipede Found; Shows How Bioluminescence Evolved' (*National Geographic*, 4 May 2015), accessed 3 May 2017 at http://news.nationalgeographic.com/2015/05/150504-glowing-millipedes-evolution-insects-animals-california/; CQ Choi, 'Strange Glowing Millipedes Ooze Cyanide to Foil Predators' (*Live Science*, 26 September 2011), accessed 3 May 2017 at www.livescience.com/16221-glowing-millipedes-toxic-warning.html.

[2] Criminal Law Act 1967 (UK) s 3(1) 'A person may use such force as is reasonable in the circumstances in the prevention of crime'; D Ormerod and K Laird, *Smith and Hogan's Criminal Law* (Oxford University Press 2015) 427.

[3] Charter of the United Nations (24 October 1945) 1 UNTS XVI (UNC) Article 51.

Nation has, but it is a duty, and one of its most sacred duties'.[4] It follows that states have a duty to protect those who belong to it, which is part of the basic agreement underlying social and legal structures.[5] A state cannot be sovereign, independent, or expect to continue to exist if it does not have the right to defend itself and its citizens from aggression emanating from an external party: if a state cannot protect its citizens from 'the subjection of a foreign power', then it has breached the social contract and so can no longer exercise a monopoly over the use of coercive force.[6]

This chapter addresses the concept of self-defence through an analysis of a drone strike conducted by the UK government in August 2015 against an individual British citizen residing in ISIL-held territory within Syria. In this case, self-defence is posited in a way that reveals a new reliance on an old understanding of self-defence, based on self-defence as sovereignty and less as a distinct and confined rule of the jus ad bellum. In a way, although individual terrorists cannot normally be seen as an existential threat to a state, their antipathy towards the status quo has the potential to erode the system of sovereign states, by challenging their continuation, but also by provoking increasingly draconian responses by states that will potentially destroy them from within.

---

[4] E de Vattel, *The Law of Nations, or the Principles of Natural Law Applied to the Conduct and to the Affairs of Nations and of Sovereigns* (trans T Nugent, Liberty 2008) 246.

[5] A fact explicitly acknowledged by national leaders, especially when justifying a use of force and framing it as being in defence of the nation, e.g. Prime Minister David Cameron speaking to the House of Commons, 'My first duty as a Prime Minister is to keep the British people safe.' HC Deb 7 September 2015, col 27, accessed 3 May 2017 at www.publications.parliament.uk/pa/cm201516/cmhansrd/cm150907/debtext/150907-0001.htm; UK Secretary of State for Defence Michael Fallon MP in his oral evidence to the Joint Committee on Human Rights asserting that preventing loss of life is the 'primary duty of government', Joint Committee on Human Rights, *Oral Evidence: The UK Government's Policy on Use of Drones for Targeted Killing* (HC 2015–16, 574) 7, accessed 3 May 2017 at http://data.parliament.uk/writtenevidence/committeeevidence.svc/evidencedocument/human-rights-committee/the-uk-governments-policy-on-the-use-of-drones-for-targeted-killing/oral/27633.pdf.

[6] J Locke, *Two Treatises of Government. In the Former, the False Principles and Foundation of Sir Robert Filmer, and his Followers, are Detected and Overthrown: the Latter, is an Essay Concerning the Original, Extent, and End, of Civil Government* (prepared by R Hay for the McMaster University Archive of the History of Economic Thought, Thomas Tegg, W Sharpe and Son 1823) 200 para 217, accessed 3 May 2017 at http://socserv2.socsci.mcmaster.ca/econ/ugcm/3113/locke/government.pdf.

The chapter examines the contribution of technology to legal change, but concludes that technology is also contributing to changes to the overall structure of international relations by facilitating the breakdown in sovereignty by its persistent erosion of borders, provoking, in response, desperate efforts to shore it up by returning to primordial understandings of sovereignty based on preservation of the nation state. Technology has sped up the escalating and apparently never-ending cycle of blows and counter-blows[7] and means that individual terrorists can orchestrate attacks on states from afar as well as being targeted by remotely operated drones, but the return to absolute forms of sovereignty by technologically advanced states is something more profound and alarming. The reversion to ancient notions of sovereignty is in contrast to progress in technology.

2. SELF-DEFENCE AND THE SOCIAL CONTRACT

The frequent targeted strikes by drones on individual (suspected) terrorists demonstrate the modern phenomenon of 'legally saturated violence',[8] namely that international relations is now characterised by multiple, every-day uses of force by states that are not simply cynically justified in legal terms, but are seen as the exercise of the essence of sovereignty as states seek to defend themselves and their citizens from violence emanating from other states, so-called states (for example, Islamic State of Iraq and the Levant, or ISIL), non-state armed groups and individual terrorists. In the current debate about drone usage, law is not peripheral to the exercise of state power but is at the heart of it, a return, in a way, to the basic social contract at the heart of a state—to provide security in return for its monopoly on the use of force. However, the monopoly of the state over force cannot take away an individual's right to self-defence. In Western legal traditions, an individual is permitted to defend their self and also others as an extension of this right.[9] Aggression by an external actor compromises an individual's freedom, who is then entitled to 'vindicate [their] freedom by repelling the aggressor'.[10]

---

[7] T J Farer, 'Beyond the Charter Frame: Unilateralism or Condominium?' (2002) 96 *AJIL* 359.

[8] A-C Martineau, 'Concerning Violence: A Post-Colonial Reading of the Debate on the Use of Force' (2016) 29 LJIL 95, 112.

[9] G P Fletcher and J D Ohlin, *Defending Humanity* (Oxford University Press 2008) 51.

[10] Ibid 48.

The endemic global violence unleashed after 9/11 in 2001 has generated numerous explanations under the jus ad bellum and jus in bello, but increasingly also by a 'meta' concept of 'self-defence' that has hitherto largely been confined to the jus ad bellum, and now seems to have been unchained from its shackles in Article 51 of the UN Charter to become the overriding norm. Its apparent transformation from a right that justified the defence of one state from an attack by another, to an overweening inherent exercise of sovereign duty to protect all aspects of a state (its territory, its government and its population, even its way of life[11]) has had a profound effect on the conduct of international relations, facilitated by advances in technology and the securitisation of many aspects of everyday life.

The advent of 'human security', broadly defined as a freedom from want and a freedom from fear, was promoted as a value to balance against the narrow state-centric focus of 'security of territory from external aggression, or as protection of national interests'.[12] However, the attempt to focus security concerns on the micro level—on the lives of ordinary people—has now been merged with 'state security' to justify the return to a focus on the macro level by framing uses of force as being against targets that present a general threat to a state's citizens and their daily lives. The reconceptualising of security as people-centric, 'a concern with human life and dignity' rather than with weapons,[13] has been accepted by politicians and commentators and also, perhaps, by history. The world is not currently locked in an ideological battle for global dominance manifested in a nuclear arms race, the Cold War has been won by the West. Accordingly, the current threat is also framed as being centred on people—the enemy is not a state, it is people who seek to

[11] See, e.g. the statement by President Hollande following the terrorist attacks Paris on 13 November 2015 declaring that the attacks were 'committed by a terrorist army, the Islamic State group, a jihadist army, against France, against the values we defend everywhere in the world, against what we are: a free country that means something to the whole planet … What we are defending is our country, but more than that, it is our values'. L Dearden, 'Paris terror attack: Francois Hollande vows merciless response to Isis "barbarity"' *The Independent* (14 November 2015).

[12] UNDP, *Human Development Report 1994* (Oxford University Press 1994) (HDR 1994) 22, accessed 3 May 2017 at http://hdr.undp.org/sites/default/files/reports/255/hdr_1994_en_complete_nostats.pdf.

[13] HDR 1994 (n 12) 22.

destroy the every-day lives of ordinary people.[14] Thus, the requisite response, the self-defence to this threat, is also people centred.

Self-defence has an instinctive, almost visceral quality to it and, when contemplated, the question asked is often not *if* violence in the face of danger is acceptable, but rather *how far* is a person or state permitted to go when repelling an apparent threat. For Grotius, defence is one of the 'just' causes of war and lies at the heart of sovereign statehood.[15] Similarly, many philosophers have connected the nature of the state with self-defence. John Locke viewed protection as being the core justification for the state's monopoly on the use of force. In order to preserve their wealth, life, liberty and general well-being, men united together to form states and made a pact (the social contract) with government that it would have coercive power, in exchange for protection. It follows that, if the government becomes unable to protect the people against threats, the people are justified in throwing over the government as the pact has been broken—the people have returned to the state of nature.[16]

By legitimising the use of force through citing self-defence, governments are able to maintain their position and the status quo—the people accept the use of force as the government is fulfilling its duty. Indeed, as mentioned above, this duty is often cited by politicians when justifying a particular incident of force.[17] Thus, the continuance of the current situation relies on a threat that people need protecting from and the ability of the state to provide that protection. Hobbes conceptualised a social order in which people join together in a social covenant that provides them security from an external threat.[18] Their motivation is to escape the state of nature in which they live in perpetual fear and under continual threat and so they agree to form communities, common laws and mechanisms to keep and enforce the laws. Although Hobbes argued that the sovereign had to be given absolute authority in order to ensure the preservation of society, he permitted the people to disobey the

---

14 The HDR 1994 (n 12) includes terrorism as being a security threat that concerns ordinary people, 22.

15 'Hugo Grotius' (*Stanford Encyclopedia of Philosophy*, 28 July 2011), accessed 3 May 2017 at http://plato.stanford.edu/entries/grotius/#JusWarDoc.

16 Locke (n 6).

17 E.g. Prime Minister David Cameron (n 5).

18 T Hobbes, *Leviathan or the Matter, Forme, and Power of a Commonwealth Ecclesiasticall and Civill* (prepared by R Hay for the McMaster University Archive of the History of Economic Thought, Andrew Crooke 1651) 106, accessed 3 May 2017 at http://socserv2.socsci.mcmaster.ca/econ/ugcm/3ll3/hobbes/Leviathan.pdf.

sovereign in the event that it failed to provide adequate protection.[19] Again, the legitimacy of the sovereign is connected to its ability to sufficiently safeguard its people.

Similarly, Jean-Jacques Rousseau also conceived of a social pact in which people agree to form a collective and to accept restrictions on their individual liberties in order to receive the benefits of state protection. The sovereign is committed to the good of the individuals that constitute it and each individual is committed to the good of the whole.[20] On this conception, the idea that an individual may break 'the contract' and so be denied membership to the collective[21] has at least an appearance of validity. An individual does not have the liberty to decide whether or not they wish to fulfil their duties to the sovereign power and collective, whilst still receiving the benefits of citizenship. However, the sovereign power has the monopoly on power in order to maintain order and to ensure compliance with the agreed rules. Expelling every individual who dissented would restrict and limit the form of direct democracy that Rousseau espoused and so is arguably not in keeping with his philosophy.

To legitimately evoke the right to self-defence, a state is required to demonstrate that it has suffered an intentional armed attack.[22] Traditionally, this was understood as being an attack on the territory or flagged ship of a state. In the modern era, it is not necessarily so straightforward to determine when an armed attack has begun and so when it is legitimate to utilise armed force in self-defence. In the aftermath of 9/11,

---

19 'The obligation of subjects to the sovereign is understood to last as long, and no longer, than the power lasteth by which he is able to protect them.' Hobbes (n 18) 136.

20 J-J Rousseau, *The Social Contract* (trans M Cranston, Penguin 2004).

21 A Nossiter, 'French Proposal to Strip Citizenship over Terrorism Sets off Alarms' *New York Times* (8 January 2016): Prime Minister Manuel Valls 'insisted in a television interview ... that "you are French because you adhere to a community. This strict measure applies to terrorists who have been convicted of especially grave crimes, and it is because they have broken the contract ... it is a way of consolidating the national pact."'

22 *Case Concerning Oil Platforms (Islamic Republic of Iran v United States of America)* (Judgment) [2003] ICJ Rep 161, 187 para 51: 'Therefore, in order to establish that it was legally justified in attacking the Iranian platforms in exercise of the right of individual self-defence, the United States has to show that attacks had been made upon it for which Iran was responsible; and that those attacks were of such a nature as to be qualified as "armed attacks" within the meaning of that expression in Article 51 of the United Nations Charter, and as understood in customary law on the use of force.'

the UN Security Council adopted Resolution 1368 and made explicit reference to 'the inherent right of individual or collective self-defence in accordance with the Charter'. Resolution 1373 reaffirmed this statement and, utilising Chapter VII powers, adopted a series of binding decisions, which included an instruction for all states to 'take the necessary steps to prevent the commission of terrorist attacks'. These resolutions designated terrorism as a threat to international peace and security and recognised or, indeed, helped to create conditions in which states would be able to exercise their inherent right to self-defence in response to terrorism or even the threat of terrorism. Thus, in this age of terrorism, non-state actors are deemed capable of mounting an armed attack that justifies and legitimises the use of force in self-defence. This developing of the principle of self-defence in extremis, without providing criteria for determining whether an act by a non-state actor is an 'armed attack' (and so a threat to international peace and security) and without thought to the consequences, has arguably undermined the prohibition on the use of armed force as now states are able to designate a fellow state or a non-state actor as terrorist and thereby justify the use of force by citing self-defence. For example, in response to the 2015 Paris shootings, France embarked upon an aerial bombing campaign against ISIL in Syria, citing self-defence, despite the lack of credible evidence that the shootings were carried out by ISIL operatives and/or explicitly co-ordinated and directed by the group and so actually attributable to them.[23] It appears that it is enough for an act of violence to be 'inspired by' ISIL for it to justify the use of force in self-defence.[24]

---

23 BBC News, 'Paris Attacks: Who were the attackers?'(BBC News, 8 March 2016), accessed 3 May 2017 at http://www.bbc.co.uk/news/world-europe-34832512. See generally K Tibori-Szabo, 'Self-Defence and the United States Policy on Drone Strikes' (2015) 20 *JCSL* 381, 401.

24 See, e.g. Prime Minister David Cameron, during the Parliamentary debate on Syrian airstrikes, claiming that the UK Security Services have foiled seven plots against the United Kingdom that were 'inspired by' the radical teachings of the 'death cult' that is ISIL in order to justify his claim that ISIL poses a credible threat and so Britain should join the airstrikes campaign. It has since emerged that there is no actual evidence of a direct link between the plots and ISIL. Thus, it appears that merely the existence of ISIL and its ideology poses a threat to international peace and security and so engages the inherent right to self-defence (and self-preservation—the Western states are arguably fighting to maintain the status quo and to preserve their existence as it is now): 'They have inspired the worst terrorist attack against British people since 7/7 on the beaches of Tunisia, and they have plotted atrocities on the streets here at home. Since November last year our security services have foiled no fewer than seven different plots against

3. DRONE TECHNOLOGY AND LEGAL CHANGE

With the odd exception, such as the regulation of outer space,[25] international law tends to develop as a reaction to change. In this way, it might be anticipated that new non-kinetic technologies that can be used to disable computer networks, or to carry mass covert surveillance of e-mail traffic, may take decades to bring within a clear legal framework, depending on how quickly states come to realise that it is in their mutual self-interest to effectively regulate cyber-space. It may, in any case, prove to be an impossible task as it raises the question of whether states can actually regulate something that has escaped the confines of sovereignty—it may simply be too late to put the genie back into the bottle. In this scenario, states will fall back on general principles of international law, such as the norm prohibiting intervention in a state's political or economic affairs, which will not prevent cyber operations but will enable selective condemnation in the General Assembly and, occasionally, executive responses to particular threats by the Security Council.

In contrast, when it comes to new technologies that seem to provide straightforward improvements in military efficacy, such as Unmanned Aerial Vehicles (UAVs), commonly known as drones, it should be expected that existing international law will be adequate. Indeed, this is quite commonly the argument made in the literature, given that drones are seen as mere 'platforms' for the launch of weapons such as missiles and not new weapons per se.[26] Furthermore, drones are portrayed by their users and supporters as upholding the value of security rather than

---

our people, so this threat is very real … do we go after these terrorists in their heartlands, from where they are plotting to kill British people, or do we sit back and wait for them to attack us?' HC Deb 2 December 2015, col 324, accessed 3 May 2017 at www.publications.parliament.uk/pa/cm201516/cmhansrd/cm151202/debtext/151202-0001.htm.

[25] See UNGA Res 1962 'Declaration of Legal Principles Governing the Activities of States in the Exploration and Use of Outer Space' (13 December 1963) UN Doc A/RES/1962 (XVIII); Treaty on Principles Governing the Activities of States in the Exploration and Use of Outer Space, including the Moon and Other Celestial Bodies (adopted 27 January 1967, entered into force 10 October 1967) 610 UNTS 205.

[26] D Turns, 'Droning on: Some International Humanitarian Law Aspects of the Use of Unmanned Aerial Vehicles in Contemporary Armed Conflicts', in C Harvey, J Summers and N D White (eds), *Contemporary Challenges to the Laws of War* (Cambridge University Press 2014) 199.

undermining it.[27] Drone-using states, in particular, argue that new law is unnecessary for the regulation of drones, since they are simply another means of delivering death and destruction. But such arguments belie the fact that existing laws have to be reinterpreted and applied to drones and that in this process the louder voice of the drone-using state tends to dominate. Such debate is not confined to the rules of humanitarian law on targeting,[28] but also include the rules on the use of lethal force found in human rights law (in the context of the right to life) and, moreover, the application of the right of self-defence both within the meaning of Article 51 and under human rights law. The essence of self-defence is action necessary to ensure the survival of a state or person under threat of imminent attack. Given that the drones themselves are not entitled to the right of self-defence since their operators are not under imminent threat of attack, and the targets are a distance away from the state using them, the dynamics of self-defence action through the use of drones are clearly different.

The increasing use of drones raises security concerns for a number of reasons. When they are used for surveillance they are potential threats to personal security and privacy. When used for targeting purposes they not only raise security concerns for civilians potentially caught in the blast (the problem of collateral losses), but they also seem to either extend the battlefield, thereby bringing the instability inherent in war, or constitute the extraterritorial application of force for the purposes of some extreme form of law enforcement. Under this model of law enforcement, capture, arrest and trial are replaced by summary execution. All of these conceptions of drone use challenge the notion that they represent a new era of clean, clinical and, above all, legitimate use of force. Perceptions and assertions of security by governments are difficult for the courts to resist, particularly in times of terrorism that are characterised by random attacks against civilians, even when government actions to protect the lives and security of its citizens may appear to tread on the very freedoms it is fighting to protect.

Governments are under a duty to provide their citizens with security, but it cannot be an absolute duty—one that it aims to achieve at all costs.

---

[27] For critical evaluation, see C Gray, 'Targeted Killings: Recent US Attempts to Create a Legal Framework' (Legal Studies Research Paper Series, University of Cambridge Paper No 52/2013, November 2013).

[28] W H Boothby, *The Law of Targeting* (Oxford University Press 2012) 593.

Due diligence obligations upon governments are obligations of conduct,[29] rather than result, and so a failure by government to prevent specific acts of terrorism is not necessarily an indication that the state has failed to fulfil its duties to protect life and security. The random nature of many terrorist actions means that it is very difficult to prevent each and every one. When considering how these obligations have been interpreted by the Human Rights Committee in the context of the rights to life and security under the International Covenant on Civil and Political Rights,[30] it is clear that states must take reasonable and appropriate measures to protect individuals within their jurisdiction who are subject to known threats to their lives.[31] The European Court of Human Rights has similar jurisprudence, stating in one judgment that a government that 'knew or ought to have known … of a real and immediate risk to the life of an identified individual or individuals from the criminal acts of a third party', must take 'measures within the scope of their powers, which, judged reasonably, might' be 'expected to avoid that risk'.[32] As has been stated by Bates:

> Applying this jurisprudence by analogy to terrorist attacks creates some challenges: the bombing of civilians on aircraft or commuter trains and the hijacking of aircraft suggests a random choice of victims, rather than the selection of an 'identified individual or individuals' as victims.[33]

When drones are used outside of a state's jurisdiction, whether for surveillance or for targeting purposes, and when lethal force is used against individuals, the human rights issues become more complex. While human rights obligations apply to individuals within a state's territory, there is considerable debate about when they apply to individuals outside its territory but, arguably, within its jurisdiction.[34] When considering the use of armed force from a drone against a terrorist

---

[29] S Marks and F Azizi, 'Responsibility for Violations of Human Rights Obligations: International Mechanism' in J Crawford, A Pellet and S Olleson (eds), *The Law of State Responsibility* (Oxford University Press 2010) 729.

[30] (Adopted 16 December 1966, entered into force 23 March 1976) 999 UNTS 171 (ICCPR) Articles 6 and 9.

[31] *Delgado Paez v Columbia* (12 July 1990) Human Rights Committee Communication No 195/1985, para 5.5.

[32] *Osman v United Kingdom* (1998) 29 EHRR 245.

[33] E S Bates, *Terrorism and International Law: Accountability, Remedies and Reform* (Oxford University Press 2011) 83–4.

[34] See generally M Milanovic, 'Human Rights Treaties and Foreign Surveillance: Privacy in the Digital Age' (2015) 56 Harv Intl L J 81, 111ff.

suspect, the question is whether the individual is within the jurisdiction of the state using force. Although there is some Inter-American case law that supports the application of the right to life in these circumstances,[35] there is contrary European jurisprudence.[36] Rather than considering whether the state using force has enough control over the targeted individual for the purposes of evaluating whether there is an assertion of jurisdiction in these circumstances, it might be better for the courts to focus on the fact that the operator of the drone, often a distance away from the target, is clearly under the control of the state using force.[37]

If jurisdiction is established, such uses of targeted force from drones, when taken outside of armed conflict, appear to be violations of the right to life as there is usually no imminent threat to the state to justify its use of force as a last resort.[38] Indeed, the use of lethal force from drones seems to be an extreme and unlawful version of law enforcement where it is easier to kill suspects than to capture them (particularly as capturing suspects would put them within the capturing state's jurisdiction).[39] Furthermore, the use of drones for targeting suspected terrorists appears to be an attempt to externalise a state's security measures to counter terrorism by taking out targets in another state's territory before they have the chance to hit the drone state's territory or nationals. The United States has tried to justify this by arguing what is the ultimate justification for using lethal force—that there is a global armed conflict against terrorists or, at the very least, a transnational armed conflict against Al Qaeda and its associates. This argument is an attempt to justify a lower standard for the use of lethal force for, in simple terms, a use of lethal force is allowed in an armed conflict if the target is either a military objective, a combatant, or a civilian who is directly participating in hostilities, and the anticipated collateral damage ('incidental loss of civilian life') is not excessive in relation to the expected military

---

[35] *Armando Alejandre Jr, Carlos Costa, Mario de la Peña and Pablo Morales v Cuba (Brothers to the Rescue)*, Case 11.589, Report No 86/99 (28 September 1999) para 25.

[36] *Banković and others v 17 NATO States* Admissibility Decision (Grand Chamber) App no 52207/99 (ECtHR, 12 December 2001) paras 52–53.

[37] F Hampson, 'The Scope of the Extra-Territorial Applicability of International Human Rights Law' in G Gilbert, F Hampson and C Sandoval (eds), *The Delivery of Human Rights: Essays in Honour of Sir Nigel Rodley* (Routledge 2011) 181–2.

[38] UNHRC 'P Alston: 'Study on Targeted Killings, Report to the Human Rights Council' (2010) UN Doc A/HRC/14.24/Add.6, paras 85–86.

[39] *Ocalan v Turkey* App No 46221/99 (ECtHR, 12 March 2003) para 125.

advantage.[40] The United States has interpreted these rules liberally: to carry out 'signature' strikes on the basis that the targeted individual is performing suspicious activities; to target funerals where there is a concentration of Taliban leaders; to target drug lords (who are criminals not combatants); and sometimes to order strikes outside of a conflict zone, for example, in Yemen in 2002 and again in 2011.[41] Under President Obama the 'war on terror' rhetoric was abandoned in favour of a mixture of jus ad bellum and jus in bello justifications, according to which a targeted killing is lawful 'if the targeted individual posed an imminent threat of violent attack against the United States, capture was not feasible, and the operation was conducted in line with law of war principles',[42] with a presumption that a known terrorist located anywhere in the world constitutes an imminent threat to the United States and its citizens.[43] Rather than determining whether that individual is a specific imminent threat (that is, is about to launch an attack), their membership of a group such as Al Qaeda or ISIL is sufficient per se. In this way, jus in bello reasoning in the form of identification of a 'combatant' is used to justify triggering a right of self-defence under the jus ad bellum. The confusion of legal concepts is a deliberate manipulation of the law to justify drones, which are used to hunt down their targets rather than respond to imminent attacks.

It seems that after the devastating attacks on the United States of 11 September 2001, governments (and not just the United States) have re-assessed their security priorities, have reasserted national security (often on the basis that this is the best way to protect human security) and have acted in violation of basic norms governing when coercion can be used by the state against individuals in order to protect the majority of its citizens. This has either been as a result of the extension of the battlefield or the extension of law enforcement. While the majority of states may support this, or, more accurately, remain supine in the face of these erosions, the securitisation of post-9/11 life has meant that (the right to) security has been elevated to a pre-eminent position in political

---

40 Article 55, Additional Protocol I 1977; Turns (n 26) 207.

41 S Casey-Maslen, 'The Use of Armed Drones', in S Casey-Maslen (ed), *Weapons under International Human Rights Law* (Cambridge University Press 2014) 400–403.

42 Tibori-Szabo (n 23) 383.

43 Ibid 402.

rhetoric and action in contradistinction to its position as one of a number of human rights and protections provided by international law.[44]

Thus, while there are international norms applicable to drone use, a great deal of the law is underdeveloped, indeterminate or ineffectual, and furthermore, has been subject to artful manipulation of the boundaries between the jus ad bellum and jus in bello, with little regard to the right to life of the target. The UN itself has not tackled the legality of drone usage in any meaningful way. Although this is probably to be expected in the executive body, it is disappointing to see that the plenary body has also failed to fulfil its functions as a security community with the ability to shape normative frameworks, confining itself instead to exhortation in general resolutions to the effect that counter-terrorism efforts by states should be undertaken in conformity with international human rights law, refugee law and international humanitarian law.[45] This simply begs the question of how these norms should be applied to drone strikes.

4. THE KILLING OF AN INDIVIDUAL IN SELF-DEFENCE OF A STATE

Given the discussion above, it seems that technology may have outstripped the law, for drones are not just new platforms for delivering weapons, they actually change the dynamics of both the battlefield and of law enforcement outside the battlefield. In the former, they enable the enemy to be taken out without any risk to the drone-state's soldiers. They change the idea of war from the clash of armies, towards asymmetrical warfare often characterised by a technologically rich state using force against a technologically poor state or non-state actor. While international humanitarian law seems to be more readily interpreted to allow the calculated killings of soldiers and other combatants, outside of that law enforcement generally requires that the state (through its police or other agents) acts out of self-defence or defence of others, sometimes also when absolutely necessary in attempting to carry out arrests or quell

[44] See generally L Lazarus, 'The Right to Security – Securing Rights or Securitising Rights?' in R Dickinson, E Katselli, C Murray and O W Pederson (eds), *Examining Critical Perspectives on Human Rights* (Cambridge University Press 2012) 87.

[45] E.g. UNGA Res 68/178 (18 December 2013) UN Doc A/RES/68/178.

riots.[46] The UN Declaration on the Basic Principles on the Use of Force and Firearms by Law Enforcement Officials 1990 states:

> 9. Law enforcement officials shall not use firearms against persons except in self-defence or defence of others against the imminent threat of death or serious injury, to prevent the perpetration of a particularly serious crime involving grave threat to life, to arrest a person presenting such a danger and resisting their authority, or to prevent his or her escape, and only when less extreme means are insufficient to achieve these objectives. In any event, intentional lethal use of firearms may only be made when strictly unavoidable in order to protect life.[47]

It is to the use of lethal force delivered by drones outside of armed conflict, and whether the above-stated standard is applicable, that this chapter now turns by focusing on one particular strike by a drone operated by the United Kingdom that resulted in the death of Reyaad Khan in Syria on 21 August 2015.

A. The Killing of Reyaad Khan

In the House of Commons, Prime Minister David Cameron justified this action as one of self-defence. He began by contextualising the United Kingdom's action:

> Turning to our national security, I would like to update the House on action taken this summer to protect our country from a terrorist attack. With the rise of ISIL, we know terrorist threats to our country are growing. In 2014, there were 15 ISIL-related attacks around the world. This year, there have already been 150 such attacks, including the appalling tragedies in Tunisia in which 31 Britons lost their lives. I can tell the House that our police and security services have stopped at least six different attempts to attack the UK in the past 12 months alone.[48]

The Prime Minister provided no more detail on these alleged plots—not even an indication of the stage at which the plots were stopped or the

---

[46] Convention for the Protection of Human Rights and Fundamental Freedoms (European Convention on Human Rights, as amended) (adopted 4 November 1950, entered into force 3 September 1953) ETS 5 (ECHR) Article 2.

[47] 'Basic Principles on the Use of Force and Firearms by Law Enforcement Officials' (Adopted by the Eighth United Nations Congress on the Prevention of Crime and the Treatment of Offenders, Havana, Cuba, 27 August to 7 September 1990) para 9, accessed 3 May 2017 at www.ohchr.org/EN/ProfessionalInterest/Pages/UseOfForceAndFirearms.aspx.

[48] HC Deb (n 5) col 25.

nature of the attempted attacks. There was also no indication as to whether or not the plots were such that those involved faced (successful) criminal prosecution under counter-terrorism legislation, or whether the security services relied on more conventional criminal law provisions. The lack of detailed information is interesting as it allows the Prime Minister to build an image of a country under siege from terrorist attacks, without being restricted by or bogged-down in the details. He continued:

> The threat picture facing Britain in terms of Islamist extremist violence is more acute today than ever before. In stepping up our response to meet this threat, we have developed a comprehensive counter-terrorism strategy that seeks to prevent and disrupt plots against this country *at every stage*. It includes new powers to stop suspects travelling. It includes powers to enable our police and security services to apply for stronger locational constraints on those in the UK who pose a risk. It addresses the root cause of the threat—the poisonous ideology of Islamist extremism—by taking on all forms of extremism, not just violent extremism.[49]

Here Mr Cameron continues in the construction of a United Kingdom facing an unprecedented threat (it has never before been as acute), which requires an even stronger response, by explaining a counter-terrorist strategy that ultimately requires the extinction of threats at their source. The vagueness of his language here is concerning, especially when viewed in the light of the government's over-broad definition of extremism in its Counter-Extremism Strategy,[50] which is potentially capable of covering any form of opposition to the current status quo. The definition has simultaneously shifted and narrowed from its literal meaning of driving something to the limit, to the extreme (or edge)[51] to a term that indicates political and/or religious views that lie outside of the (acceptable) mainstream attitudes of a given society. Given that 'Extremism is a

---

[49] Ibid, emphasis added.

[50] Home Office, *Counter-Extremism Strategy* (Cm 9148, Counter-Extremism Directorate: UK Home Office 2015) 9: 'Extremism is the vocal or active opposition to our fundamental values, including democracy, the rule of law, individual liberty and the mutual respect and tolerance of different faiths and beliefs.'

[51] Random House Kernerman Webster's College Dictionary (*KDictionaries*), accessed 4 May 2017 at www.kdictionaries-online.com/DictionaryPage.aspx?ApplicationCode=18#&&DictionaryEntry=extremism&SearchMode=Entry; Merriam-Webster Dictionary (*Merriam-Webster.com*), accessed 3 May 2017 at www.merriam-webster.com/dictionary/extremism.

relational concept'[52] and '... the labelling of activities, people, and groups as "extremist", and the defining of what is "ordinary" in any setting is always a subjective and political matter'.[53] For something to be considered 'extreme', there must be a mainstream against which it can be measured—extremism is now used to label views and opinions as 'bad' in an apparent attempt to create an objective standard to which all must adhere. The Prime Minister continued by outlining the UK government's response to the threat:

> We have pursued Islamist terrorists through the courts and the criminal justice system. Since 2010, more than 800 people have been arrested and 140 successfully prosecuted. Our approach includes acting overseas to tackle the threat at source, with British aircraft delivering nearly 300 air strikes over Iraq. Our airborne intelligence and surveillance assets have assisted our coalition partners with their operations over Syria. As part of this counter-terrorism strategy, as I have said before, if there is a direct threat to the British people and we are able to stop it by taking immediate action, then, as Prime Minister, I will always be prepared to take that action. That is the case whether the threat is emanating from Libya, from Syria or from anywhere else.[54]

The Prime Minister then turned to the targeted drone strike in question, explaining it as a precise use of lethal force taken under the UK's inherent right to self-defence in order to eliminate the threat caused by the terrorist activities of the targeted individual:

> In recent weeks it has been reported that two ISIL fighters of British nationality, who had been plotting attacks against the UK and other countries, have been killed in air strikes. Both Junaid Hussain and Reyaad Khan were British nationals based in Syria and were involved in actively recruiting ISIL sympathisers and seeking to orchestrate specific and barbaric attacks against the west, including directing a number of planned terrorist attacks right here in Britain, such as plots to attack high-profile public commemorations, including those taking place this summer.[55]

Interestingly, the Prime Minister does not explain why the drone strike was still considered to be necessary in August when the commemorative

---

[52] M Malik, 'Engaging with Extremists' (2008) 22 *International Relations* 85, 88.

[53] P T Coleman and A Bartoli, 'Addressing Extremism' (*International Center for Cooperation and Conflict Resolution*, nd) 2, accessed 3 May 2017 at www.fpamed.com/wp-content/uploads/2015/12/WhitePaper-on-Extremism.pdf.

[54] HC Deb (n 5) col 25.

[55] HC Deb (n 5) col 25.

events had already occurred without incident.[56] Lethal force used too late is an illegal reprisal or punishment, not a form of self-defence in the face of an imminent attack. Action taken too early because the pattern of behaviour suggested further attacks were being planned is illegal force taken pre-emptively in anticipation of a future attack.[57] The Prime Minister's justification comes close to an explicit recognition that this was a form of capital punishment:

> We should be under no illusion; their intention was the murder of British citizens, so on this occasion we ourselves took action.[58]

Here Prime Minister Cameron asserts knowledge of Reyaad Khan's intention regarding British citizens. It is not clear how he came to be able to state this so emphatically and with such certainty, especially given that Reyaad Khan was not interviewed by any security agency or asked about his intentions directly. This statement is more a rhetorical device encouraging the listener to accept Khan as a particularly dangerous threat that necessitated a response from the UK government than a statement of fact.

The justifications for lethal uses of force in self-defence often involve mentioning the illegitimacy of the victim due to their perceived wrongdoing.[59] This approach links the act of self-defence with the concept of punishment and promotes the concept that some states are in a position to

---

[56] 'It is understood that the two events were the VE Day commemorations, presided over by the Queen at Westminster Abbey on *10 May*, and a ceremony to mark the murder of Lee Rigby in Woolwich on Armed Forces Day on *27 June*' (emphasis added), N Watt, P Wintour and V Dodd, 'David Cameron Faces Scrutiny Over Drone Strikes Against Britons in Syria' *The Guardian* (8 September 2015), accessed 3 May 2017 at www.theguardian.com/world/2015/sep/07/david-cameron-justifies-drone-strikes-in-syria-against-britons-fighting-for-isis.

[57] G Fletcher, *Basic Concepts in Criminal Law* (Oxford University Press 1998) 133.

[58] HC Deb (n 5) col 25.

[59] See e.g. Israeli Spokesperson Mr Gillerman speaking at the UN Security Council 2004 debates on Israel's targeted killings on the death of Sheikh Yassin at the hands of Israeli security forces, 'He was an arch-terrorist with international aims and international ties … This is the man whom the Council is asked to defend.' (23 March 2004) UN Doc S/PV/4929; Michael Fallon in evidence before the Joint Committee on Human Rights stating that, 'These were people fighting for ISIL. They were not innocent civilians' when discussing the deaths of those with Khan, Fallon (n 5) 15; Liam Fox MP stating that those killed in the United Kingdom's drone strike 'were part of a barbaric organisation involved in systematic gang rape, torture and beheadings' in his article defending the Government's action and describing scrutiny of the legality of the strike as

discipline individuals in other states. However, the 'purpose of a defensive act is not to inflict harm according to the desert of the aggressor; its purpose is to repeal the attack'.[60] Thus, self-defence needs to be just that – the *defence* of the self.

B. Killing an Individual as an Act of Self-Defence of the United Kingdom

The Prime Minister relayed to the House the circumstances of Reyaad Khan's death:

> Today, I can inform the House that in an act of self-defence and after meticulous planning, Reyaad Khan was killed in a precision airstrike carried out on 21 August by an RAF remotely piloted aircraft while he was travelling in a vehicle in the area of Raqqa in Syria. In addition to Reyaad Khan, who was the target of the strike, two ISIL associates were also killed, one of whom, Ruhul Amin, has been identified as a UK national. They were ISIL fighters, and I can confirm that there were no civilian casualties.[61]

In this statement Mr Cameron unequivocally asserts that the United Kingdom was acting in self-defence. For a state to be able to assert its inherent right to self-defence under the jus ad bellum against Reyaad Khan, there needs to be an 'armed attack'.[62] In the *Nicaragua* case of 1986 the International Court of Justice stated that not every use of force would amount to an armed attack justifying the use of serious retaliatory force.[63] A substantial imminent attack by *ISIL* may well meet the threshold of 'scale and effects' specified by the Court.[64] However, the Prime Minister is asserting the right to self-defence against an *individual*. Given that the only information regarding potential targets was in relation to events that had already passed at the time of the airstrike and, in the absence of any details regarding what exactly Khan was planning, it is

---

spurious. L Fox, 'Drone Strikes in Syria are not just Legally Justified' *Independent* (9 September 2015), accessed 3 May 2017 at www.independent.co.uk/voices/comment/drone-strikes-in-syria-are-not-just-legally-but-morally-justified-10493619.html.

[60] Fletcher and Ohlin (n 9) 57.

[61] HC Deb (n 5) col 25.

[62] UNC Article 51.

[63] *Military and Paramilitary Activities in and against Nicaragua (Nicaragua v United States of America)* (Judgment) [1986] ICJ Rep 14, para 195.

[64] Ibid.

hard to see how an individual would be able to remotely launch an armed attack in a jus ad bellum sense.[65]

The point is that self-defence, by itself, is not a sufficient justification, especially when considering a use of force by a state against a specific individual. While not confining self-defence at the international jus ad bellum level to force used against an attacking state, the attacker must still represent an imminent threat to the *state*, as per the scale and effects test in the *Nicaragua* case. The attacks of 9/11 conducted by a non-state actor, or imminent attacks of a similar nature, may cross the threshold of such an attack justifying self-defence under Article 51 of the Charter, but care must be taken that the imminent threat posed by one individual is of sufficient gravity to trigger the Article 51 right, and if not, the standard must be the more precise one as to when life can be taken under human rights law, where lethal force can only be used when absolutely necessary to defend oneself or others under imminent threat, unless the state using force is engaged in an armed conflict under which there are more generous rules on the use of lethal force.[66] The United Kingdom was not engaged in an armed conflict in Syria at the time of Reyaad Khan's killing in August 2015 and so the law of armed conflict was not applicable. It was not until December 2015 that the House of Commons approved airstrikes in Syria.[67]

C. Self-Defence as a Legal Principle

That states have the right to self-defence is undisputed; what is potentially problematic here is the proposition that states have the right to self-defence under the jus ad bellum against an individual person outside of their territory. The importance to a government of being able to bring a use of force under the self-defence banner is that it renders a potentially wrongful act lawful and legitimate. Article 21 of the International Law Commission's Articles on State Responsibility 2001 provides that: 'The

---

[65] Of course, if Khan was in possession of a WMD then he may have been able to commit an act of sufficient magnitude to pass the threshold of armed attack. However, there was no suggestion from the UK Government that this was the case.

[66] UK Ministry of Defence, *The Manual on the Law of Armed Conflict* (Oxford University Press 2004) 54–7.

[67] The UK House of Commons resolved that it 'supports Her Majesty's Government in taking military action, specifically airstrikes, exclusively against ISIL in Syria'. It also noted the 'clear legal basis to defend the UK and our allies in accordance with the UN Charter' HC Deb (n 24) col 499.

wrongfulness of an act of a state is precluded if the act constitutes a lawful measure of self-defence taken in conformity with the Charter of the United Nations'. The government seems to be deliberately conflating the right of individual self-defence that a state has under the UN Charter and customary law, with the right that individuals have to defend themselves when attacked. Essentially these are separate rights exercised within different legal orders (international law and domestic criminal law), although international human rights law also recognises the right of individuals to self-defence in delineating the right to life. The human rights standard is applicable to state agents when they use force to defend themselves or others, or to prevent a serious crime from being committed.

(i) In criminal law

In UK domestic law, self-defence is the use of reasonable force in defence of the self, another person, or property. Imminence of threat is a necessary element.[68] UK domestic criminal law generally does not apply extraterritorially, although for certain serious crimes it does—those offences include murder and manslaughter.[69]

(ii) In human rights law

The United Kingdom did not have control or authority over the area in which Khan resided. According to the European Court of Human Rights, for human rights to have extra-territorial application, the state in question must be exercising control and authority over the relevant individual and/or the territory in which they are in.[70] As the United Kingdom was not involved in military action in Syria at that time, this was prima facie not the case and so the European Convention on Human Rights (ECHR) would not appear to apply. If the ECHR were held to apply, on the basis either that the use of force against an individual should be construed as an assertion of jurisdiction by the United Kingdom, and/or that the drone operator situated in the United Kingdom should trigger jurisdiction as he or she is clearly under the control (and is within the territory) of the United Kingdom, Article 2(2) ECHR permits killing in defence of others where absolutely necessary to protect life, which is one way of appraising the UK government's claim to self-defence in this case, although it

---

[68] Above n 2.

[69] D J Harris and S Sivakumaran, *Cases and Materials on International Law* (8th edn, Sweet & Maxwell 2015) 226.

[70] *Al Skeini and Others v United Kingdom* App no 55721/07 (ECtHR, 7 July 2011) paras 130–142.

presented no evidence that the lives of specific individuals in the United Kingdom or elsewhere were under imminent threat of existential attack.

(iii) In international law

States have a right to self-defence that pre-dates the UN Charter. In the *Nicaragua* case the International Court established that the right to self-defence was an 'inherent' right under customary international law that was 'confirmed and influenced by' the UN Charter.[71] The articulation of self-defence as an international legal principle developed through the *Caroline* case in the 19th century. In that case, 'self-defence was changed from a political excuse to a legal doctrine'.[72]

The *Caroline* case was later referred to in the Nuremberg Trials where it was stated that, 'preventative action in foreign territory is justified only in case of "an instant and overwhelming necessity for self-defense, leaving no choice of means, and no moment of deliberation"'.[73] The trial also made it clear, when discussing Germany's invasion of Norway, that the claim that an act was one of self-defence is subject to objective scrutiny.[74] Thus, a state's judgement of its own actions is not final.[75] It is therefore appropriate that the actions of any state claiming self-defence as a justification for a use of force be placed under proper scrutiny, especially in the current climate of broadening legal claims to a right of self-defence.

During correspondence between the British Foreign Office and US Secretary of State discussing the *Caroline* incident the British laid out three arguments to justify its destruction of the steamer *Caroline*, the latter of these being self-defence. It was clear that this was not regarded as 'strictly legal',[76] that the only criterion was sufficient provocation and that the British Law Officers regarded the action as being fully justified under 'self-defence and self-preservation'[77] as it was performed with the

---

[71] *Nicaragua* (n 63) 94.

[72] R Y Jennings, 'The Caroline and McLeod Cases' (1938) 32 *AJIL* 82.

[73] Nuremberg Trial (1947) 41 *AJIL* 172, 205.

[74] Ibid.

[75] C W Jenks, *A New World of Law? A Study of the Creative Imagination in International Law* (Longmans 1969) 203.

[76] Letter from British Foreign Office to US Secretary of State, 13 January 1838 (Record Office FO5 322) cited in Jennings (n 72) fn 13.

[77] There is a significant difference between these two concepts that needs to be examined: self-defence is the defence of the self against an imminent threat or attack; self-preservation is to secure the preservation of the self—in this situation, the self is the state. Thus, a state may justify the use of force against any enemy that seeks to destroy its very existence. In the case of terrorism, this

aim of guarding against future hostile activities.[78] The British had launched a surprise midnight attack on the vessel and in his correspondence US Secretary of State Webster stated that the British carried the burden of demonstrating that threats or a warning would have been 'impracticable, or … unavailing,' that 'day-light could not be waited for', and that seizing and detaining the vessel 'would not have been enough'.[79] The US Secretary of State also provided the fundamentals of self-defence, which were accepted by the British Foreign Office and are now acknowledged as being part of customary international law. It is for the state claiming self-defence to demonstrate that there existed a 'necessity of self-defence, instant, overwhelming, leaving no choice of means, and no moment for deliberation'. Furthermore, the state must also evidence that it 'did nothing unreasonable or excessive; since the act, justified by the necessity of self-defence, must be limited by that necessity, and kept clearly within it'.[80] Thus, the essential elements can be summarised as: (1) necessity (or imminence); and (2) proportionality.

(1) *Necessity*: The right of self-defence 'can be invoked only against a danger which is serious and actual or imminent'.[81] Necessity is a fundamental principle of the doctrine of self-defence in both domestic and international law. The *Caroline* case made it clear that the situation must be such that no practical alternative exists. The International Court of Justice held that necessity was a criterion for self-defence in the *Nicaragua* case,[82] and also in its *Nuclear Weapons* Advisory Opinion.[83] Additionally, the Chatham House Principles state that:

---

is precisely what is asserted. Prime Minister David Cameron asserts that ISIL/Daesh pose a very real and existential threat to the United Kingdom as they seek to 'destroy our way of life' HC Deb (n 24). J L Brierly describes self-preservation as being 'an instinct' rather than a 'legal right', *The Law of Nations* (2nd edn, Clarendon Press 1936), which means that it would be limited by law (not all acts of self-preservation are permissible) and it is now self-defence that is referred to.

[78] Jennings (n 72) 87.

[79] Jennings (n 72) 89.

[80] Webster's letter to Lord Ashburton, Parliamentary Papers (1843) Vol LXI; British and Foreign State Papers, Vol 30, 193 cited in Jennings (n 72).

[81] Jenks (n 75).

[82] *Nicaragua* (n 63) para 176.

[83] *Legality of the Threat or Use of Nuclear Weapons* (Advisory Opinion) [1996] ICJ Rep 226, para 41.

> … there should be no practical non-military alternative to the proposed course of action that would be likely to be effective in averting the threat or bringing an end to an attack.[84]

In the strike against Reyaad Khan, it was revealed that the 'preparations took place over a period of months after the intelligence agencies briefed ministers',[85] strongly indicating that this was a premeditated killing rather than an act forced upon the government as the only means of preventing an imminent attack.

(2) *Proportionality*: According to a classic statement of self-defence by de Vitoria writing in the 16th century: 'In war everything is lawful which the defence of the common weal requires … the end and aim of war is the defence and preservation of the State'[86] as without the existence of the state, all international law and the international system would be redundant. However, in the modern context the 'defensive measure must be limited to what is necessary to avert the attack or bring it to an end',[87] and the measures taken must be reasonably limited to the necessity of protection and proportionate to the danger.[88] From the Prime Minister's speech, it appears that although Khan was allegedly planning the attacks from his safe-haven in Syria, the actual attacks would have emanated from *within* the United Kingdom. Thus, it would seem that the UK security forces would have had the opportunity to intervene within the United Kingdom in order to prevent the attacks. Additionally, depending upon the stage of the plans, it may be that the attacks could still occur, notwithstanding the death of Khan—especially if those working in concert with Khan remain at large. The Prime Minister gave no details regarding arrests of those with whom Khan was working and upon whom he was relying for the implementation of his plans.

Applying these principles of necessity and proportionality to the present case: the killing of Reyaad Khan was 'meticulously planned' over a

---

84 'The Chatham House Principles of International Law on the Use of Force in Self-Defence' (2006) 55 ICLQ 963, 967.

85 Watt, Wintour and Dodd (n 56).

86 Francisco de Vittoria quoted in J Brown Scott, *The Spanish Origin of International Law: Francisco de Vittoria and his Law of Nations* (The Lawbook Exchange 2000) 430.

87 Chatham House Principles (n 84) 967.

88 Jenks (n 75) 29.

course of months, suggesting that the drone strike was not a reaction to an imminent threat. Furthermore, the events identified as potential targets had passed. Khan was not going to perform the attack himself and so his co-conspirators may have retained the capacity to continue with the plot despite his death. Thus, it is asserted that whatever nefarious plots Reyaad Khan may or may not have been involved in, they do not appear to have been such as to leave 'no moment for deliberation' and that the UK security services would have had the potential capability to prevent the attacks from occurring without the death of Khan.

(iv) Pre-emptive strikes

The concept of anticipatory self-defence is not uncontroversial and is not accepted by the majority of states.[89] It seems that it is generally prohibited in law, although there may be circumstances in which anticipatory self-defence is morally or politically justified.[90] However, pre-emptively repelling an identified and imminent attack (however that is defined and recognising that the 'battleground for this debate is the correct definition of imminence'[91]) is not the same as launching a preventative war, which would, in all likelihood, lack the necessary element of imminence as it would be aiming to prevent at attack from manifesting at all—unless the notion of imminence is stretched beyond common sense limits. Where the line is drawn is crucial in determining whether or not a defensive act by a state was justified. In his evidence to the United Kingdom's Joint Committee on Human Rights, Michael Fallon MP made it clear that he did not consider the *Caroline* definition to be current and that it was not 'possible to have a hard and fast rule about how you would define "imminent"'.[92] This is a problematic proposition as a flexible approach to imminence would allow states to justify almost any use of force against a perceived enemy by claiming that an attack was imminent, notwithstanding the seeming reassurance of Mr Fallon's assertion that he has 'to be absolutely satisfied that there is simply no other way of preventing an attack that is imminent'.[93]

According to the Chatham House principles:

---

[89] C Gray, *International Law and the Use of Force* (Oxford University Press 2008) 112–18.

[90] A Cassese, *International Law in a Divided World* (2nd edn, Oxford University Press 2005) 362.

[91] Fletcher and Ohlin (n 9) 156.

[92] Fallon (n 5) 9.

[93] Ibid 4.

> The requirements set out in the *Caroline* case must be met in relation to a threatened attack. A threatened attack must be 'imminent' and this requirement rules out any claim to use force to *prevent* a threat emerging. Force may be used in self-defence only when it is necessary to do so, and the force used must be proportionate.[94]

The UN's High Level Panel of 2004 opined that 'a threatened state, according to long established international law, can take military action as long as the threatened attack is *imminent*, no other means would deflect it and the action is proportionate'.[95] The UN Secretary General has also declared that imminent threats are also covered by the right to self-defence.[96] The United Kingdom holds the view that states have the right to pre-emptively strike against an imminent attack.[97]

Thus, it seems that a pre-emptive strike can be acceptable when the defender perceives an attack is about to occur. It would be to go against natural instinct and would be 'unrealistic in practice to suppose that self-defence must in all cases await an actual attack'.[98] When considering violence between individuals, the legitimacy of pre-emption is understandable as an individual faces possible extinction from an attack, especially one involving a weapon. A state does not face the same degree of threat in that it is hard to see how one individual can sufficiently threaten a state with extinction. However, to allow terrorist threats to materialise would potentially undermine the social contract between the state and its citizens as the monopoly on force the state enjoys is quid pro quo for the security each citizen enjoys. This would only occur, however, if the level of terrorist force was allowed to be such as to have the 'scale and effects' spoken about in the *Nicaragua* case.[99] Below that threshold, terrorist violence should be dealt with in the same manner as the serious threat of violent crime arising within the United Kingdom from organised

---

94 Chatham House Principles (n 84) 965.

95 UNGA 'Report of the UN High Panel on Threats, Challenges and Change' (2004) UN Doc A/59/565, para 188.

96 UNSG 'In Larger Freedom' UN Doc A/59/2005, para 124.

97 See e.g. the British proposition in the *Caroline* case, Jennings (n 72); in answer to a question in the House of Commons on the legitimacy of pre-emptive armed attack, the Attorney-General replied that '… it has been the consistent position of successive United Kingdom Governments over many years that the right of self-defence under international law includes the right to use force where an armed attack is imminent'. United Kingdom Materials of International Law (2004) *BYIL* 595, 822–3.

98 Chatham House Principles (n 84) 964.

99 *Nicaragua* (n 63).

crime, drug-related crime, vigilantism, and other similar challenges to the state monopoly on force.

D. An Additional Criterion

In continuing his speech to the House, the Prime Minister appeared to add another criterion to the assessment of when the use of force in self-defence is legitimate:

> We took this action because there was no alternative. In this area, there is no Government we can work with; we have no military on the ground to detain those preparing plots; and there was nothing to suggest that Reyaad Khan would ever leave Syria or desist from his desire to murder us at home, so we had no way of preventing his planned attacks on our country without taking direct action.[100]

The United Kingdom appears to be adding an additional criterion of 'unable or unwilling', on the part of the host state where the terrorists are found, to the assessment of when the use of self-defence is legitimate. This is a problematic additional criterion as it is not clear how that is to be assessed and who is qualified to make that determination.[101] It is not clear where this criterion has come from in relation to self-defence and it appears to be a mixing of different legal principles (principally borrowing from international criminal law). Although it has gained traction in the literature,[102] it has not been expressly accepted by states.[103]

E. The United Kingdom's Legal Basis for the Strike

The Prime Minister continued by outlining the legal basis for the strike:

> With these issues of national security and with current prosecutions ongoing, the House will appreciate that there are limits on the details I can provide. However, let me set out for the House the legal basis for the action we took,

---

[100] HC Deb (n 5) col 5.

[101] C Gray (n 27), 11.

[102] 'Where a State is unable or unwilling to assert control over a terrorist organization located in its territory, the State which is a victim of the terrorist attacks would, as a last resort, be permitted to act in self-defence against the terrorist organization in the State in which it is located' (fn omitted), Chatham House Principles (n 84) 970.

[103] Gray (n 27) 11.

> the processes we followed and the implications of this action for our wider strategy in countering the threat from ISIL. First, I am clear that the action we took was entirely lawful. The Attorney General was consulted and was clear that there would be a clear legal basis for action in international law. We were exercising the UK's inherent right to self-defence. There was clear evidence of these individuals planning and directing armed attacks against the UK. These were part of a series of actual and foiled attempts to attack the UK and our allies, and given the prevailing circumstances in Syria, the airstrike was the only feasible means of effectively disrupting the attacks that had been planned and directed. It was therefore necessary and proportionate for the individual self-defence of the United Kingdom. The United Nations Charter requires members to inform the President of the Security Council of activity conducted in self-defence, and today the UK permanent representative will write to the President to do just that.[104]

The Prime Minister went on to explain that this was not action undertaken as part of an armed conflict in Syria in which the United Kingdom was involved, something that did not happen until December 2015 when the government won a vote in the House of Commons in favour of UK airstrikes in Syria.[105] Despite this, however, the Prime Minister did allude to principles of humanitarian law—minimising civilian casualties, proportionality and military necessity. These are referenced as also framing the conduct of the operation, but the overriding claim was that the government had no other choice but to use lethal force in defence of the United Kingdom.

> Our intelligence agencies identified the direct threat to the UK from this individual and informed me and other senior Ministers of that threat. At a meeting of the most senior members of the National Security Council, we agreed that should the right opportunity arise, military action should be taken. The Attorney General attended the meeting and confirmed that there was a legal basis for action. On that basis, the Defence Secretary authorised the operation. The strike was conducted according to specific military rules of engagement, which always comply with international law and the principles of proportionality and military necessity. The military assessed the target location and chose the optimum time to minimise the risk of civilian casualties. This was a very sensitive operation to prevent a very real threat to our country, and I have come to the House today to explain in detail what has happened and to answer questions about it.

---

[104] UNSC 'Letter dated 7 September 2015 from the Permanent Representative of the United Kingdom of Great Britain and Northern Ireland to the United Nations addressed to the President of the Security Council' (8 September 2015) UN Doc S/2015/688.

[105] Above n 24.

> I want to be clear that the strike was not part of coalition military action against ISIL in Syria; it was a targeted strike to deal with a clear, credible and specific terrorist threat to our country at home. The position with regard to the wider conflict with ISIL in Syria has not changed. As the House knows, I believe there is a strong case for the UK taking part in airstrikes as part of the international coalition to target ISIL in Syria, as well as Iraq, and I believe that that case only grows stronger with the growing number of terrorist plots being directed or inspired by ISIL's core leadership in Raqqa. However, I have been absolutely clear that the Government will return to the House for a separate vote if we propose to join coalition strikes in Syria.
>
> My first duty as Prime Minister is to keep the British people safe. That is what I will always do. There was a terrorist directing murder on our streets and no other means to stop him. The Government do not for one minute take these decisions lightly, but I am not prepared to stand here in the aftermath of a terrorist attack on our streets and have to explain to the House why I did not take the chance to prevent it when I could have done. That is why I believe our approach is right. I commend this statement to the House.[106]

The government has declined to publish the Attorney-General's advice and it is not exactly clear whether or not the advice was specific to the killing of Khan or was more in principle.[107]

In response to a question from Harriet Harman MP, the Prime Minister stated:

> She asked: is this the first time in modern times that a British asset has been used to conduct a strike in a country where we are not involved in a war? The answer to that is yes. Of course, Britain has used remotely piloted aircraft in Iraq and Afghanistan, but this is a new departure, and that is why I thought it was important to come to the House and explain why I think it is necessary and justified … If it is necessary to safeguard the United Kingdom and to act in self-defence, and there are no other ways of doing that, then yes, I would [do it again].[108]

Debates about the extra-territorial application of human rights can tend to obscure the central problem with drone strikes taken outside of armed conflict, namely the claim that they are justified actions of self-defence of a state under the international law governing the use of force by states. Rather than this jus ad bellum standard, cases of threats posed by

---

[106] HC Deb (n 5) cols 26–7.

[107] O Bowcott, 'Syria Drone Strikes: UK Attorney General Refuses to Disclose Advice' *The Guardian* (15 September 2016), accessed 3 May 2017 at www.theguardian.com/politics/2015/sep/15/syria-drone-strikes-uk-attorney-general-refuses-to-disclose-advice.

[108] HC Deb (n 5) col 30.

individual terrorists should be assessed at the level of the individual right to life, at least in principle, where lethal force is only permitted for state agents acting in self-defence when absolutely necessary to protect the lives of those using force or other individuals under attack or in danger of imminent attack. It is clear that the UK government is deliberately using a self-defence standard that does not recognise this divide on the basis that the average citizen will not recognise 'fine' legal distinctions. There is also a deliberate attempt by the UK government to mould the concept of self-defence to capture any state use of force against state, non-state actor or individual, who is presented as a threat to the nation, understood not only as the state but also every UK national. This is not justifiable under the concept of self-defence as 'military action, even in national-defense, remains morally problematic in a profound and troubling way'.[109] What may be used to justify legitimate self-defence by an individual person, 'fails to do so for national defense'.[110] The use of the self-defence argument is predicated on the notion that it is possible to determine which actor has behaved illegitimately and in a manner that makes the use of force against them justifiable.[111]

However, the interpretation of self-defence at both national levels and international levels is much disputed, creating the conditions for drone-using states to exploit and operate what appears to be a new form of remote self-defence in the shape of drone strikes undertaken by operators thousands of miles away from the targets. Even within federal states, there are different understandings of what is allowed in self-defence.[112] There are differences in criminal law standard between states, while at the international level self-defence means different things in different contexts. For example, in the case of peacekeeping, self-defence has been extended from a narrow conception of a peacekeeper defending himself and his colleagues from attack, to include defending third parties under imminent threat, to extend to defending the mandate and the peace

---

[109] D Rodin, *War and Self-Defense* (Oxford University Press 2002) 199.

[110] Ibid 198.

[111] Ibid 107 and 193.

[112] For an overview of the differences in self-defence laws within the United States, see e.g. C Currier, 'The 24 States that have Sweeping Self-Defence Laws Just Like Florida's' (*ProPublica*, 22 March 2012), accessed 3 May 2017 at www.propublica.org/article/the-23-states-that-have-sweeping-self-defense-laws-just-like-floridas.

process.[113] In peacekeeping, self-defence straddles a national criminal law/human rights standard and what could be called a jus ad bellum standard of self-defence, when a state is defending itself from attack. Normally a peacekeeper is concerned with defending himself from attack and increasingly defending civilians under imminent existential threat, but occasionally, especially under mandates given to forces in the 21st century, he or she is concerned with defending the state in which they have been deployed. However, that expanded concept of self-defence of the state can only be justified under a Chapter VII mandate given to the peacekeeping force by the Security Council under which certain 'necessary measures' have been authorised. In any case, a peacekeeper acting in defence of a civilian under attack is to be judged by narrower and specific standards of imminence, proportionality and necessity, than when he or she is acting to defend the state from non-state actors who threaten the peace. Similarly, when a state authorises a drone operator to use lethal force against an individual target in defence of potential victims, as the United Kingdom did against Reyaad Khan, it should be judged by the narrower and specific standards of a state agent coming to the defence of threatened citizens in the United Kingdom, which focuses on when an individual's life can be taken by a state that owes him a duty not to violate his right to life, unless a 'use of force is no more than is absolutely necessary ... in defence of any person from unlawful violence',[114] and not by standards applicable to self-defence of a state under Article 51 of the UN Charter and under customary international law.

Doubts about sufficiency of the claim to self-defence made by the Prime Minister in relation to the lethal strike against Reyaad Khan may explain why, in the letter to the Security Council explaining the defensive nature of the action, referred to by the Prime Minister, the British ambassador widened the threat to include ISIL:

> On 21 August 2015 armed forces of the United Kingdom of Great Britain and Northern Ireland carried out a precision airstrike against an ISIL vehicle in which a target known to be actively engaged in planning and directing imminent armed attacks against the United Kingdom was travelling. This airstrike was a necessary and proportionate exercise of the individual right of self-defence of the United Kingdom.

---

[113] N Tsagourias, 'Consent, Neutrality/Impartiality and the Use of Force in Peacekeeping Operations: Their Constitutional Dimension' (2006) 11 *JCSL* 465, 473.

[114] ECHR Article 2(2)(a).

> As reported in our letter of 25 November 2014, ISIL is engaged in an ongoing armed attack against Iraq, and therefore action against ISIL in Syria is lawful in the collective self-defence of Iraq.[115]

The justification becomes an even broader jus ad bellum one, namely that Reyaad Khan's killing was both an act of individual self-defence of the United Kingdom and an action in collective self-defence of Iraq, and, moreover, was a strike against ISIL, an armed group that is more likely to be able to mount attacks of the scale and effects required to trigger the United Kingdom's right of self-defence of the state under Article 51 of the Charter. This effectively became the position of the United Kingdom following the terrorist attacks on Paris in November 2015, when Parliament voted for airstrikes against ISIL, after gaining a resolution in the Security Council that lent support to states using force against that organisation.[116] It might be argued that this renders the debate about the killing of Reyaad Khan an academic one as ISIL and its members became a legitimate target after Paris, but that would leave the standalone killing of an individual by a drone strike as a precedent for future terrorist scares and threats that are not overtaken by events.

Further obfuscation of the legal basis of the strike is found in evidence given on 16 December 2015 to the Joint Committee on Human Rights Inquiry into UK government's policy on the use of drones for targeted killings, when the Secretary of State for Defence, Michael Fallon MP, stated that:

> I think that compliance with international humanitarian law discharges any obligation that we have under international human rights law, if I can put it that way. If any of those obligations might be thought to apply, they are discharged by our general conformity with international humanitarian law.[117]

Although by this time the United Kingdom's use of force in Syria had been approved by the House of Commons (on 3 December 2015), the discussion focused on the drone strike of August 2015 in which Reyaad Khan was killed. Furthermore, the above was an answer given to a specific question: 'The human rights law standard says that lethal force outside an armed conflict situation is justified only if it is absolutely necessary to protect life. Is that the standard?'[118] The most relevant

115 UN Doc (n 104).
116 UNSC Res 2249 (20 November 2015) UN Doc S/RES/2249, op para 5.
117 Fallon (n 5) 4.
118 Ibid.

question was not answered; instead the government fell back on arguments of humanitarian law that only apply in an armed conflict to which the United Kingdom is a party. The introduction of *in bello* standards to displace the stricter human rights ones is yet another example of the government playing fast and loose with international law.

5. CONCLUSION

There will be drone strikes in the future where there is no link to an armed conflict, where the standard against which the action should be measured is one of self-defence and then care must be taken to assess whether the claim can be founded under the UN Charter, essentially as a defence of state, or under human rights law, as a defence of individuals. Self-defence at both levels shares common features, such as imminence, that drone strikes, like the one against Reyaad Khan, struggle to match, but killing individuals in defence of individuals is properly assessed at the more precise level of the right to life. The fact that human rights law has lagged behind the technology reality of drone strikes, by failing to recognise whether the use of lethal force against an individual by a state agent firing a weapon from a drone flying over another state is an assertion of jurisdiction over the target, should not detract from the conclusion that the targeted killing of an individual is an act that should be judged by human rights standards. There are two exceptions to this: first, if the targeted individual can be said to be part of an imminent armed attack on the United Kingdom that has such scale and effects that it triggers the right of self-defence of the state itself under Article 51 of the Charter, when proportionate force can be used to eliminate the attack or imminent threat of it; secondly, where the United Kingdom is engaged in an ongoing armed conflict against a non-state actor such as ISIL, and the individual is a legitimate target under the law of armed conflict (international humanitarian law). It is submitted that Reyaad Khan's killing did not fit either of the exceptions and, moreover, on the evidence presented by the Prime Minister, did not represent such an existential threat to individuals within the United Kingdom to justify the use of lethal force against him in violation of his right to life.

New technology, whereby an operator, sitting thousands of miles away, can in real time and with great precision kill an individual, means that individual drone strikes outside of an armed conflict challenge our conceptions of when force is legally justifiable. The surgical killing of an individual by a drone operator who is not under imminent threat of existential violence or physically close to others who are under such a threat, does not seem to fit our definition of self-defence as either captured in criminal law or human rights law, but it is argued that it should be these standards that are applicable. Evidence of an existential imminent threat to individuals in the United Kingdom or to its citizens abroad must be given for such strikes to be justified. Claiming the right to defend the state by targeted killings of individuals cannot be accepted per se without evidence that the individual was part of an imminent orchestrated attack that is of sufficient scale as to elevate the attack to one against the state and not only against individuals within it. To blithely accept that the United Kingdom has the right to defend itself against Reyaad Khan is to grossly exaggerate the threat one individual can pose, but it also represents a reversion to a very primitive view of the state whereby its promise to protect its citizens at all costs is used to circumvent the basic rights of individuals. Moreover, the portrayal of individuals like Khan as dangerous and evil serves the purpose of justifying their demise. Precise and clinical summary execution of individuals suspected of terrorist activities in a country far away from the United Kingdom is technologically possible but it clearly violates human rights *standards*. It is time, however, that human rights *laws* caught up with technology.

8. Drone warfare and the erosion of traditional limits on war powers

Geoffrey Corn[*]

I. INTRODUCTION

Drones. There are few words that symbolize more things to more people. For a military commander, it symbolizes precision lethality that can prove decisive against an enemy. For the enemy, it symbolizes a terrifying silent killer, necessitating constant caution to avoid detection and attack. For legal, national security, and social science scholars it symbolizes everything from the inherent illegitimacy of expansive notions of war and authority to kill, to the decisive tool for disrupting international terror organizations, to simply a tool of war, no different than any other weapon. For political leaders, it symbolizes flexibility and risk avoidance in the scheme of leveraging national power to destroy or disrupt national and international threats.

The debate about the legality and legitimacy of drone operations has raged since the United States began to conduct lethal drone operations as a staple of military and paramilitary operations. This debate has progressed along two primary vectors. First, whether use of lethal drone attacks outside 'hot' or 'active' areas of combat operations comply with international law. Second, whether employing deadly force as a measure of first resort violates international law. These two lines of inquiry and debate have, to a significant extent, conflated the nature of the weapon system with broader questions related to the controlling international legal framework for counter-terror operations, and the international legal authority to conduct military operations in the territory or airspace of a sovereign state absent that state's consent.

There are no easy answers to these questions, but one thing is clear: the ability to conduct highly precise lethal attacks with minimal risk to

* I am indebted to the assistance of my friend Brigadier General (Retired) Kenneth Watkin for his insights into these complex issues, and to the outstanding efforts of my research assistant, Andrew Culliver, JD candidate, South Texas College of Law Houston.

friendly forces has incentivized the use of drones, even amidst the fog of legal uncertainty. While this trend stresses traditional understandings of international humanitarian law and international human rights law, it also has a significant influence on the willingness of national leaders to employ military force as a tool of national security. Like all national decisions to use combat power to advance national security objectives, the decision to employ lethal drones must be founded upon assessments of international and domestic legal authority. These assessments must ensure compliance with both domestic and international law. While there will be times when both these legal regimes empower national leaders to unleash the tools of war on an enemy, in most situations it is quite the opposite, and international and domestic law actually constrain such actions.

One important question related to the increasing availability and efficacy of drone capability is whether it dilutes these traditional legal barriers or constrains them to the use of military force. This chapter will explore this question. Section II considers how, at least from a functional standpoint, drones offer national level decision-makers a combat capability that is really different from the other tools in the military force arsenal. Section III considers how this capability has influenced the assessment of when a threat triggers the law of armed conflict (LOAC), and more specifically the international legal authority to use lethal force as a measure of first resort against a threat. This section also explains why the impact of drones does not extend across the so-called spectrum of conflict, but instead is limited to the assessment of non-international armed conflict. Section IV then considers how drone capability impacts the assessment of constitutional war powers, specifically focused on the dilution of political risk associated with drone-dominated military action.

II. ARE DRONES DIFFERENT?

Drones, or remotely piloted vehicles armed with lethal combat power, are a relatively new capability. However, at the fundamental level, a drone is just another weapon system—a combination of capabilities that enables commanders to employ lethal combat power against an enemy. Indeed, proponents of drones frequently assert that vilifying drones distorts the legal and policy debate because drones are just weapons. Reality, however, probably lies between the two extreme ends of this argumentative spectrum.

Drones are highly effective weapon systems. They are lethal, precise, and situationally-aware. They afford significant stand-off capability while maintaining cost effectiveness, and are largely immune from enemy counter-measures. Unsurprisingly, they have become the weapon of choice for conducting precision strikes against individual enemy targets whose conduct complicates distinguishing them from the general civilian populations in which they operate.

Individually, none of these attributes are unique to drones. What is unique is the combination of these attributes in one weapon system. From the inception of armed drones, no other existing weapon system has offered national and operational-level leaders an analogous capability—the ability to seek out, identify, and engage a target with a high degree of precision, all while posing little to no risk to friendly forces. This capability has proven especially valuable in the post-11 September era, largely due to the nature of the non-state enemies that the United States has placed within its war-making crosshairs. Like any other conflict, synchronizing available resources to maximize the effects of combat power is essential to disrupt this enemy. However, in this type of ongoing, asymmetric conflict, the capability provided by drones has become highly coveted. Intelligence accuracy and precision engagement are essential when facing an enemy who makes no effort to distinguish himself from the civilian population, who exploits the presence of civilians to seek functional immunity from attack, and who exploits any civilian casualty for strategic information gain.

Of course, other tools in the combat arsenal offer some of the capabilities of drones. Aircraft, cruise missiles, and even platforms as basic as a sniper, all offer precision engagement through the employment of smart munitions. Of these examples, only the sniper offers anything close to the real-time surveillance capability of the drone, although how close is a matter of degree based on any specific tactical situation. In contrast, the drone can linger for extended periods of time over a suspected target, gathering highly precise information to support both target verification and identification of the ideal attack options and situations. Significantly, unlike the human operative, the drone can provide this package of capabilities with virtually no risk to friendly forces. Even in the rare situation where an enemy is armed with a counter-measure effective against the drone, the worst-case scenario is loss of the physical asset; the operator remains immune from the effects of any such attack.

Being safe, accurate, and precise, the drone obviously offers strategic and operational leaders tremendous advantages over the many other tools at their disposal. What makes this attack option even more appealing is the nature of the enemy's center of gravity in asymmetrical warfare: *command and control.* Unlike more conventional opponents, it is this ability to disrupt enemy leadership that proves so decisive in achieving the objective of disruption and dispersal. For a conventional opponent, this might involve using a range of capabilities to attack enemy command, control, and communication structures. But for the terrorist organization, it is the individual leaders who are the focal point of such attacks. Furthermore, because these leaders routinely co-mingle with the civilian population, the accuracy, precision, and lethality of drones are all the more decisive.

At the strategic level, drones offer one additional advantage: a minimal footprint. In the aftermath of the 11 September attacks, the United States adopted a clear position that the struggle against al Qaeda and associated groups qualified as an armed conflict, with an accordant assertion of authority to strike this enemy when he presented himself.[1] This led to invocation of what is commonly referred to as the 'unable or unwilling' test to justify projecting US military power into the sovereign territory of other states to conduct lethal attacks on high value enemy targets, even without the state's consent.[2] Because drones provide the capability to conduct attacks in such locations with minimal physical intrusion into the state's territory with virtually no risk of mission compromise or loss to US personnel, the drone option fits ideally within this legal paradigm.

All of these attributes and considerations point to two almost indisputable conclusions. First, like any other weapon system, drones are just one of the many tools within a mosaic of lethal and non-lethal options available for strategic and operational leaders to leverage in achieving a desired effect against an enemy. Second, the nature of this weapon system is uniquely suited for producing these effects. Therefore, it is no surprise that drones have become so central to both the conduct and criticism of the so-called US 'war on terror'. But there are other unique consequences of the rise of drone warfare, consequences that transcend military or operational considerations and almost certainly also explain why drones have become the symbolic focal point for debates over the

---

1 See Authorization for the Use of Military Force, Pub. L. No. 107–40, 115 Stat. 224 (2001); Int'l Comm. of the Red Cross, Commentary on Article 3: Conflicts Not of an International Character, of the Geneva Convention of 12 August 1949, para. 400 (2016), accessed 4 May 2017 at https://www.icrc.org/applic/ihl/ihl.nsf/Comment.xsp?action=openDocument&documentId=59F6CDFA490736C1C1257F7D004BA0EC#_Toc452041779 [hereinafter ICRC Art. 3 Commentary]. (In 2016, the ICRC released an updated commentary to common Article 3, which is used throughout this chapter.)

2 Geoffrey S Corn et al, *U.S. Military Operations: Law Policy and Practice* (Oxford University Press 2016) 108.

legitimacy of the US assertion of an armed conflict against transnational non-state enemies, such as al Qaeda. Most notable among these is the impact drones seem to have had on how law is perceived as a limitation or constraint on the use of military force to advance national security objectives.

III. DRONES AND THE ASSERTION OF ARMED CONFLICT

One of the most fundamental obligations of any state is to protect itself and its population from internal and external threats. When necessary, the state authorizes its agents to use lethal force to achieve this objective. When and under what conditions this authority is properly exercised is, however, dictated by law. International law establishes limitations on the state's power to use force in response to threats in both peacetime and during armed conflicts.[3] Peace is the normal condition of national and international affairs, and therefore it is the peacetime legal framework that should be applied as the 'default' rule.[4] That legal framework is provided by international human rights law (IHRL).[5]

IHRL protects individuals from the arbitrary deprivation of life at the hands of state agents.[6] Accordingly, such agents are legally permitted to use lethal force in response to a threat only where there exists actual

[3] Kenneth Watkin, 'Controlling the Use of Force: A Role for Human Rights Norms in Contemporary Armed Conflict' (2004) 98 Am J Intl L 1, 2; see generally Kenneth Watkin, 'The Humanitarian Law and Human Rights Interface' in *Fighting at the Legal Boundaries: Controlling the Use of Force in Contemporary Conflict* (Oxford University Press 2016) 121, 121–58.

[4] Watkin (n 3) 2; Gábor Kardos, 'The Relationship Between International Humanitarian Law and International Human Rights Law: A Legal Essay' (1993) 34 Annales U Sci Budapestinensis Rolando Eotvos 49, 50. Though, IHRL is not exclusively applicable to solely 'peacetime' situations. IHRL applies at all times; however, it may be derogable where permitted by treaty. Derogations must still remain proportional, and are still limited by IHL. Ibid; see also ICRC Advisory Service on Int'l Humanitarian Law, *Int'l Humanitarian Law and Int'l Human Rights Law: Similarities and Differences* (January 2003), available 4 May 2017 at https://www.icrc.org/en/download/file/1402/ihl-and-ihrl.pdf [hereinafter ICRC Advisory Service].

[5] ICRC Advisory Service (n 4).

[6] International & Operational Law Department, The Judge Advocate General's Legal Center & School, US Army, Judge Advocate 422, *Operational Law Handbook* 45 (2013) 45; see also ICRC Advisory Service (n 4).

necessity, and when resort to deadly force is a measure of last resort.[7] Use of military forces to achieve state security objectives does not automatically alter this fundamental IHRL legal equation, even when those forces operate outside national territory. Unless operating within the alternative international humanitarian law legal framework, military forces, like police forces, are subject to IHRL-based obligations and legal limitations on the use of lethal capabilities.[8]

International humanitarian law (IHL) fundamentally alters the use of force legal equation applicable to state agents. Unlike IHRL, IHL does not restrict the use of lethal force to a measure of last resort based on individualized assessments of necessity.[9] Instead, IHL permits use of lethal force as a measure of first resort based on status determinations.[10] Because armed conflict is defined fundamentally as a contest between organized belligerent groups, use of force authority is not triggered by individualized assessments of actual threat, but instead by the presumptive threat resulting from an assessment that an individual is a member of an enemy belligerent group. Once that status determination is made, state agents may employ lethal force as a measure of first resort, limited only by a conclusion that the enemy is rendered incapable of continued participation in hostilities (*hors de combat*) as the result of wounds, sickness or capture.[11]

The line between peacetime response to security threats and armed conflict is therefore profoundly significant. While both IHRL and IHL impose important limits on the state's authority to implement measures to incapacitate such threats, the existence of armed conflict substantially expands the scope of authority available to the state and its agents. Historically, that line was defined as the line between war and peace—the laws and customs of war applied only during war. However, until

---

7 *Operational Law Handbook* (n 6) 51. (While IHL may provide for expressed derogations, IHRL already contemplates the balance between military necessity and humanity.)

8 Geoffrey Corn, 'Mixing Apples and Hand Grenades: The Logical Limit of Applying Human Rights Norms to Armed Conflict' 44 (Journal of Int'l Humanitarian Legal Studies, Working Paper), accessed 4 May 2017 at http://ssrn.com/abstract=1511954.

9 Gary D Solis, *The Law of Armed Conflict: International Humanitarian Law in War* (2nd edn, Cambridge University Press 2016) 301.

10 Ibid ('[U]nder the law of war [IHL], deadly force may be the lawful *first resort*; under human rights law [IHRL], deadly force is the *last resort*'); Corn (n 8) 44.

11 Solis (n 9) 301–2; Corn (n 8) 30.

1949, international law, or more specifically treaties codifying international law, did not define what constituted 'war' for purposes of bringing the laws of war, or IHL, into force. Furthermore, prior to 1949, it was unclear whether a situation of domestic instability and/or violence could ever qualify as a war for purposes of legal regulation. It was true that some civil wars might fall within the scope of the doctrine of belligerency, thereby bringing into force the laws and customs of war applicable to inter-state wars. However, brutal and bloody 'internal' conflicts of the early 20th century—like those in Spain and Russia[12]—indicated a gap in international law, wherein major conflicts might rage within the borders of a state with no consensus on the applicability of international legal regulation.[13]

The international community sought to fill this gap by including articles in the four Geneva Conventions of 1949 indicating when the obligations established in the treaties become applicable.[14] Common Article 2 of the treaties defines what is today known as international armed conflict (IAC).[15] Common Article 3 defines non-international armed conflict (NIAC).[16] By adopting the notion of armed conflict as the trigger for treaty application, the Conventions fundamentally altered the law applicability equation. The existence of a war was no longer the decisive question. Instead, a more pragmatic and fact-oriented assessment of armed conflict became decisive. Furthermore, common Article 3 extended baseline treaty-based regulation to conflicts between a state and

---

[12] Specifically, reference being made to the Russian Civil War that ensued after the 1917 Bolshevik October Revolution; and also the Spanish Civil War between democratic Republicans, and Nationalists led by General Francisco Franco, among others. Each multi-year conflict resulted in the death of hundreds of thousands, and national regime change.

[13] Corn et al (n 2) 77.

[14] See generally Geneva Convention for the Amelioration of the Condition of the Wounded and Sick in Armed Forces in the Field, 12 August 1949, 75 UNTS 31 [hereinafter GWS]; Geneva Convention for the Amelioration of the Condition of the Wounded, Sick and Shipwrecked Members of Armed Forces at Sea, 12 August 1949, 75 UNTS 85 [hereinafter GWS-Sea]; Geneva Convention Relative to the Treatment of Prisoners of War, 12 August 1949, 75 UNTS 135 [hereinafter GPW]; Geneva Convention Relative to the Protection of Civilian Persons in Time of War, 12 August 1949, 75 UNTS 287 [hereinafter GC]. Collectively, each of these treaties contains a set of articles that are common to each, which are referred to as the Common Articles.

[15] GWS (n 14) Article 2; GWS-Sea (n 14) Article 2; GPW (n 14) Article 2; GC (n 14) Article 2.

[16] GWS (n 14) Article 3; GWS-Sea (n 14) Article 3; GPW (n 14) Article 3; GC (n 14) Article 3.

a non-state group, or even between multiple non-state groups.[17] Thus, after 1949, armed conflict, no matter the location or nature of the contestants, fell under the scope of humanitarian regulation.

Of course, as a matter of treaty obligation, common Articles 2 and 3 dictate applicability of the Geneva Conventions only, treaties that are almost exclusively devoted to humanitarian protection. Nothing in these treaties provides authority to employ lethal force, even against an enemy during an armed conflict. However, these law-triggering provisions of the Conventions have evolved to be considered the definitive standard for assessing when the entire corpus of IHL, to include so-called 'conduct of hostilities' rules, become applicable.[18] Accordingly, the IAC and NIAC definitions in these common articles evolved into a customary international law standard for assessing if and when a state is engaged in an armed conflict. Importantly for this discussion, once that line is crossed, it triggers not only humanitarian protection, but also the expanded scope of authority to employ force to bring the enemy into submission.

The two types of armed conflicts coined and defined by the Geneva Conventions, IAC and NIAC, are assessed quite differently.[19] Hostilities between two or more opposing organized belligerent groups is the common element of both IAC and NIAC. This is only logical, as the entire notion of armed conflict, as noted above, is a contest between organized belligerent groups. But assessing when such hostilities exist is relatively apparent in the context of IAC. This is because such conflicts require some hostile action between state armed forces, action that is usually not difficult to identify, as state armed forces rarely interact with violence in anything other than such a contest, even if brief and limited in scope.

In contrast, state police authorities—and in some cases even military authorities—constantly interact with internal and even external non-state threats across a broad 'spectrum of conflict'. In many situations, this interaction is insufficient to qualify as an armed conflict within the meaning of IHL. Thus, in a very real sense, when comparing IAC with NIAC, there is an inverse relationship between the use of military force

---

[17] GWS (n 14) Article 3; GWS-Sea (n 14) Article 3; GPW (n 14) Article 3; GC (n 14) Article 3.

[18] See ICRC Art 3 Commentary (n 1) paras 351–356; see also Geoffrey S Corn, 'Hamdan, Lebanon, and the Regulation of Hostilities: The Need to Recognize a Hybrid Category of Armed Conflict' (2007) 40 Vand J Transnatl L 295, 300–301.

[19] See GWS (n 14) arts 2, 3; GWS-Sea (n 14) arts 2, 3; GPW (n 14) arts 2, 3; GC (n 14) arts 2, 3.

and what that use indicates in terms of the legal status of a conflict. In interstate relations, confrontations between state armed forces that result in the use of force almost always qualify as armed conflicts, even if brief in duration. This is because there are few situations where such hostilities will occur below the armed conflict threshold. Thus, it is the exception, and not the rule, that such confrontations fall within the scope of a law enforcement or non-armed conflict legal framework.

In contrast, it is common for states to utilize armed forces in response to internal disturbances that challenge the response capacity of domestic law enforcement, or even to use armed forces to augment extraterritorial law enforcement activities. Accordingly, such use does not necessarily, or even normally, indicate the existence of an armed conflict against a non-state threat. Instead, it is necessary to focus on the nature of the threat demanding the use of military force in assessing when that use qualifies as an armed conflict.

When common Article 3 was first proposed, states were hesitant to consent to application of IHL to purely internal conflicts.[20] As the ICRC Commentary to common Article 3 indicates, not all civil or internal disturbances are to be considered armed conflicts.[21] However, the Commentary also indicates that recognition of an armed conflict has no impact on the legal or political status of a non-state opposition group.[22] Nonetheless, the Commentary also indicates that the states that agreed to common Article 3 expressed concerns over the impact of acknowledging when a situation of armed conflict existed within their borders.[23] In response, the Commentary not only challenges the validity of such a concern, but emphasizes the de facto nature of the armed conflict

---

[20] ICRC Art 3 Commentary (n 1) paras 361–362.

[21] Ibid paras 387–392 (a situation of violence crosses the threshold of becoming an armed conflict only when a requisite level of violence of a certain degree of intensity, which is a factual determination); see also *Operational Law Handbook* (n 6) 15.

[22] GWS (n 14) Article 3; GWS-Sea (n 14) Article 3; GPW (n 14) Article 3; GC (n 14) Article 3; ICRC Art 3 Commentary (n 1) paras 861, 864–869.

[23] ICRC Art 3 Commentary (n 1) para 417. The 2016 Commentary cites notable discussion from Pictet's 1952 Commentary on the First Geneva Convention, stating: '[M]any of the delegations feared that it might be taken to cover any act committed by force of arms—any form of anarchy, rebellion, or even plain banditry. For example, if a handful of individuals were to rise in rebellion against the State and attack a police station, would that suffice to bring into being an armed conflict within the meaning of the Article?' Jean S Pictet, *Geneva Convention for the Amelioration of the Condition of the Wounded and Sick in Armed Forces in the Field: Commentary* (ICRC 1952).

assessment and the purely humanitarian consequence of crossing this threshold.[24]

Ultimately, the line between internal disturbances that do not qualify as NIACs and situations justifying, or perhaps more importantly necessitating that characterization, was blurry from inception. Consistent with the Commentary discussion, no single factor was, or is, dispositive in assessing the existence of a NIAC.[25] Not even use of military force is dispositive, as it is common for states to use such forces to augment civil law enforcement capabilities, or even to even assume law enforcement functions in situations that do not objectively qualify as armed conflicts. Instead, this determination must be based on an assessment of the totality of the circumstances, to include, the nature of the threat, threat capabilities, and the nature of the government response.[26]

But this assessment methodology inevitably led to, and will continue to lead to, disparate conclusions between states, non-state groups, and external organizations like the ICRC or the United Nations. And, because prior to the US response to the 11 September terrorist attacks, the NIAC was almost universally considered synonymous with 'internal' armed conflicts, more simply stated to be conflicts between states and internal opposition groups, these disparate interpretations almost always focused on the point at which states must comply with IHL in response to such internal challenges. In this context, the concerns expressed during the discussions of Common Article 3 seem to be manifested by state practice: it is almost axiomatic that states resist acknowledging that an internal challenge qualifies as an armed conflict.

Why do states resist acknowledging when an internal threat crosses the threshold into the realm of armed conflict? Inverting the question may reveal the most obvious answer: why would states want to acknowledge such a state of affairs? According to the Commentary to common Article 3, the answer is that it advances humanitarian protection for all victims of the armed conflict.[27] However, the reality is that states seem to continue to view such acknowledgment as carrying with it a host of negative consequences. These include providing some level of legitimacy or credibility to the non-state opposition group (even though the Commentary to common Article 3 clearly indicates that no legal consequence

---

24 ICRC Art 3 Commentary (n 1) paras 414–421.

25 Ibid paras 419–421; see also Corn et al (n 2) 74. ('These "convenient criteria" are merely indicative ... Nonetheless, if met, the "convenient criteria" may certainly indicate the existence of a non-international armed conflict'.)

26 Corn et al (n 2) 74.

27 ICRC Art 3 Commentary (n 1) para 388.

derives from acknowledging the existence of a NIAC), signaling a loss of control or authority by the state, and opening the proverbial door to increased international legal regulation and involvement of international actors in domestic affairs.[28]

Of course, the applicability of humanitarian protection is not the only consideration relevant to NIAC recognition. Responding to non-state groups that threaten state authority necessitates the use of state power, and the characterization of the threat will significantly impact that response authority, at least in theory. For states fully committed to compliance with both IHRL and IHL, acknowledging the existence of a NIAC results in an expansion of response authorities through the conduit of customary international law. While, as noted above, common Article 3 only addresses humanitarian constraints applicable during the NIAC,[29] the existence of the NIAC also brings into effect the fundamental principles of IHL related to methods and means of warfare, most notably status-based targeting authority and preventive detention authority. Thus, this expansion of state response authority would seem to provide an important incentive for NIAC acknowledgment.

In practice, however, this incentive has produced a relatively insignificant influence on states, as most states confronting internal threats seem to simply expand response authorities without acknowledging the existence of a NIAC. Instead, a pattern of legal and operational fictions seems to define state response to internal armed threats: states refuse to acknowledge the existence of a NIAC, but nonetheless employ military power in a manner that cannot be squared with a law enforcement legal framework. While there is some diplomatic and supra-national judicial risk associated with such practices, this seems to be a relatively consistent state practice, a pattern that continues to this day. The unwillingness of the Syrian government to acknowledge the existence of a NIAC while it was pummeling Syrian communities with indirect fire and air attacks,[30] or the extensive violence among armed groups in Mexico, illustrate how states invoke IHL-type authority without acknowledging the existence of armed conflicts.

---

[28] GWS (n 14) Article 3(4); GWS-Sea (n 14) Article 3(4); GPW (n 14) Article 3(4); GC (n 14) Article 3(4); see also ICRC Art 3 Commentary (n 1) paras 861–869.

[29] Infra, p 5, fnn 9–10.

[30] Cf, e.g., SC Res 2139, para 10, UN Doc S/RES/2139 (22 February 2014) (illustrating the UN Security Council's acknowledgment of an armed conflict in Syria).

What happens, however, when the non-state threat is not confined to the territory of the threatened state—when a state faces a non-state threat operating internationally? Between 1949 and 2001, such situations rarely arose, or at least if they did, states rarely (if ever) considered them to qualify as NIACs. While there are examples of states acting against extraterritorial non-state threats before 2001, such as the Israeli hostage rescue raid against Palestinian terrorists in Entebbe,[31] or the US cruise missile attack against al Qaeda training camps in Afghanistan, it is unclear how these operations were legally classified. Are they considered NIACs? Extraterritorial law enforcement actions executed by military forces? Or short duration IACs against the states allowing their territory to be used by non-state groups?

The US decision to characterize its military response to the 11 September terrorist attacks as a NIAC opened a new chapter in conflict characterization. For the first time since the advent of common Article 3, a state unequivocally asserted it was engaged in a NIAC with a transnational non-state group. This characterization triggered widespread criticism, but also initiated a process of conflict classification reassessment. While it would be an exaggeration to assert that NIAC is today understood to include NIACs of international scope—what is often referred to as 'transnational' armed conflicts—there does seem to be growing support for this interpretation.[32]

The assertion of transnational NIAC was significantly influenced by a number of factors. First among these was the assessment of the non-state threat capability and the resulting conclusion that law enforcement authority and capability was insufficient to effectively address this threat. This led to the conclusion that only an expanded invocation of the nation's military power would be effective in addressing this threat. Reliance on law enforcement authority would not allow these forces to

---

[31] On 4 July 1976, the Israeli Defense Force conducted a hostage rescue operation (Operation Thunderbolt) at Entebbe Airport in Uganda. One hundred and two of the hostages, passengers of an Air France flight hijacked by members of the People's Liberation Front for the liberation of Palestine, were rescued by the IDF. Aside from the success of the raid in the midst of a mixed reaction from the international community, the incident has gained notoriety for the sole death on the IDF task force—Lt Col Yonatan Netanyahu, the brother of current Israeli Prime Minister Benjamin Netanyahu.

[32] See generally Watkin (n 3) ch 2.4.5; Geoffrey Corn and Eric Talbot Jensen, 'Transnational Armed Conflict: A "Principled" Approach to the Regulation of Counter-Terror Combat Operations' (2009) 42 Israel L Rev 1, 5; Monica Hakimi, 'International Standards for Detaining Terrorism Suspects: Moving Beyond the Armed Conflict-Criminal Divide' (2008) 33 Yale J Intl L 369.

fully leverage their combat power, power that could only be employed within an armed conflict legal framework. And, while there was tremendous uncertainty at the inception of these operations as to the true nature of the applicable legal authority, over time the proverbial legal dust settled to reveal an unavoidable conclusion: the US viewed the struggle against al Qaeda as an armed conflict.

The authority to employ lethal combat power as a measure of first resort—quite often through an armed drone—is perhaps the most significant consequence of this armed conflict determination. Drones were and are certainly not the only tool for employing such power, but they are often considered ideal, for all the reasons discussed above.[33] The rapid evolution of lethal drone capability in many ways complemented, or perhaps responded to, the assertion of transnational NIAC. And this dual evolution has produced both an expanded scope of legal authority to attack non-state opponents and an expanded capability to do so.

The international legal standards for assessing the existence of armed conflict is not generally understood as a constraint on the state's authority to engage in armed conflict, but rather as addressing the distinct question of the law applicable to such conflicts. From inception, the law of NIAC and the 'test' for assessing the existence of a NIAC has had a constraining effect, if not de jure, then at least de facto. This de facto constraint flowed from the perceived second and third-order consequences of treating an internal threat to state authority as an 'enemy' in an armed conflict. As noted above, states have been and remain reticent to acknowledge the existence of internal NIAC based on these concerns, which inevitably constrain the invocation of armed conflict authority to address internal threats.

The de facto consequence of the transnational NIAC interpretation is arguably the exact opposite. Unlike an internal domestic threat, what the US response to al Qaeda indicates is that states may often perceive a powerful political benefit from asserting a more aggressive response to transnational non-state threats. Confining responses to such threats to the more limited authority permitted outside the context of armed conflict risks a perception of national weakness. Indeed, in the realm of US political and policy discourse, even strategically and operationally motivated restrictions on armed conflict authority in the form of rules of engagement are condemned as manifestations of national weakness.

Thus, when confronting an external non-state threat, the state will not perceive recognition of armed conflict as an indication of national

---

[33] See section II.

weakness, but instead one of national strength. And because of the NIAC classification, the responding state is much less constrained by jus ad bellum considerations; the NIAC classification allows the state to disavow any intention to act aggressively against another state. Instead, by invoking 'failed state' or 'unable or unwilling' theories, the state using force against the transnational non-state opponent minimizes concerns related to jus ad bellum constraints.

Ultimately, unlike in the domestic context, there seem to be powerful incentives for an aggressive assertion of transnational NIAC, with few disincentives. From a tactical and operational perspective, the classification opens the door to assert otherwise unavailable national military power: the power to kill as a measure of first resort, the power to detain preventively without charge or trial, and the power to use extraordinary criminal tribunals to punish captives. From a strategic perspective, it facilitates the use of national military power largely free of the legal constraints that flow from the jus ad bellum and the practical risks of escalation resulting from attacking another state. And, from a political perspective, the invocation of armed conflict signals a message of strength and resolve, not of loss of control.

Drones substantially contribute to this incentive equation. Indeed, almost all the potential incentives associated with an aggressive NIAC characterization are enhanced by the availability of drones. The tactical and operational benefit of drones is almost self-evident, in that they are highly precise and effective weapons ideally suited to strike the non-state enemy's center of gravity—*leadership*. Strategically, drones facilitate the use of deadly combat power within the sovereign territory of another state. It is true that any attack will inevitably implicate jus ad bellum considerations and the risk of a military response by that state. However, the minimal sense of physical intrusion into that territory, coupled with precision engagement, mitigates these considerations.

Perhaps the most significant impact of drones on this incentive equation is political. Drone operations against non-state actors seem to offer national political leaders a windfall of benefits. First, drone operations are routinely marketed as highly effective at striking at the enemy's center of gravity, producing a powerful disruptive effect.[34] Whether these assertions are justified or exaggerated is almost impossible to assess. Critics of drone warfare routinely argue that they produce more

[34] James Igoe Walsh, *The Effectiveness of Drone Strikes in Counterinsurgency and Counterterrorism Campaigns* 14, accessed 4 May 2017 at http://www.strategicstudiesinstitute.army.mil/pdffiles/pub1167.pdf.

harm than good, driving up support for the very enemy we seek to undermine.[35] But because the nature of this enemy necessitates minimal public disclosure of the threat identification characteristics that justify attack, it is almost impossible to determine the true efficacy of these operations. Thus, even if critics are correct, the political benefit resulting from the perception of aggressive and decisive action cannot be ignored.

Drones also offer another political advantage: avoiding complex issues related to non-lethal incapacitation efforts.[36] Few issues have generated more legal and political complexity than the indefinite preventive detention of captured al Qaeda and Taliban operatives. While the common thread that runs through these detentions and drone operations is the assertion of a transnational NIAC, detention is simply more complicated than lethal attacks. A lethal drone attack avoids this complexity.

Drones are, of course, not uniquely responsible for incentivizing aggressive assertions of armed conflict authority to deal with transnational non-state threats. However, it is difficult to ignore how the intersection of an expanded notion of NIAC and drone capability have influenced the perceived political incentives for treating counter-terror operations as an armed conflict. In this sense, drones truly are 'different', as they offer a specialized attack capability that allows low risk/high payoff action against the transnational terrorist or non-state threat. This perceived risk/reward imbalance may often result in situations where failing to authorize attack is perceived as producing unacceptably high risk, both strategically and politically. Strategically, foregoing an opportunity to strike an elusive enemy with a highly lethal and accurate capability will almost certainly enhance the pressure to exploit windows of attack opportunity. Furthermore, failing to do so will generate concerns over potential political blowback in the event some future harm to the United States or its interests is attributed, even in part, to an alleged failed opportunity to neutralize a particular threat.

The significance of this pressure is confirmed by the debates that continue to this day over the consequence of alleged presidential hesitation to employ decisive military force against high level al Qaeda operatives before and shortly after the 11 September attacks.[37] The

---

[35] Ibid v.

[36] Some opine that the Obama administration has resorted to the use of drones as a practical alternative to staving off the issues the Bush Administration encountered pertaining to Gitmo detainees. See John B Bellinger III, 'Will drone strikes become Obama's Guantanamo?', *Washington Post* (2 October 2011).

[37] See, e.g., Kurt Eichenwald, 'The Deafness Before the Storm' *New York Times* (10 September 2012).

political risks for any president who appears hesitant to exploit potentially high payoff attack opportunities against threats to national interests are immense. This risk is substantially increased when the means available to take decisive action produce minimal risk to either US military personnel or of triggering an escalation of military violence with another nation. Drones are the tool that has created this dilemma for US presidents, a reality candidly acknowledged by the former legal adviser to the Secretary of State, John Bellinger, in a presentation made at the University of Texas.[38] To paraphrase a well-known parable, 'when the best tool in your toolbox is a hammer, almost every problem starts looking like a nail'.

IV. DRONES AND CONSTITUTIONAL WAR POWERS

The decision to use military force in response to national security threats is influenced not only by international law considerations, but also by considerations related to constitutional war powers. Since the inception of our nation, war powers have been exercised pursuant to a complex balance of power among our three branches of government (and occasionally even state governments). Indeed, few issues have implicated greater national strategic, political, and social significance than decisions of when, where, how, and for how long US military forces should be committed into hostilities.

An extensive discussion of presidential war powers is well beyond the scope of this chapter. Quite generally, the president's constitutional authority to commit US armed forces into hostilities lies somewhere between two extremes. One end of the spectrum posits that the president's vested authority as commander in chief is purely operational, and that he possesses constitutional authority to direct congressionally authorized military operations.[39] Under this theory, the president has no constitutional authority to *initiate* hostilities, and must always seek

---

[38] John B Bellinger, Former Legal Advisor of the Dept of State, Address at the Texas International Law Journal Symposium (14 April 2016).

[39] See Richard Brust, 'Constitutional Dilemma: The Power to Declare War is Deeply Rooted in American History', *ABA Journal* (1 February 2012), accessed 4 May 2017 at http://www.abajournal.com/magazine/article/constitutional_dilemma_the_power_to_declare_war_is_deeply_rooted_in_america, citing Harvey Rishikof et al, *Patriot Debate: Contemporary Issues in National Security Law* (American Bar Association 2012) (this position being a stance taken by Louis Fisher); see also Louis Fisher, 'Lost Constitutional Moorings: Recovering the War Power' (2006) 81 Ind L J 1199 (notably, see part I).

congressional authorization for any military action that goes beyond peaceful, 'military diplomacy'.[40] The other end of the spectrum posits that, absent enactment of statutory authority to legally prohibit the president from initiating or continuing hostilities, the president may authorize such hostilities.[41] This view treats congressional war powers as primarily faciliatory in nature: Congress provides the sinew of war, and may also choose to legally perfect wars through declaration or other statutory endorsement.[42] But short of express prohibition, the president is free to act when, where, and how he determines it is necessary.[43]

Neither of these extremes has been manifested by war-making practice. Instead, a much more complex equation evolved, one in which presidents exercise a broad range of war-making initiatives based on indicators of implicit congressional consent, or perhaps more accurately, an absence of congressional opposition. Furthermore, based on the seminal Supreme Court decision in the Civil War-era *Prize Cases*, there is widespread support for the proposition that the president is vested with inherent constitutional authority to use military force in response to attacks, either ongoing or imminent, against the nation or its armed forces.[44] Protection of nationals abroad is also generally considered to fall within the scope of this inherent presidential authority,[45] although the level of consensus is not as strong as that associated with defensive war powers.

Historical practice and the rare forays into war powers by the judicial branch call into question the validity of either extreme view of presidential war powers. Perhaps more importantly, these sources of authority have armed presidents with powerful support for asserting authority to

---

[40] Ibid.

[41] Burst (n 39) (this position being a stance taken by John Yoo); John Yoo, 'War Powers Belong to the President', *ABA Journal* (2 February 2012), accessed 4 May 2017 at http://www.abajournal.com/magazine/article/war_powers_belong_to_the_president.

[42] Ibid.

[43] Ibid.

[44] The *Prize Cases*, 67 US 635, 669 (1862) (When war is thrust upon the nation, the president needs no congressional authorization to use appropriate measures to quell the invasion or insurrection); Furthermore, the War Powers Resolution of 1973 arguably expanded the president's power to this end by acknowledging the existence of such ability and by explicitly stating that it extended to US territories and soldiers stations abroad. See The War Powers Resolution of 1973, 50 USC §§ 1541 (c)(2–3) (1973).

[45] See *In re Neagle*, 135 US 1, 63–4 (also, notably, the *Neagle* court's discussion of the Koszta Affair).

engage in war-making initiatives.[46] Nonetheless, the cryptic nature of this shared constitutional power almost always injects a certain degree of uncertainty into the validity of these assertions. Ironically, this uncertainty was increased by the 1973 War Powers Resolution (WPR), a law Congress enacted (over President Nixon's veto) for the express purpose of defining the extent and limits of presidential war-making power.[47] In an overt effort to prevent presidents from initiating hostilities without express congressional authorization, the WPR provides that:

> The constitutional powers of the President as Commander-in-Chief to introduce United States Armed Forces into hostilities, or into situations where imminent involvement in hostilities is clearly indicated by the circumstances, are exercised only pursuant to (1) a declaration of war, (2) specific statutory authorization, or (3) a national emergency created by attack upon the United States, its territories or possessions, or its armed forces.[48]

The apparent clarity of this provision of the WPR was, however, substantially eroded by the mechanisms incorporated into the statute to implement the obvious congressional objective of restraining presidential war-making initiatives. Most notable among these provisions is the so-called 60-day clock. According to § 1544 (a):

> Within sixty calendar days after a report is submitted or is required to be submitted pursuant to section 1543(a)(1) of this title, whichever is earlier, the President shall terminate any use of United States Armed Forces with respect to which such report was submitted (or required to be submitted), unless the Congress (1) has declared war or has enacted a specific authorization for such use of United States Armed Forces, (2) has extended by law such sixty-day period, or (3) is physically unable to meet as a result of an armed attack upon the United States.[49]

The WPR also provides, in § 1547(d)(2), that nothing in the WPR, 'shall be construed as granting any authority to the President with respect to the introduction of United States Armed Forces into hostilities or into situations wherein involvement in hostilities is clearly indicated by the

---

[46] See generally *Holtzman v Schlesinger*, 484 F 2d 1307 (2d Cir 1973) (once Congress has acted to allow the president to conduct war-making initiatives, the judiciary will likely not interfere unless the president is subsequently acting contrary to an expressed revocation of such authority by Congress).

[47] 50 USC §§ 1541–48.

[48] Ibid § 1541(c).

[49] Ibid § 1544(a).

circumstances which authority he would not have had in the absence of this chapter'.[50]

It is therefore clear that the 60-day clock cannot properly be interpreted as a congressional authorization for presidents to initiate and conduct hostilities for up to 60 days. Nonetheless, taken has a whole, the WPR produces that precise effect. This effect was exacerbated when the Supreme Court struck down the constitutionality of the so-called legislative veto in *INS v Chadha*.[51] This is because the WPR also provides that Congress may, at *any* time prior to the termination of the 60-day period, order termination of hostilities by a majority vote of both houses.[52] Because such a concurrent resolution is considered a legislative veto, this provision was effectively nullified by *Chadha*. As a result, only by enacting a law—ostensibly requiring the requisite super-majority to overcome a presidential veto—will Congress be able to direct termination of hostilities already initiated (or even contemplated) by the president.[53]

The combined effect of the WPR, post-enactment practice, and the invalidation of the legislative veto, has arguably strengthened the presidential war-making hand. This is especially the case when hostilities are expected to be of short duration. Of course, nothing in the WPR nor its legislative history suggests that Congress intended the law to apply only to hostilities that extend beyond 60 days, or which are expected to involve a magnitude likely to produce such a duration.[54] The best evidence of this is found in the text of the law itself, which includes no 'intensity', 'gravity', or 'magnitude' qualifiers, and also includes a provision that expressly prohibits treating anything in the statute as a source of authority to initiate or continue hostilities. Indeed, the clear purpose of the WPR was to prevent presidents from committing the nation to conflicts based on an expectation of quick or limited involvement, precisely because of the inherent risk of escalation associated with such military ventures.

But a presidential judgment that a military objective can be achieved within a 60-day period substantially mitigates both the risk of congressional opposition, and judicial action in response to a challenge to the action. An effective congressional challenge would require Congress to muster the political will to enact a law to restrict the president, a daunting task at any point during a conflict, but especially so at the outset. Nor

---

50 Ibid § 1547(d)(2).

51 See *INS v Chadha*, 462 US 919, 959 (1983).

52 50 USC § 1544(c).

53 *Chadha*, 462 US at 951–9.

54 Cf 50 USC §§ 1541–48.

could opposition members of Congress turn to the courts to enforce the WPR, as the doctrine of legislative standing would almost certainly function as an impenetrable barrier to judicial action absent enactment of a prohibitory statute.

Where hostilities can be conducted without subjecting US forces to significant risk, it further mitigates the risk of political opposition. Hence, the capability offered by drones produces a potentially significant advantage for presidents who seek to employ US military power without congressional authorization. The ability to achieve the strategic and operational objectives without ever subjecting US forces to the mortal risks of combat may, like drones themselves, make such operations almost unnoticed. And when military operations are not noticed, it is unlikely congress will seriously question the president's authority to conduct them.

The limited US risk associated with drone operations is not just a practical political consideration; it actually is at the core of what may be an emerging theory of presidential war powers. In 2011, President Barack Obama ordered US armed forces to participate in the 2011 military action against Libya, Operation Odyssey Dawn.[55] President Obama emphasized that the US role in the broader coalition effort would be limited, focused primarily on suppression of enemy air defense, surveillance, intelligence, and logistics.[56] However, there was no question that the President had authorized initiation of hostilities. Without either congressional authorization or an assertion that the mission was ordered in response to an attack on the nation or its armed forces, his action seemed to clearly violate the WPR.[57]

From inception, President Obama asserted inherent executive authority as the constitutional basis for ordering the US participation.[58] However,

---

[55] President Barak Obama, Remarks by the President on Libya (29 March 2011) (transcript accessed 4 May 2017 at https://www.whitehouse.gov/the-press-office/2011/03/19/remarks-president-libya) [hereinafter Libya Remarks].

[56] Ibid.

[57] Jeremiah Gertler, Congressional Research Service, R41725, Operation Odyssey Dawn (Libya): Background and Issues for Congress 3–4 (2011), accessed 4 May 2017 at https://www.fas.org/sgp/crs/natsec/R41725.pdf (in President Obama's remarks from 18 March 2011, he makes no reference to an attack on, or a threat to, the United States. Instead, President Obama alludes only to the United States' commitment to a broader, international coalition tasked with enforcing a cease-fire between Libyan troops and civilians).

[58] *Libya and War Powers: Hearing before the Committee on Foreign Relations*, 112th Cong 8 (2011) (statement of Harold Koh, Legal Adviser, US Dept of State) ('[F]rom the outset, we noted that the situation in Libya does not

he also emphasized the extremely limited exposure of US personnel involved in the operation.[59] Like the air campaign against Serbia in 1998, the expectation of short duration proved erroneous, and the operation dragged on for several months. And, also like the Serbia air campaign, once operations exceeded 60 days, addressing compliance with the WPR became a significant issue. In the case of US operations against Serbia (the first US military campaign subsequent to enactment of the WPR involving ongoing hostilities that exceeded 60 days without express statutory authorization), the extended duration led to litigation between the President and members of Congress.[60] President Clinton had, like prior presidents, asserted he was not bound by the WPR due to its impermissive intrusion into his inherent constitutional authority.[61] However, his Justice Department asserted, and the DC Circuit Court relied on, justiciability considerations to terminate the litigation without ever reaching the constitutional question.[62]

---

constitute a war requiring specific congressional approval under the Declaration of War Clause of the Constitution'.) [hereinafter Koh Report]; Libya Remarks (n 55) (President Obama states that 'after consulting the bipartisan leadership of Congress, I authorized military action to stop the killing and enforce U.S. Security Council Resolution 1973'. This language is carefully articulated to be consistent with the WPR, however still reverent of inherent executive war power authority, chiefly in the words, 'I authorized'.).

[59] Libya Remarks (n 55).

[60] *Campbell v Clinton*, 203 F 3d 19, 19, 24 (DC Cir 2000). (Members of congress filed suit seeking declaratory relief against the President in an effort to force congressional action. The Court, however, held that no vote of Congress was being nullified by the President, and that the legislators lacked standing; therefore, essentially, so long as the legislature has options at their disposal, they lack standing in Court.)

[61] 'Letter to Congressional leaders reporting on airstrikes against Serbian targets in the Federal Republic of Yugoslavia (Serbia and Montenegro)' (1999) 1 Published Papers of William Jefferson Clinton 459, 459 ('United States and NATO forces have targeted the [Yugoslavian] government's integrated air defense system, military and security police command and control elements, and military and security police facilities and infrastructure … I have taken these actions pursuant to my constitutional authority to conduct U.S. foreign relations and as Commander in Chief and Chief Executive'.).

[62] *Campbell*, 203 F 3d at 19, 24 (The government challenged the jurisdiction of the federal courts to adjudicate this claim on three separate grounds: the case is moot, appellants lack standing (as the district court concluded), and the case is non-justiciable. The Court never reached the mootness and political question assertions, as they agreed with the government in that the congressmen lacked standing to bring the claim.).

Unlike President Clinton, President Obama was confronted with almost no congressional opposition when the Libya campaign crossed the same temporal phase-line. Nonetheless, President Obama directly addressed the issue of WPR applicability and compliance. In a controversial speech, Harold Koh, the Legal Adviser to the Secretary of State, presented the administration's new theory of WPR inapplicability: de minimis risk.[63] According to Koh, the WPR trigger of situations where commitment of US armed forces into hostilities or situations where hostilities were imminent was intended to prevent the type of incrementally escalating quagmires typified by the Vietnam conflict—the conflict that originally motivated enactment of the law.[64] Koh posited that where the nature of the hostilities posed little to no risk of escalation, especially an escalation that would require putting US ground forces at risk, the law was inapplicable.[65]

The limited nature of the US involvement in the Libya campaign nullified the type of risks the administration concluded were necessary predicates for applicability of the WPR.[66] The bulk of the US forces participating in the operation performed combat support functions, and were not directly engaged in confrontation with Libyan forces.[67] Furthermore, even when US forces did conduct combat operations against Libyan forces, only air and missile assets were being used, with no 'boots on the ground'.[68] As a result, the administration concluded the commitment of US forces did not fall within the intended scope of the WPR.[69]

---

63 See Koh Report (n 58) 7–11, 11–40.

64 Ibid 9–10.

65 Ibid 9.

66 Ibid; see also Office of the President, United States Activities in Libya (2011), accessed 4 May 2017 at https://www.scribd.com/fullscreen/57965200?access_key=key-1u10mi6mo7qaatybceao.

67 Ibid. For example, President Obama's report states the following:

> The overwhelming majority of strike sorties are now being flown by our European allies while American strikes are limited to the suppression of enemy air defense and occasional strikes by unmanned Predator UAVs against a specific set of targets, all within the UN authorization, in order to minimize collateral damage in urban areas ... The United States provides nearly 70 percent of the coalition's intelligence capabilities and a majority of its refueling assets, enabling coalition aircraft to stay in the air longer and undertake more strikes.

68 Koh Report (n 58) 9–10.

69 Ibid 8.

In-depth analysis of the relative merit of this interpretation is beyond the scope of this chapter. Suffice to say that it seems difficult to reconcile this 'de minimis risk' or 'no boots on the ground' theory with a statute that evolved from the difficult US experience in Vietnam, obviously intended to prevent presidents from what might best be called 'incremental escalation'. The mere fact that the WPR addresses not only commitment of US armed forces into hostilities, but also into situations indicating an imminence of hostilities—which is defined, inter alia, to include a substantial increase in the presence in any given area of US armed forces equipped for combat[70]—seems to contradict this interpretation. Ultimately, no matter how credible or incredible the interpretation, the bottom line remains that it opened a new theory of WPR avoidance.

The assertion is, therefore, in itself a significant development in the evolution of constitutional war powers. And, the fact that there was little congressional resistance to this interpretation of the WPR, much less any effort to enforce its terms, increases this significance. It is not difficult to imagine that subsequent presidents will look back on this campaign and President Obama's assertion of constitutional authority as an example of how to frame their own assertions of executive war powers.

Drones may very well be central to any future assertion of this de minimis risk theory of executive war power. Almost no other weapon system capable of employing analogous lethal and destructive power with virtually no risk to US forces is currently available in the US military arsenal. While long-range weaponry such as cruise missiles and other, 'beyond the horizon' strike assets may present virtually no risk to US forces, they lack the analogous real-time information dominance and precision engagement offered by drones. This capability is, therefore, an ideal fit within this theory of virtually unconstrained presidential war power.

This raises serious concerns. It may be true that the constitutionality of the WPR remains uncertain. It is certainly true that all presidents have aligned themselves with President Nixon's initial conclusion that the law unconstitutionally infringed on Article II authority.[71] However, it would

---

[70] 50 USC § 1543 (a)(3).

[71] See, e.g., Letter (Regarding Cameroon) from Barak Obama, President, United States, to John Boehner, Speaker of the House of Representatives, United States, and Orrin Hatch, President Pro Tempore of the Senate, United States (14 October 2015), accessed 4 May 2017 at https://www.whitehouse.gov/the-press-office/2015/10/14/letter-from-president-war-powers-resolution-cameroon; Letter (Regarding Iraq) from Barak Obama, President, United States, to John Boehner, Speaker of the House of Representatives, United States, and Patrick Leahy,

be misleading to conclude the WPR has been a complete failure. To the contrary, perhaps because its constitutionality and ultimate impact on national security policy remains uncertain for *both* presidents *and* Congress—it has generated greater war powers interaction between these branches of government. As noted above, only twice since enactment of the law has the United States conducted a military campaign beyond 60 days without express congressional authorization. Furthermore, even when presidents believe they are initiating hostilities that fall within the scope of their constitutional authority, they have routinely provided notice to Congress, 'consistent' with the notification provisions of the WPR.

Perhaps the WPR is responsible for presidents seeking express statutory authorizations for the conflicts they intend to initiate. But even if the WPR has accomplished nothing more than stimulating more dialogue between presidents and Congress, it has been a success. WPR notifications and the discussions they stimulate provide Congress with an early opportunity to endorse or constrain presidential war-making initiatives. And, where Congress is silent or ambivalent in response, presidents may consider even this as acquiescence that may be treated as a constitutional 'green light' to move forward with the military action. In fact, the importance of robust inter-branch war powers dialogue was recognized by the Baker-Warren proposal to replace the WPR with a new law titled the War Powers Consultation Act, which would focus exclusively on ensuring such dialogue.[72]

The capabilities provided by drones may stifle the positive trend of increased inter-branch war powers dialogue. When presidents are armed with a capability that allows for the rapid and highly effective use of combat power with almost no perceived risk to US personnel, assertions

---

President Pro Tempore of the Senate, United States (23 September 2014) accessed 4 May 2017 at https://www.whitehouse.gov/the-press-office/2014/09/23/letter-president-war-powers-resolution-regarding-iraq. From these letters, note the redundancy in language used to show consistent action with the WPR: 'I am providing this report as part of my effort to keep the Congress fully informed, consistent with the War Powers Resolution …' Compare this 'consistent with' language to 'in accordance with', for example, and reflect this as an indication that US presidents do not find the WPR obligatory. This language, or variations thereof, has been used by presidents since the enactment of the WPR during the Nixon administration.

[72] See generally The War Powers Consultation Act of 2014, S 1939, 113th Cong. (2014); James A Baker III et al, *National War Powers Commission Report* (Miller Center Public Affairs 2009), accessed 4 May 2017 at http://web1.millercenter.org/reports/warpowers/report.pdf.

of unilateral executive power will almost certainly become more likely. Presidents confronted with threats vulnerable to drone attacks will perceive significant pressure for swift action; action that will be perceived as in the national interest because of its speed and decisiveness. Where an attack option offers a high probability of neutralizing a target in a short period of time with no risk to US forces, the risk of congressional backlash for failing to provide notification will likely be considered minimal. In such situations, it is more likely a president will assess the risk of congressional backlash to actually be more significant in response to a lost attack opportunity resulting from efforts to involve Congress or even congressional leaders in the discourse.

V. CONCLUSION

The military, diplomatic, political, legal, and moral consequences of lethal drone capability have been a central focus of discourse since drones emerged as a weapon of choice for US national security decision-makers. From a purely tactical perspective, a drone is just a weapon system, offering many beneficial attributes, most notably accuracy. Indeed, as Professor Oren Gross has noted, the efficacy of drones implicates international humanitarian law obligations related to civilian risk mitigation, and may result in obligatory use in certain situations.[73]

But should drones be considered simply as just another weapon system? Or is something inherent in this capability that distinguishes drones from other weapons? This question has been a constant focal point of debate, one laced throughout the other chapters of this book. While perhaps indistinct from a tactical or operational perspective, it does seem that drones present unique strategic implications.

The strategic impact of drones cannot be assessed in a vacuum. Instead, it is essential to consider how drone capability interacts with international and domestic law. In this context, drones are indeed different than other weapon systems. For the United States, and an increasing number of like-minded states, once it is determined that a threat is of sufficient magnitude to trigger the international legal right of self-defense, the inability or unwillingness of the 'host' state to eliminate

[73] See generally Oren Gross, 'The New Way of War: Is There a Duty to Use Drones?' (2016) 67 Florida L Rev 1.

that threat will open the door to the use of military force in self-defense.[74] Ironically, the capability provided by drones does not significantly impact this ad bellum legality assessment, which focuses primarily on the capacity of the host state to address the threat and not so on the capacity of the state to neutralize the threat.

In contrast, the intersection of 'conflict classification' law and drone capability seems to be far more significant. Ironically, unlike the jus ad bellum, this law was never intended to function as a limitation on the use of military force. Nonetheless, state practice suggests that since the concept of NIAC emerged in 1949,[75] states have been extremely conservative in their willingness to acknowledge that a non-state threat has risen to the level of NIAC.[76] The law of NIAC is clear that such acknowledgment does not impact the legal status of the non-state opposition forces.[77] However, perception is often more powerful than reality, and it seems relatively clear that states prefer to avoid the perception that the magnitude of an internal non-state threat necessitates characterization as an armed conflict. But once the notion of NIAC was extended internationally, and recognized as the so-called transnational armed conflict, the incentives and disincentives for an armed conflict characterization were flipped. In this context, a NIAC characterization not only opens the door to more robust response authority, but also involves little risk that it will be perceived as somehow validating or legitimizing the threat.

Drones, therefore, provide the ideal tool to exploit this expansion of military response authority. This advent of the transnational armed conflict theory, coupled with the capacity to conduct virtually risk-free attacks with decisive lethal force, has arguably incentivized an aggressive invocation of armed conflict. If this is true, how should international law respond? Ultimately, drones are central to significant evolutions of law and practice related to state response to non-state threats: dilution of the

---

[74] See *Operational Law Handbook* (n 6) 7; Yoram Dinstein, *War Aggression and Self Defence* (5th edn, Cambridge University Press 2011) 226.

[75] See generally GWS (n 14) Article 3; GWS-Sea (n 14) Article 3; GPW (n 14) Article 3; GC (n 14) Article 3; see also *Operational Law Handbook* (n 6) 37 ('With respect to NIACs, Common Article 3 of the Geneva Conventions recognizes the prerogative of the ICRC or other impartial humanitarian organizations to offer its services to the parties to the conflict').

[76] ICRC Art 3 Commentary (n 1) paras 357–383.

[77] GWS (n 14) Article 3(4); GWS-Sea (n 14) Article 3(4); GPW (n 14) Article 3(4); GC (n 14) Article 3(4); ICRC Art 3 Commentary (n 1) paras 861–869.

restraining effect of the jus ad bellum by the tendency of states to invoke self-defense in response to non-state transnational threats, and the incentives associated with characterizing the response to such threats as an armed conflict. This evolution can only be effectively managed if it is accurately assessed.

Drones are also at the center-point of a similar evolution of constitutional war powers. The capability to employ decisive combat power with virtually no US military 'footprint' and equally minimal risk to US personnel has incentivized assertions of unilateral presidential war powers. The 'de minimis' risk, or 'no boots on the ground' theory invoked by President Obama to avoid compliance with the WPR during the extended US involvement in the military campaign against Libya, may forecast an emerging trend. Drones enable presidents to employ military force that offers high strategic payoff and almost no political risk. As a result, this weapon system may dilute the post-WPR trend for more, rather than less inter-branch war powers interaction.

Drones are not going away. No nation would abandon such a highly effective combat capability, especially one that offers relatively unique strategic, political, and diplomatic advantages. But the impact of this capability, especially in an era of external non-state threats that will often push states to consider a military response, must be carefully assessed. Both international law and our Constitution create an expectation that war will be an exceptional situation, and that our nation will cross this profoundly significant threshold only when doing so is legitimately necessary. In such situations, drones will often quite appropriately be a weapon of choice. But if the weapon drives the decision to cross that threshold, what should be exceptional may become the norm.

9. Developing norms for cyber conflict

William C Banks[*]

1. INTRODUCTION

The prospect of cyber war has evolved from science fiction and doomsday depictions on television, in films and novels to reality and front page news. As early as 1982, a little-noticed but massive explosion of the trans-Siberian pipeline was caused by malware apparently inserted into Canadian software by the CIA. The CIA and Canadians knew that the software would be illegally acquired by Soviet agents. Although the incident greatly embarrassed the KGB, the Soviets never disclosed the incident or accused the United States of causing it. If a US missile had struck the pipeline, the Soviets would have expressed their outrage publicly and almost surely would have retaliated.[1]

As the Internet grew exponentially over the next quarter century, so did the frequency and variety of cyber intrusions. By 2012, reports confirmed that the Stuxnet malware attack on the computers that ran Iran's nuclear enrichment program was carried out as part of a larger Olympic Games campaign of cyber war against Iran begun in 2006 by the United States and perhaps Israel. This use of cyber-weapons to attack a state's infrastructure became the second (following the Siberia explosion in 1982) known use of computer code to affect physical destruction of equipment—in this case Iranian centrifuges—instead of disabling computers or stealing data.[2] Like the Soviet Union in 1982, Iran did not acknowledge the cyber-attack. However, in 2012 Iran released the Shamoon virus in a major cyber-attack on US ally Saudi Arabia's state-owned oil company, Aramco. Shamoon replicated itself inside

* The author is grateful to Kyle Lundin for excellent research assistance.

1 Brett Stephens, 'Long before There Was the Stuxnet Computer Worm, There Was the 'Farewell' Spy Dossier', *Asian Wall Street Journal*, 19 January 2010.

2 See David E Sanger, 'Obama Order Sped Up Wave of Cyberattacks Against Iran', *New York Times* (2012).

30,000 Aramco computers, destroying the computers and disrupting operations for nearly two weeks.[3]

In 2013, the US Director of National Intelligence named cyber as the number one strategic threat to the United States, ahead of terrorism.[4] Increasingly frequent and intense cyber-attacks are mounted at military and intelligence targets, as Edward Snowden demonstrated in 2013.[5] In the United States, our electric grid, municipal water and sewer systems, air traffic and railway control, banking system, and even military operations are persistently subject to cyber penetration. At least as costly and sometimes more destructive, cyber intruders attack businesses and industry. Because industrial control systems in most of the world are connected to the Internet, all of them are vulnerable. In 2009, President Barack Obama said that 'cyberintruders have penetrated our electric grid', and that 'in other countries cyberattacks have plunged entire cities into darkness'.[6]

Nor has it been only nation states that carry out the cyber-attacks. Profit-seeking criminals, ideological hackers, extremists, and terrorists have also directed attacks towards state-owned facilities and infrastructure, and against the private sector. At the same time, there are increasing signs that cyber techniques are now an integral part of heretofore violent conflicts between terrorist groups and states. In October 2015, the US government arrested Kosovar Ardit Ferizi in Malaysia. Ferizi was charged with providing material support to terrorism on the basis of his hacks of a private US company for the purpose of gaining access to personally identifiable information of US military and federal employees. Ferizi allegedly released the information on behalf of the terrorist group the Islamic State (IS).[7] Meanwhile, state and non-state cyber threats now often blend and merge, as privateers operate as surrogates for states and provide cover for state-based actors.[8]

---

[3] See Fergus Hanson, '*Norms of cyberwar in peacetime*', Brookings Institution, 17 November 2015.

[4] Office of the Director of National Intelligence, *Worldwide Threat Assessment of the US Intelligence Community* (2013).

[5] Joel F Brenner, 'Eyes Wide Shut: The Growing Threat of Cyber Attacks on Industrial Control Systems' (2013) 69 Bulletin of Atomic Scientists 15–20, 16.

[6] Barack Obama, 'Remarks by the President on Security Our Nation's Cyber Infrastructure' 29 May 2009.

[7] Ellen Nakashima, 'At least 60 people charged with terrorism-linked crimes this year—a record', *Washington Post*, 25 December 2015.

[8] US Dept of Defense, *The Department of Defense Cyber Strategy* (2015) 9.

Hostile cyber penetration is now a daily occurrence. The perpetrators are too numerous to count, and the targets continue to expand in number and by type. The frequency, breadth and persistence of an ever wider set of cyber exploitations reflects bad actors racing to the bottom of every data repository that might generate profit or impose costs, inflict pain, instill fear, create inconvenience, or disrupt operations. Some states and private actors justify cyber intrusions on the grounds that their adversaries are pursuing cyber operations against them. Others simply attack, for financial or some other gain or to inflict harm, for whatever reasons.

Despite the growing prominence of cyber threats, international law does relatively little to regulate cyber conflict. For the most part, treaty-based and customary international law provide limits on state but not private actions, and only in conflict that has kinetic consequences. Even as experts recognize that terrorists may engage in cyber war, the international community continues to rely on a legal conception that limits terrorism to 'acts of violence committed in time of peace',[9] a categorization that excludes most cyber-attacks. Despite the growing role of the cyber domain in the security sectors of many governments over the last decade, the legal architecture for cyber pays little attention to cyber-attacks that do not produce harmful effects equivalent to kinetic attacks.

A distinguished International Group of Experts was invited by NATO in 2009 to produce a manual on the law governing cyber warfare.[10] The resulting *Tallinn Manual on the International Law Applicable to Cyber Warfare* (*Tallinn Manual*) defines the scope of their project to include only those forms of cyber-attack that meet the UN Charter and IHL conceptions of 'use of force' or 'armed attack'.[11] Beyond limiting their inquiry into state-on-state cyber conflict to these traditional conceptions, the *Tallinn Manual* restates the consensus view that prohibits 'cyber-attacks, or the threat thereof, the primary purpose of which is to spread terror among the civilian population'.[12] The *Tallinn Manual* experts concluded that cyber-attacks can constitute terrorism, but only where the attack has been conducted through 'acts of violence'.[13] In other words,

---

[9] Jelena Pejic, 'Armed Conflict and Terrorism: There Is a (Big) Difference', in A-M Salinas De Frias, Katja L H Samuel and N D White (eds), *Counter-Terrorism: International Law and Practice* (Oxford University Press 2012) 203.

[10] *Tallinn Manual on the International Law Applicable to Cyber Warfare* (Cambridge University Press 2013).

[11] Ibid rule 18.

[12] Ibid rule 36.

[13] Ibid rules 30, 36.

the *Tallinn Manual* concludes that international law proscribes only kinetic harm by states and violent terrorism and thus leaves unregulated an entire range of disruptive cyber intrusions.[14] To the great credit of the NATO Cyber Centre of Excellence, the organizers and Group of Experts are finishing *Tallinn Manual 2.0*, which will consider the application of various customary international law doctrines and principles that could apply to govern cyber conflict where the intrusions do not meet the traditional kinetic thresholds.

As reflected in the *Tallinn Manual*, there is international legal clarity in some cyber conflict situations. In instances where a cyber-attack causes physical destruction and/or casualties at a significant level, a cyber-intrusion may constitute a 'use of force' or an 'armed attack' under the UN Charter. In these extreme circumstances, even where the attacker is a state-sponsored non-state actor, customary law permits a forceful response in self-defense, assuming attribution of the attacker.[15] In addition, whether the Charter criteria have been met is most likely a function of the consequences of the cyber event, and is not dependent on the instrument used in the attack.[16] Apart from this relatively small subset of cyber-intrusions, however, the legal regime remains clouded and ambiguous.

Developing a more fully-formed international law of cyber conflict is complicated by a few unique attributes of the cyber domain. Prompt attribution of an attack and even threat identification can be very difficult. As a result, setting the critical normative starting point for invoking international law is elusive—which is the offending state, and what is the line between offense and defense? Preliminary questions include: Is it lawful to anticipate cyber-attacks by implementing countermeasures in advance of the intrusion? How disruptive or destructive a response does the law permit once a source of the incoming intrusions is identified, even plausibly? If victim states cannot reliably attribute incoming attacks, must they delay all but the most passive responses until the threat can be reliably identified? Beyond challenging threshold questions like these, because cyber-attacks will likely originate from multiple sources in many states, using geography as a proxy for a battle space may not be realistic or useful in the cyber context. Even assuming attribution of incoming attacks, which if any geographic borders should define the scope of a victim state's responses?

[14] Ibid rule 30.
[15] Ibid rule 13.
[16] Ibid rules 11–12.

International law scholars and operational lawyers have struggled in recent years to accommodate IHL and the UN Charter system to asymmetric warfare waged by non-state actors, including terrorist groups. The language and structure of IHL—the regulation of 'armed conflict'—and of the Charter—focusing on 'use of force' and 'armed attack'—present considerable analytic challenges and even incongruities in attempting to fit cyber into the conventional framework for armed conflict, even for state-on-state cyber conflict. Because cyber-attacks may be carried out by states or non-state actors and may occur continuously or in stages with no overt hostility and range from low-level harassment to potentially catastrophic harms to a state's infrastructure, the either/or dichotomies of war and peace and armed conflict/no armed conflict are not in most instances well suited to the cyber domain. Over time, the ongoing struggle to fit cyber into the IHL and Charter categories may threaten their normative integrity and their basic commitment to collective security and restraints on unilateral uses of force.

The core component of the framework for regulating the use of force—the UN Charter—is less important in developing future prescriptions for cyber conflict than customary international law, developed over time through state practice. Most cyber-intrusions for the foreseeable future will take place beyond the traditional consensus normative framework for uses of force supplied by international law. For the myriad and multi-faceted cyber-attacks that disrupt but do not destroy, whether state-sponsored or perpetrated by organized private groups or single hacktivists, much work remains to be done to build a normative architecture that will set enforceable limits on cyber intrusions and provide guidelines for responses to disruptive cyber-intrusions. The next two sections of the chapter first review and assess the historical and contemporary normative justifications for cyber conflict, and then outline the components of future cyber conflict norms.

2. EXAMINING HOW AD BELLUM PRINCIPLES MAY APPLY TO CYBER CONFLICT

Cyber-weapons are adaptable and relatively easy to use. One common view is that because the collective law of war does not reach most cyber conflict, states enjoy relatively new non-kinetic options for achieving their conflict objectives, untethered by law. A state's security objective that may have required the use of military force in the past may now be accomplished through the use of cyber techniques. Better still, a state may be able to act in cyberspace without acknowledging responsibility for what it has done. In order to place the international legal issues in context, consider these scenarios:

Assume that fictional State A launches a massive malware attack at fictional State B. The botnets and sophisticated software unleashed by the malware cause power failures when generators are shut down by the malware. Train derailments and airplane crashes with hundreds of casualties soon follow, as traffic control and communications systems that rely on the Internet are made to issue false signals to pilots and conductors. Dozens of motorists die when traffic lights and signals malfunction at the height of an urban rush hour. State A acknowledges its responsibility for the cyber-attacks, and it says that more are on the way. Clearly there is an international armed conflict (IAC) between states A and B, and pending Security Council action, B is lawfully permitted by Article 51 of the Charter to use self-defense to respond to the 'armed attack' by A. The Charter and IHL norms provide sufficient ad bellum authority for B to respond to these cyber-attacks.

Assume instead that unknown assailants have launched a series of cyber-attacks on the banking system of a state. The malware is sophisticated; large and small customers' accounts are targeted and account balances are reduced by hundreds of millions of dollars. For the time being the attacks cannot be attributed, but non-state terrorists are suspected in light of intelligence reports. No one has been injured or killed. There is no international armed conflict (IAC), either because there is no known state adversary and/or because there has been no 'attack' as contemplated by Article 49 of Additional Protocol I. (Additional Protocol I was added to the 1949 Geneva Conventions in 1977, and Article 49 expands on the definition of 'attack' contained in the Fourth Geneva Convention in 1949.) There is no non-international armed conflict (NIAC) because the conflict is not sufficiently intense, or because the likely culprit is not an organized armed group. It is far from clear that there has been a 'use of force' as contemplated by Article 2(4) of the Charter, or an 'armed attack' within the meaning of Article 51. Even if the incoming attacks could be attributed to a state, the conflict likely is not an armed conflict. Surely the state must respond to deflect and/or dismantle the sources of the malware, and delaying responses until attribution is certain will greatly exacerbate the crisis.

Although these scenarios do not fully represent the wide range of possible cyber-intrusions that occur now on a daily basis, they do underscore that only the most destructive cyber-attacks fall clearly within the existing international law framework for cyber conflict. What international law principles offer the best options for extending their application more broadly to cyber conflict?

One of the most challenging aspects of regulating cyber war is timely attribution. As Joel Brenner reminds us, 'the Internet is one big masquerade ball. You can hide behind aliases, you can hide behind proxy servers, and you can surreptitiously enslave other computers to do your dirty work'.[17] Cyber-attacks also often occur in stages, over time. Infiltration of a system by computers operated by different people in different places may be followed by delivery of the payload and, perhaps at a later time, manifestation of the harmful effects. At what stage has the cyber-attack occurred? Attribution difficulties also reduce the disincentives to cyber-attack and further level the playing field for cyber war waged by terrorists and other non-state actors. Although identifying a cyber-intruder can be aided by a growing set of digital forensic tools, attribution is not always fast or certain, making judgments about who was responsible for the cyber intrusion that harmed the victim state probabilistic.[18] Even where the most sophisticated forensics can reliably determine the source of an attack, the secrecy of those methods may make it difficult to demonstrate attribution in a publicly convincing way. Because the ad bellum justifications for responding to a cyber-attack are tied to attribution of the attack and thus identification of the enemy, the legal requirements for attribution may at least delay effective defenses or responses.

The 'use of force' rubric from Article 2(4) establishes the benchmark standard for determining a violation of international law in the world of kinetic conflict. Once a use of force occurs, permissible responses are determined by the law of state responsibility, potential Security Council resolutions, and the law of self-defense.[19] The traditional and dominant view among member states is that the prohibition on the use of force and right of self-defense apply to armed violence, such as military attacks,[20] and only to interventions that produce physical damage. As such, most

---

[17] Joel F Brenner, *America the Vulnerable: Inside the New Threat Matrix of Digital Espionage, Crime, and Warfare* (Penguin Press 2011).

[18] W A Owens, K W Dam and H S Lin (eds), *Technology, Policy, Law, and Ethics Regarding U.S. Acquisition and Use of Cyberattack Capabilities* (National Research Council 2009) §2.4.2: 33–4, 245, 253, 261, 263; S E Goodman and H S Lin (eds), *Toward a Safer and More Secure Cyberspace* (National Research Council 2007).

[19] See Michael N Schmitt, 'Cyber Operations and the *Jus Ad Bellum* Revisited (2011) 56 Vill L Rev 573–80.

[20] *Technology, Policy, Law, and Ethics Regarding U.S. Acquisition and Use of Cyberattack Capabilities* (n 18) 253.

cyber-attacks will not violate Article 2(4).[21] Throughout the Cold War, some states argued that the Article 2(4) 'use of force' prohibition should focus not so much on the instrument as the effects of an intrusion and thus forbids coercion, by whatever means, or violations of sovereign boundaries, however carried out.[22] The United States opposed these efforts to broaden the interpretation of 'use of force' by developing states, and by the end of the Cold War Charter interpretation had settled on the traditional and narrower focus on armed violence.[23]

An interpretation of Article 2(4) could evolve to include cyber intrusions, depending on the severity of their impact. State practice may in the future recognize cyber intrusions as 'uses of force', at least when cyber-attacks deliver consequences that resemble those of conventional armed attacks.[24] Public statements by the United States in recent years suggest that the US government is moving toward this sort of effects-based interpretation of the Charter's use of force norm in shaping its cyber-defense policies, a position at odds with the US government's history of resisting flexible standards for interpreting Article 2(4).[25] As historically interpreted, however, the Charter purposefully imposes an additional barrier to a forceful response to a use of force. The response to such a use of force cannot itself rise to the level of use of force unless authorized by the Security Council or is a lawful action in self-defense.[26] In other words, unilateral responses to a use of force are permitted only if the intrusion constitutes an armed attack recognized by Article 51.

---

[21] Jason Barkham, 'Information Warfare and International Law on the "Use of Force"' (2001) 34 NYU J Intl L & Pol 56.

[22] Matthew C Waxman, 'Cyber-Attacks and the Use of Force: Back to the Future of Article 2(4)' (2011) 36 Yale J Intl L 421.

[23] Ibid 431.

[24] *Technology, Policy, Law, and Ethics Regarding U.S. Acquisition and Use of Cyberattack Capabilities* (n 18) 33–4; Waxman (n 22) 438, citing Abraham D Sofaer et al, 'Cyber Security and International Agreements' in *Proceedings of a Workshop on Deterring Cyberattacks: Informing Strategies and Developing Options for U.S. Policy* (2010) 179, 185; Michael N Schmitt, 'Computer Network Attack and the Use of Force in International Law: Thoughts on a Normative Framework' (1999) 37 Colum J Transnatl L 914–15; Oona A Hathaway, 'The Law of Cyber-Attack' (2012) 100 Cal L Rev 848; US White House, *The National Security Strategy of the United States of America* (2010) 22; *Tallinn Manual* (n 10) rule 11.

[25] See Waxman (n 22) 463–7; Ellen Nakashima, 'U.S. Official Says Cyberattacks Can Trigger Self-Defense Rule' *Washington Post*, 18 September 2012.

[26] Vida M Antolin-Jenkins, 'Defining the Parameters of Cyberwar Operations: Looking for Law in all the Wrong Places?' (2005) 51 Naval L Rev 172–4.

Some scholars have argued that cyber-attacks that are especially disruptive but have not been traditionally considered as armed attacks under Article 51 might give rise to the Article 51 right of self-defense.[27] But no international tribunal has so held. In a case involving conventional armed violence, but on a smaller scale, the United States argued unsuccessfully before the ICJ that its naval attacks on Iranian oil platforms was justified by the right of self-defense following low-level Iranian attacks on US vessels in the Persian Gulf.[28] Although the separate opinion of Judge Simma in the *Oil Platforms* case argued that self-defense should permit more forceful countermeasures where the 'armed attack' threshold has not been met,[29] this more flexible approach has not been accepted by the ICJ or any court, and only state practice is likely to change the prevailing traditional interpretation.

In addition, the 'use of force' framework has little value in developing responses to terrorists. By the terms of the Charter, non-state actors cannot violate Article 2(4), and responses to uses of force are limited to actions carried out by or otherwise the responsibility of states.[30] Guidance on the degree of state control that must exist to establish state liability for a non-state group's actions was supplied by the ICJ in the *Nicaragua* case, where the Court limited US responsibility for actions of the Nicaraguan Contras to actions where the United States exercised 'effective control of the military or paramilitary operations [of the Contras] in the course of which the alleged violations were committed'.[31] Only if the state admits its collaboration with terrorists or is otherwise found responsible for the terrorists' actions may the victim state use force against the terrorists and sponsoring state.[32]

The law of self-defense remains unsettled. The text of Article 51—'armed attack'—is not as amenable as 'use of force' to a flexible interpretation. Nor did the Charter drafters consider the possibility that very harmful consequences could follow from a non-kinetic cyber-attack.

---

27 See Eric Talbot Jensen, 'Computer Attacks on Critical National Infrastructure: A Use of Force Invoking the Right of Self-Defense' (2002) 38 Stan J Intl L 207, 233–9; Schmitt (n 24) 930–4; *Tallinn Manual* (n 10) rule 13.

28 *Iran v US*, 161 ICJ Rep paras 12, 46–7 (2003).

29 Ibid.

30 UN International Law Commission, *Draft Articles on Responsibility of States for Internationally Wrongful Acts, with Commentaries* (2001), UN GAOR, 53rd Sess. Supp. No. 10, at 80, UN Doc A/56/20, Article 8.

31 *Nicaragua v US*, 14 ICJ Rep (1986) paras 115, 109; *Prosecutor v Tadić*, ICTY Appeals Chamber Judgment (1999) para 145.

32 *Draft Articles on Responsibility of States for Internationally Wrongful Acts, with Commentaries* (n 30).

Nonetheless, outside the cyber realm, state practice has evolved toward accepting that attacks by terrorists may constitute an armed attack that triggers Article 51 self-defense.[33] The text of Article 51 does not limit armed attacks to actions carried out by states, although the state-centric model of the Charter strongly suggests that the drafters contemplated only those armed attacks by non-state actors that could be attributed to a state as Article 51 armed attacks.

The dramatic development that made it clear that armed attacks may occur by non-state terrorists regardless of the role of a state was 9/11. Within days of the attacks, the Security Council unanimously passed Resolutions 1368 and 1373 and recognized 'the inherent right of individual or collective self-defense in accordance with the Charter' in responding to the attacks.[34] NATO adopted a similarly worded resolution.[35] Unlike prior instances where non-state attackers were closely linked to state support, the Taliban merely provided sanctuary to Al-Qaeda and did not exercise control and were not substantially involved in Al-Qaeda operations.[36]

State practice in the international community supported extending self-defense as the ad bellum justification for countering Al-Qaeda on a number of occasions since 2001.[37] While the ICJ has not ratified the evolving state practice, and even seemed to repudiate it in at least three decisions—twice since 9/11 (*Nicaragua v US* in 1986, *Democratic Republic of the Congo v Uganda* in 2005, and *Wall Advisory Opinion* in

---

[33] Steven R Ratner, 'Self-Defense Against Terrorists: The Meaning of Armed Attack' in N Schrijver and L van den Herik (eds), *The Leiden Policy Recommendations on Counter-Terrorism and International Law* (2012) 5–6, 8–9; Michael N Schmitt, 'Cyber Operations in International Law: The Use of Force, Collective Security, Self-Defense, and Armed Conflicts' in *Proceedings of a Workshop on Deterring Cyberattacks: Informing Strategies and Developing Options for U.S. Policy* (National Academies Press 2010) 151, 163–4; Michael N Schmitt, 'Responding to Transnational Terrorism under the *Jus Ad Bellum*: A Normative Framework' (2008) 56 Naval L Rev 18-19; Sean Watts, 'Low-Intensity Computer Network Attack and Self-Defense' (2011) 87 Intl L Stud 60-61; Dept. of Defense Office of Gen. Counsel 1999; *Tallinn Manual* (n 10) rule 13.

[34] Security Council Resolution 1368 (2001), UN Doc S/RES/1368; Security Council Resolution 1373 (2001), UN Doc S/RES/1373.

[35] North Atlantic Treaty Organization (2001), *Statement by the North Atlantic Council*, accessed 4 May 2017 at http://www.nato.int/docu/pr/2001/p01-124e.htm.

[36] See Derek Jinks, 'State Responsibility for the Act of Private Armed Groups' (2003) 4 Chi J Intl L 89.

[37] Ratner (n 33).

2004)—the trend is to accept the extension of armed attack self-defense authorities when non-state groups are responsible, provided the armed attack predicate is met and the group is organized and not an isolated set of individuals. In general, states that were victimized by non-state terrorist attacks were more likely to advocate the more expansive conception of self-defense. Unsurprisingly, the US Department of Defense supports the same position.[38] Thus, despite the apparent gulf between the text of the Charter as interpreted by the ICJ and state practice, whether an 'armed attack' is kinetic or cyber-based, armed force may be used in response to an imminent attack if it reasonably appears that a failure to act promptly will deprive the victim state of the opportunity to defend itself.[39]

The legal bases for self-defense may also be extended to anticipatory self-defense in the cyber context. As evolved from Secretary of State Daniel Webster's famous formulation in response to the *Caroline* incident that self-defense applies in advance of an actual attack when 'the necessity of that self-defense is instant, overwhelming, and leaving no moment for deliberation',[40] contemporary anticipatory self-defense permits the use of force in anticipation of attacks that are imminent, even if the exact time and place of attack are not known.[41] Imminence in contemporary contexts is measured by reference to a point in time where the state must act defensively before it becomes too late.[42] In addition to imminence or immediacy, the use of force in self-defense must be necessary—law enforcement or other non-use of force means will not suffice—and the attacking group must be shown to have the intent and means to carry out the attack.[43]

In contemporary state practice, nearly every use of force around the world is justified as an exercise of self-defense.[44] As Sean Watts has observed, 'in the post-Charter world ... states have resurrected pre-Charter notions that self-defense includes all means necessary for self-preservation against all threats'.[45] So interpreted, the legal parameters of self-defense law may be adapted to cyber-attacks, subject to meeting the

---

[38] Ibid.

[39] Schmitt (n 19) 593.

[40] Daniel Webster, 'Letter', reprinted in H Miller (ed), *Treaties and Other International Acts of the United States of America, Vol 4* (1934) (1842).

[41] *The National Security Strategy of the United States of America* (n 24).

[42] See Schmitt 2008 (n 33) 18–19; *Tallinn Manual* (n 10) rule 15.

[43] See Schmitt 2008 (n 33) 18–19.

[44] Watts (n 33).

[45] Ibid 76.

formidable Article 51 threshold of armed attack. Thus, if a cyber-attack by a non-state actor constitutes an armed attack as contemplated by the Charter, self-defense allows the victim state to conduct forceful operations in the state where the terrorist perpetrators are located if that state is unwilling or unable to police its territory.[46] In the sphere of anticipatory self-defense, the fact that cyber-attacks arrive unattributed and without warning provide strong analogs to the challenges of counter-terrorism law that give rise to the contemporary interpretation. At the same time, even though reliance on self-defense arguments is and will remain tempting in the cyber arena, the Charter system remains subject to the 'armed attack' qualification.

What does international law say about cyber-attacks that do not meet the armed attack threshold? One potentially important rule distilled from the Charter and state practice is that a number of small cyber attacks that do not individually qualify as armed attacks might do so when aggregated, provided there is convincing evidence that the same intruder is responsible for all of the attacks.[47] The so-called 'pin-prick' theory could have emerging importance in supporting cyber self-defense, especially if technical advances aid in attribution. Otherwise, distilling the conclusions developed in this section so far, the international law of self-defense may only justify responses to cyber-attacks that are sufficiently destructive to meet the armed attack threshold.

What international law determines the permissible responses to a cyber-attack that causes significant economic harm but no physical damage? Is the loss or destruction of property sufficient to trigger a kinetic response? The answer turns in part on whether the state wishes to use force in response. For non-forceful responses, customary international law has long allowed countermeasures—temporarily lawful actions undertaken by an injured state in response to another state's internationally unlawful conduct.[48] The state that places malware inside the cyber systems in another state has violated the victim state's sovereignty. In the cyber context, sovereignty intrusions that fall short of armed attacks as defined by the Charter are nonetheless in violation of the international law norm of non-intervention and thus permit the reciprocal form of violation by the victimized state. As codified by the

---

[46] Ashley S Deeks, 'Unwilling or Unable: Toward a Normative Framework for Extraterritorial Self-Defense' (2012) 52 Va J Intl L 483.

[47] *Tallinn Manual* (n 10) rule 13.

[48] *Draft Articles on Responsibility of States for Internationally Wrongful Acts, with Commentaries* (n 30).

UN International Law Commission's Draft Articles on State Responsibility for Internationally Wrongful Acts, countermeasures must be targeted at *the state* responsible for the prior wrongful act, and must be temporary and instrumentally directed to induce the responsible *state* to cease its violation.[49]

In the cyber arena, one important question is whether countermeasures include active defenses, 'hack backs' which attempt through an in-kind response to disable the source of an attack while it is underway.[50] Whatever active defense technique pursued by the victim state thus has a reciprocal relationship with the original cyber-intrusion, and like the original intrusion the active defense presumptively breaches state sovereignty and violates the international law norm of non-intervention. (Passive defenses, such as firewalls, attempt to repel an incoming cyber-attack.) Active defenses may be pre-set to deploy automatically in the event of a cyber-attack, or they may be managed manually.[51] Computer programs that relay destructive viruses to the original intruder's computer or packet-flood the computer have been publicly discussed.[52] Although descriptions of most active defenses are classified, the United States has publicly stated that it employs 'active cyber defense' to 'detect and stop malicious activity before it can affect [Department of Defense] networks and systems'.[53]

In theory, countermeasures provide a potentially effective defensive counter to many cyber-attacks. In practice, a few problems significantly limit their effectiveness. First, the Draft Articles codify customary law requirements that before a state may use active defense countermeasures it must find that an internationally wrongful act caused the state harm, identify the state responsible, and follow various procedural requirements, delaying execution of the active defense.[54] The delay may be exacerbated by the problems in determining attribution. Second, countermeasures customarily are available in state-on-state conflicts, not in response to intrusions by a non-state actor. A non-state actor's actions may be attributable to a state when the state knows of the non-state actors' actions and aids them in some way,[55] or possibly when the state

---

[49] Ibid Article 49.

[50] Jensen (n 27) 230.

[51] Ibid 231.

[52] Ibid.

[53] US Dept of Defense, *Strategy for Operating in Cyberspace 7* (2011) 230.

[54] *Draft Articles on Responsibility of States for Internationally Wrongful Acts, with Commentaries* (n 30) Articles 49–52.

[55] Ibid Article 16.

merely knowingly lets its territory be used for unlawful acts.[56] In most instances, however, international law supplies no guidance on countermeasures that respond to intrusions by non-state actors. Third, the normative principle that justifies countermeasures is that the initial attacker must find the countermeasure sufficiently costly to incentivize lawful behavior. For non-state groups that act independently of any state, a fairly simple relocation of their servers or other equipment may evade or overcome the countermeasures and remove any incentives to stop the attacks. In sum, although the countermeasures doctrine is well-suited to non-kinetic responses to cyber-attacks by states, attribution delays may limit their availability, and the line between permitted countermeasures and a countermeasure that constitutes a forbidden 'use of force' is not clear. Nor do countermeasures apply in responding to an attacker unaffiliated with any state.

Even if each of these limitations is overcome, the prevailing view is that active defenses may only be employed when the intrusion suffered by a victim state involves a 'use of force' as interpreted at international law.[57] Taken together, the promise of countermeasures in responding to cyber-attacks is significantly compromised by problems of attribution, timing, efficacy and logic. However, if active defense countermeasures do not involve a 'use of force', the attribution problem loses its urgency. There is no clear international barrier to non-use of force countermeasures, and attribution may be determined when feasible since no force is being used. Finally, the International Group of Experts that prepared the *Tallinn Manual* acknowledged that while victim states may not continue countermeasures after the initial intrusion had ended, state practice 'is not fully in accord … States sometimes appear motivated by punitive considerations … after the other State's violation of international law had ended'.[58] In other words, customary law on cyber countermeasures is in flux.

Whether the development of cyber-law so removed from the text of the Charter represents the optimal path forward for the law of cyber-conflict is unclear. On the one hand, the Charter's traditional self-defense doctrine may not leave states sufficient authority to respond to the full range of cyber threats they face. On the other hand, the development of customary

---

[56] *UK v Albania*, 4 ICJ Rep 22 (1949); Matthew J Sklerov, 'Solving the Dilemma of State Responses to Cyberattacks: A Justification for the Use of Active Defenses against States Who Neglect their Duty to Prevent' (2009) 210 Mil L Rev 1, 43.

[57] Jensen (n 27) 231.

[58] *Tallinn Manual* (n 10) rule 9.

law through state practice is the ultimate flexible vehicle for making new law to confront emerging problems. As with other aspects of norm development in international law, many states with vested interests in applying norms from the kinetic warfare realm to cyber tend to favor retaining core Charter principles, while states more often victimized by terrorism have looked to state practice to develop customary law norms. In any case, even Charter law interpreted at degrees of separation from the Charter is preferable to a legal vacuum.[59]

3. DEVELOPING CONTEMPORARY AD BELLUM PRINCIPLES FOR CYBER CONFLICT BELOW THE ARMED CONFLICT THRESHOLD

Particularly for cyber-attacks that are especially disruptive but not destructive—intrusions that may be increasingly pervasive, operating beneath the radar of many states' existing defensive mechanisms, and capable of fairly easily and cheaply being perpetrated by virtually any state or non-state actor—the Charter provides an incomplete normative blueprint. The asymmetric opportunities for state and non-state adversaries abound, and under the Charter norms victim states may have to choose between defending themselves unlawfully and absorbing continuing cyber-attacks.[60] Alternately, arguing that the measure of compliance with the gateway articles of the Charter should be practical, based on the effects of a cross-border intrusion and not on the nature of the instruments that cause the effects, Michael Schmitt and other scholars have argued that cyber-attacks that cause significant harm should count as uses of force and, less plausibly, armed attacks. Their view is that once the gateway determinations are made for the Charter to reach the cyber domain, international law supplies at least a serviceable roadmap for limiting cyber-war. The debates continue as the daily tally of cyber-attacks escalates.

This chapter has shown that arguments to apply the 'use of force' and 'armed attack' Charter categories to cyber may be based on a tautology; if the incoming cyber intrusion is construed as an armed attack, the victim state may respond in kind. If not so construed, the same or a

---

59 Watts (n 33) 66.
60 Ibid 60–61.

similar response may not be considered an armed attack.[61] The fact that it may be possible simply to characterize a new form of intrusion as a use of force or armed attack is not satisfying analytically and, over time, such tautological reasoning may diminish the normative values embedded in these critical cornerstones of the Charter. In a similar vein, state practice in shaping responses to cyber-intrusions has been characterized as applying a 'know it when you see it'[62] approach to deciding when the intrusion constitutes a 'use of force' or 'armed attack' that would trigger IHL requirements. Such ad hoc reasoning does little to build confidence that the international community may arrive at acceptable norms for protecting critical infrastructure from cyber threats.

Meanwhile, the dynamic growth of reliance on the Internet to support our infrastructure and national security have caused the United States to modify its longstanding views on the predicates for treating a cyber-intrusion as an 'armed attack' or 'use of force'. As Matt Waxman has noted, US government statements suggest that cyber-attacks that have especially harmful effects will be treated as armed attacks, while lower level intrusions would enable cyber countermeasures in self-defense.[63] The result is a tiered interpretation of Article 51 based on the instrument of attack—an expansive interpretation when defending against armed violence, and a narrower view with a high impact threshold for cyber-attacks.[64] Whatever precision and calibration of authorities is gained by these fresh reinterpretations of the Charter, they replace the relative clarity of an 'armed attack' criterion with fuzzier effects-based decision-making that injects ever more subjectivity and less predictability into future self-defense projections. Taking into account the characteristics of cyber conflict—uncertainty, secrecy and lack of attribution—finding consensus on international regulation through these Charter norms will be a tall order.[65]

Attribution of cyber-attacks is a technical problem, not one that the law can fix. Yet the challenges in attributing intrusions in real time with confidence should not foreclose the development of legal authorities that protect national and human security. Anonymity and surprise have long been central tenets of terrorist attacks, and international law has developed normative principles, including anticipatory self-defense, that

---

61 Dept of Defense Office of Gen Counsel, *An Assessment of International Legal Issues in Information Operations* (1999) 16–19.

62 *Jacobellis v Ohio*, 378 US 197 (1964) (Stewart J, concurring).

63 Waxman (n 22) 439.

64 Ibid.

65 Ibid 443.

accommodate these characteristics. By analogy international law may develop along similar lines to provide ad bellum bases for responding to cyber-attacks. In light of continuing attribution problems, and the likelihood that cyber-attacks will come from sources around the world, a cyber-international law could subordinate traditional legal protections that attach to national boundaries and narrowly tailor mechanisms that permit defending against the sources of the attacks, whatever their locations.

One of the difficulties of attribution is that learning that an attack comes from within a certain state does not tell us whether the attack is state-sponsored or was carried out by a non-state actor. Existing Charter and IHL law of state responsibility—heavily influenced by the United States and other western states that do not have comprehensive controls over private infrastructure—does not make the state responsible for the actions of private actors over which it has no direction or control. There is thus no clear IHL or Charter-based authority to go after the private attackers inside a state when that state was not involved in the attacks.[66] International law offers an alternative normative path. For example, criteria could be developed that indicate the circumstances where absolute attribution may be delayed in favor of immediate defensive action, when intelligence is reliable enough to authorize those actions, and under which circumstances defensive operations may invade territorial sovereignty without state permission.

The 2011 US *International Strategy for Cyberspace* asserts that 'the development of norms for state conduct in cyberspace does not require a reinvention of customary international law, nor does it render existing international norms obsolete. Long-standing international norms guiding state behavior—in times of peace and conflict—also apply in cyberspace'.[67] Because cyberspace has been around for a relatively short period of time, there is no extensive catalog of state practice that provides the basis for a body of customary cyber conflict law. Further complicating the search for evidence of customary law in cyber conflict is the secrecy that surrounds most cyber operations, and their lack of attribution.[68] In the last decade, however, several mostly public cyber-attacks occurred, including those in Estonia (2007), Georgia (2008), and the first in an escalating series of attacks on US military, Intelligence Community, and commercial networks for the purpose of transferring sensitive

---

[66] *Tallinn Manual* (n 10) rule 6.

[67] Gary Brown and Keira Poellet, 'The Customary International Law of Cyberspace' (2012) Strategic Studies Q 126, 140.

[68] Ibid.

information or stealing intellectual property. By 2010, the Stuxnet worm had targeted Iranian nuclear facilities, although the attack was not publicly revealed until 2012. Surprisingly, Iran did not blame Stuxnet or even a cyber-attack by the United States or Israel for the delays in making its nuclear plant operational. (It surely would have responded if a missile had damaged its facilities.) In any case, Iran did not allege a violation of international law by cyber means in the Stuxnet episode.

The Russian cyber-attacks against Georgia in 2008 likewise did not clearly constitute a case of a purely cyber conflict waged by one state against another. Russian troops crossed the border as an invasion force on the same day that Russian cyber actions were taken, most likely to interfere with Georgia's communications during the surprise armed attack by the Russian military. Georgia then declared it was in a state of war with Russia, but it did not single out the cyber intrusions as an attack.[69] Likewise, the Estonian intrusions by Russia in 2007 involved distributed denial of service activities, more like a series of criminal acts than a use of force. A further complication in Estonia was the inability to clearly attribute the denial of service intrusions.

As these examples show, customary international law governing cyber conflict is likely to develop unevenly over time, as state, regional and perhaps even global policies and practices evolve. Consider one example. Intelligence collection is practiced by every state. While the domestic laws of nearly every state forbid spying within its territory, neither those laws nor any international law purports to regulate espionage internationally. In the digital world, the equivalent intelligence collection activity is cyber-exploitation—espionage by computer, a keystroke monitor, for example—and nothing in the Charter, IHL, or other customary law traditionally stands in its way, except to the extent that espionage involving military weapons systems constitutes armed aggression.[70] Given the growing capabilities of digital devices to spy, exploit and steal, including military and other sensitive national secrets, the absence of international regulation is problematic. It is possible that IHL could develop customarily through state practice to recognize legal limits on one variant of cyber-exploitation where the software agent is capable of destructive action or may facilitate the same.[71] For example, malware has infiltrated and interfered with the oil and gas, freight and passenger rail

---

[69] Ibid.

[70] Roger D Scott, 'Territorially Intrusive Intelligence Collection and International Law' (1999) 46 A F L Rev 223–4.

[71] National Research Council 2009 (n 18) 261, 263.

signaling systems,[72] and the US air traffic control system is vulnerable to cyber-attack.[73]

International law for cyber-operations could evolve through something like natural law-type or just war theory reasoning, as has been the case with development of some other international law norms.[74] Just war theory and natural law reasoning or its equivalent has served as a gap-filler in international law, and could do so for cyber. The making of customary international law is often unilateral in the beginning, followed by a sort of dialectic of claims and counterclaims that eventually produce customary law that is practiced by states.[75] As some prominent US academics developed theories of 'vertical domestication'[76] to encourage greater respect and adherence to international law by the US government, in the last decade the US government sought to export its emerging counterterrorism law as international law in response to kinetic attacks on the United States and its interests. Although controversy surrounded some of the US government policies and practices, counterterrorism law has matured and developed normative content around some of its revised tenets, such as the permissible use of force against non-state terrorists inside a sovereign state.[77]

However it occurs, international law norm development for cyber might expand or contract the authorities that would otherwise govern under current interpretations of the Charter. On the one hand, an evolving international law regime may enable victim states more tools and greater flexibility in anticipating and responding to cyber-attacks. Active defense countermeasures and other kinds of responses may be permitted, through state practice, but predicated upon legal authority, where the same responses would not have been lawful under the Charter as traditionally

---

[72] Brenner (n 17) 105–10.

[73] US General Accounting Office, *Information Security: FAA Needs to Address Weaknesses in Air Traffic Control Systems,* GA0-15-221 (2015).

[74] Jeffrey L Dunoff and Mark A Pollack, 'What Can International Relations Learn From International Law?' (2012) Temp Univ Legal Stud 11.

[75] Michael W Reisman, 'Assessing Claims to Revise the Laws of War' (2003) 97 Am J Intl L 82.

[76] Harold H Koh, 'The 1998 Frankel Lecture: Bringing International Law Home' (1998) 35 Hous L Rev 626–7; Harold H Koh, 'Transnational Legal Process' (1996) 75 Neb L Rev 181, 183–4.

[77] Robert M Chesney, 'Who May Be Killed? Anwar al-Awlaki as a Case Study in the International Legal Regulation of Lethal Force' (2011) 13 Y B Intl Hum L 3; James B Steinberg and Miriam R Estrin, 'Harmonizing Policy and Principle: A Hybrid Model for Counterterrorism' (2014) 7 J Natl Sec L & Poly 161.

interpreted because the armed attack threshold was not met. On the other hand, some cyber responses that are now lawful under international law because there is no use of force or armed attack involved in the response—a small scale action designed to neutralize an incoming cyber-intrusion aimed at one system, for example—could be considered unlawful if the harmful consequences are significant.[78]

For the United States, the fact that so much of the infrastructure is privately owned makes securing the infrastructure legally and practically problematic,[79] and yet heavy reliance on networked information technology makes the United States highly vulnerable to cyber-intrusions. The government's recent posture on cyber operations has been to mark out preferred clear positions on the authority to respond to destructive cyber-attacks with armed or forceful responses, while maintaining what Matt Waxman aptly calls 'some permissive haziness'[80] concerning the norms for responding to cyber-intrusions that are less harmful but distracting. From the domestic perspective, the United States can assure itself of the authority to respond to serious intrusions while preserving the flexibility to tailor its countermeasures and develop its cyber defenses according to the nature and severity of the threat faced.

The nuanced calculations by the United States in developing its cyber doctrine are consistent with its longstanding opposition to some other states' expansive interpretations of Articles 2(4) and 51 to include economic coercion and political subversion.[81] Yet emerging cyber doctrine by the United States may be seen in the international community as just the sort of proposed expansion of the Charter norms that the United States has publicly opposed in the past. Indeed, as the evolving criteria for what triggers the Article 51 right of self-defense over the last 25 years shows, freighting fast-developing cyber-defense norms onto an already-burdened Article 51 invites controversy and may destabilize and even undermine the normative value of the Charter.

In activating the US Cyber Command in 2010, the Department of Defense confronted congressional skepticism and challenges from across the political spectrum that focused on fears of the Command's capabilities for interfering with the privacy rights of citizens, the policies and authorities that would define its mission, and its relationship to the

---

78 National Research Council 2009 (n 18) 245.

79 Waxman (n 22) 451.

80 Ibid 452.

81 Ibid 453.

nation's largely privately held critical infrastructure.[82] Against this backdrop, from the beginning Cyber Command and the Department of Defense have stated that existing Charter interpretations and the laws of war adequately provide the authorities needed to defend the United States from cyber-attack.[83] Even as President Obama in 2013 issued a classified policy directive that detailed basic principles for US responses to cyber intrusions (PPD-20 2013), including defensive and offensive cyber operations, the Legal Adviser to the State Department continued to affirm that the United States would engage in cyber-conflict according to existing understandings of international law.[84]

In 2015, the US Department of Defense publicly announced two major cyber milestones. First, in April, the *Department of Defense Cyber Strategy* stated that 'DoD must be prepared to defend the United States and its interests against cyberattacks of significant consequence ... [which] may include loss of life, significant damage to property, serious adverse U.S. foreign policy consequences, or serious economic impact on the United States'.[85] As a statement of US government policy, note the subtle but unmistakable shift away from the 'armed attack' and 'use of force' categories. Seriously adverse foreign policy or economic impacts may occur absent kinetic attacks, by cyber means. The *Strategy* also reiterates that 'the United States will always conduct cyber operations under a doctrine of restraint, as required to protect human lives and to prevent the destruction of property ... in a way that reflects enduring U.S. values, including support for the rule of law, as well as respect and protection of the freedom of expression and privacy, the free flow of information, commerce, and ideas'.[86] These additional cornerstone principles are important in limiting US cyber operations and in setting an example for other states as they shape their cyber policies. Other than a reminder that DoD cyber operations are conducted 'in accordance with the law of armed conflict', the *Strategy* does not indicate that the new characterization of the DoD cyber mission is based on legal obligation. Still, if practiced through publicly acknowledged cyber actions over some

---

82 Ellen Nakashima, 'Cyber Command Chief Says Military Computer Networks Are Vulnerable' *Washington Post* (4 June 2010).

83 Watts (n 33).

84 Harold H Koh, 'International Law in Cyberspace: Remarks Prepared for Delivery to the USCYBERCOM Inter-Agency Legal Conference' (2012) 54 Harv Intl L J Online 3.

85 *The Department of Defense Cyber Strategy* (n 8).

86 Ibid.

period of years, the DoD formulation could provide a pillar of a normative architecture for cyber conflict.

To its credit, the 2015 *Strategy* suggests that developing cyber doctrine may be more effective and more likely to be accepted internationally if it is separated from the effects-based approach relied upon by the Charter and IHL-based doctrines for cyber-operations. Not that such a legal code of conduct based in international law would be a panacea. Law must follow, not lead, particularly in an area like cyber, where policies are not yet well defined and strategies are unclear.[87]

Second, in June 2015, the Department of Defense released its long-awaited *Law of War Manual*.[88] For the first time, DoD included a chapter on cyber operations. In general, the *Manual* anticipates that cyber-attacks that cause physical damage will be subject to the rules governing kinetic attacks.[89] The *Manual* also recognizes that cyber operations may constitute 'use of force' under the Charter, based on the effects of the cyber intrusion,[90] and that the Article 51 right of self-defense applies to a use of force or armed attack,[91] whether the attack is attributed to another state or to a non-state actor.[92] In other words, the *Manual* lags behind the *Strategy* and simply superimposes ad bellum principles from kinetic armed conflict on cyber operations. Follow but not lead, indeed.

4. CONCLUSIONS

Imagine this scenario. It is summertime in the not-distant future. Just before the afternoon rush hour on a hot and steamy July day, the northeastern United States is hit with a massive blackout. The electric grid is crippled from Boston to New York, Philadelphia to Baltimore and Washington, and from there west as far as Cleveland. While back-up generators resume the most critical operations in hospitals and other critical care centers, all other activities that depend on electricity come to a sudden halt.

Government and private industrial security experts quickly discover the software and malware that has accessed supervisory control and data

---

87 Waxman (n 22) 455–7.
88 US Dept of Defense, *Law of War Manual* (2015).
89 Ibid 16.2.
90 Ibid 16.3.1.
91 Ibid 16.3.3.1.
92 Ibid 16.3.3.4.

acquisition (SCADA) controls—the industrial control system that supervises data over dispersed components of the electric grid and which are connected to the global Internet.[93] In recent years, industry reports that a few laptops containing information on how to access SCADA controls were stolen from utility companies in the Midwest. During the same period, computers seized from Al-Qaeda and IS captives abroad contained similar details about US SCADA systems. The vast majority of the affected electric grid is privately owned, and officials estimate that the cyber-attacks have done long-term damage to critical system components, and have rendered useless generators and other equipment that must be replaced where no back-up replacement equipment is standing by. Even rudimentary repairs will take weeks or months, and full system capabilities may not be restored for more than one year. Economic losses will be in the billions of dollars, and millions of Americans' lives will be disrupted for a long time.

The software and malware were set to trigger the blackout at a pre-determined time. The attacks were not attributed, and although intelligence and law enforcement experts quickly traced the original dissemination of the attacks to computers in South Asia, the only other available intelligence comes from the seized and stolen laptops. The governments of Russia, China and Iran have denied any involvement in the attacks, and no intelligence points to their involvement. Al-Qaeda and IS have shown interest in cyber capabilities, and the seized laptops suggest that some steps were taken to acquire them.

Assuming that the United States concludes that terrorists are most likely behind the attacks, what law governs the response? If, instead, we decide that the attacks were launched by Russian intelligence operatives situated in South Asia, what law applies? This chapter has helped draw attention to the incompleteness of the legal regime that will be required to provide the normative justifications in international law for responding to these intrusions.

The stakes are escalating. The United States used offensive cyber weapons to target Iran's nuclear program, and states and non-state actors are increasingly aware that cyber weapons—offensive and defensive—are available, with ever-growing sophistication. Although reports indicated the United States declined to use cyber weapons to disrupt and disable the Qaddafi government's air defense system in Libya at the start of the US/NATO military operation in 2011 because of the fear that such a cyber-attack might set a precedent for other nations to carry out their own

[93] Brenner (n 17) 96–7.

offensive cyber-attacks,[94] Stuxnet created the precedent, as did Israel's cyber-attack on Syrian air defenses when it attacked a suspected Syrian nuclear site in 2007,[95] Russia's cyber-attacks in its dispute with Georgia,[96] and the apparent use of cyber-weapons by the United States to target Al-Qaeda websites and terrorists' cell phones.[97] Now that the cyber war battlefield apparently has expanded to Beirut banks and a neutral state (Lebanon),[98] it appears that cyber weapons are being used beyond countering imminent national security and infrastructure threats.

Developing an international consensus on the norms for cyber conflict will not be easy. The state of doctrinal international law is only partly to blame. At least as important as constraints are the political differences among states and non-state actors in shaping cyber norms. In addition, the facts needed to make the normative judgments in this fast-paced realm of changing technologies are now and will be for the foreseeable future hard to come by and even more difficult to verify.[99] Law will play catch up, as it should, but the lag between evolving technologies and normative stability in cyber operations may be a long one. Legal change will occur, to be sure, but the process may be fraught.

This chapter has shown that the international community runs significant risks in continuing to build cyber-conflict law using the Charter/IHL model. One overarching concern is that categorizing cyber-attacks as a form of armed attack or use of force may enhance the chance that a cyber-exchange could escalate to a military conflict.[100] If, over time, the thresholds for what constitutes an armed attack are lowered to reach more forms of cyber-intrusion, legal barriers to military force will be lowered, leading to more military conflicts in more places. The high threshold for invoking the Charter's self-defense authorities traditionally supported by

---

[94] Eric Schmitt and Thom Shanker, 'U.S. Debated Cyberwarfare in Attack Plan on Libya' New York Times (17 October 2011).

[95] Dave A Fulghum and Robert Wall, 'Cyber-Combat's First Shot: Attack on Syria Shows Israel is Master of the High-Tech Battle' (2007) Aviation Week & Space Technology 28.

[96] John Markoff, 'Before the Gunfire, Cyberattacks' *New York Times* (12 August 2008).

[97] Markoff (n 96); Jack Goldsmith, 'Quick Thoughts on the USG's Refusal to Use Cyberattacks in Libya' Lawfare (18 October 2011).

[98] Katherine Mayer, 'Did the Bounds of Cyber War Just Expand to Banks and Neutral States?' *The Atlantic* (17 August 2012).

[99] Waxman (n 22) 448.

[100] Martin C Libicki, *Cyberdeterrence and Cyberwar* (RAND 2009) 69–70; Mary Ellen O'Connell, 'Cyber Security Without Cyber War' (2012) 17 J Conflict & Sec L 187, 190–91, 199.

the United States also offers some insurance against precipitous action in response to unattributed cyber-attacks. That such a high threshold fails to deter low-level hostilities may be a reasonable price to pay.[101]

Yet the high self-defense threshold also leaves unregulated a wide swath of cyber-intrusion techniques, those now in existence and others yet to be invented. This byproduct of the bifurcation of international law into war and peace, armed conflict or not armed conflict, armed attack and use of force or not leaves every intrusion that fails to meet the kinetic standard not subject to international law limitations, except for the limited customary authorities for countermeasures and the open-ended rule of necessity.[102] If states or the international community attempt to further expand the reach of self-defense and IHL in idiosyncratic ways to non-destructive cyber intrusions, the Charter and IHL will be compromised.

Despite the disconnect between the text of the Charter as interpreted by the ICJ and state practice, whether an attack is kinetic or cyber-based, state practice has been to enable armed force in response to an imminent attack if it reasonably appears that a failure to act promptly will deprive the victim state of the opportunity to defend itself. Article 51, or at least its self-defense shadow, has become the go-to authority for military action waged by states, whatever the context. The self-defense arguments may be adapted to cyber, but the further the analogies to responses to armed attacks stray from kinetic means, the greater the likelihood that Article 51 norms will erode. The temptation to rely on Article 51 is great, to be sure, particularly where, as in cyber, other sources of legal authority to take what is viewed as essential defensive action may not exist.

101 Waxman (n 22) 446–47.
102 *Tallinn Manual* (n 10) rule 9.

10. Some legal and operational considerations regarding remote warfare: drones and cyber warfare revisited

*Terry D Gill, Jelle van Haaster and Mark Roorda**

1. INTRODUCTION

'Remote warfare' is a term which can denote a variety of forms and techniques of warfare and which can be defined in a number of ways. It can refer to the use of long range artillery and strategic bombing in 20th century warfare, to the use of weapons systems which allow the selection and engagement of targets by an operator far removed from the traditional battlespace in contemporary and emerging military practice. It is in the latter sense that this term will be used in this chapter. Such weapons and techniques have in common that they need not be and usually are not controlled from the traditional area of operations, and are capable of being used against targets which are themselves not necessarily located on or near a 'hot battlefield'. This is a relatively new phenomenon in warfare which, until the middle of the last century, was generally characterized by a relatively clear geographical demarcation between a front line, or in any case an area of operations, where opposing forces manoeuvred and engaged each other in head on encounters and a 'rear area', which was generally far removed from actual fighting. For centuries warfare was waged between opposing armies which employed weapons which were capable of being used only over a distance of several miles at most, were dependent upon direct visual contact and involved concentrating forces and firepower at close range in order to achieve maximum effect.[1] With the advent of airpower and later

* The views presented here are expressed in a personal capacity and do not necessarily reflect the official views of the Netherlands Armed Forces or Government.

[1] Archer Jones, *The Art of War in the Western World* (Oxford University Press 1987) gives an excellent overview and analysis of the historical development of warfare and how different arms (infantry, cavalry, artillery) and weapons

of ballistic missiles capable of engaging and destroying targets that were far removed from a front line, the concept of 'remote warfare' emerged. Strategic bombing campaigns in World War II were followed by the development and deployment of ballistic missile systems on land and at sea which were designed to be used at intercontinental range and were based on strategies of mutually ensured destruction, which had the aim and effect of deterring the prospective opponent from their actual use in warfare.[2]

Contemporary techniques and weapons of remote warfare differ from their predecessors in a number of ways. They are not designed to destroy the industrial capacity and infrastructure of a state as strategic bombing was intended to do in World War II, or even the state itself as nuclear ballistic weapons were capable of. Post-industrial warfare in which remote warfare takes place is usually not between highly organized industrialized states, but more often than not between a (Western) state and a non-state entity, usually an armed group equipped with relatively unsophisticated weapons, and relying on deception, irregular warfare, often employing terrorist tactics and deliberate lack of discrimination in its operations, which are aimed at the influencing of public opinion both in the state(s) where the conflict is ongoing and further afield. It is not necessarily restricted to a particular area of operations, and some armed groups operate regularly across international borders and have 'affiliates' and imitators in a variety of countries. Consequently, the types of and use of techniques and weapons of contemporary remote warfare differ from their predecessors in World War II and the Cold War. They are not only usually operated from a great distance from the conflict area against targets which are not necessarily located in a traditional area of operations, but they are specifically designed to be used to unbalance the ability of the opponent by striking at particular individuals or groups of individuals rather than at industrial targets or massed enemy forces and to effect the adversary's command, communications and control systems and obtain detailed information relating to the adversary's capabilities, rather than inflicting large numbers of casualties. They do not involve the use of mass firepower or destructive capacity, but are rather designed to create specific effects, such as degrading the leadership of a particular

combined with mass firepower and concentration of effort dominated Western warfare over a long period from classical antiquity though the modern era.

[2] Lawrence Friedman, 'The First Two Generations of Nuclear Strategists' in Peter Paret (ed), *Makers of Modern Strategy from Machiavelli to the Nuclear Age* (Princeton University Press 1986) 735ff, gives an authoritative account of nuclear strategy in the Cold War era.

group, affecting the adversary's ability to utilize particular weapons, or its ability to communicate and to influence public opinion. They may be used as 'stand-alone' weapons, but more likely will be employed alongside other techniques of warfare and other weapons systems to prepare and shape a battlefield for more conventional operations.[3]

Two types of remote warfare techniques and weapons that are currently in use or are emerging as potential means and methods of warfare will receive attention in this contribution. These are unmanned aerial vehicles (UAVs) (widely referred to as drones) and the concept of cyber warfare, which is rapidly emerging as a technique of warfare that can be used for a wide variety of purposes. Each of these has specific operational characteristics, (potential) uses and possibilities. Each has specific (potential) drawbacks and poses specific challenges and dilemmas of a policy, ethical and legal nature. Some of these will be given attention in this chapter, the purpose of which is to give a brief overview of some of the operational characteristics of the use of both UAVs and cyber warfare and discuss some of the main legal challenges and dilemmas they have posed or are likely to pose. Much has been written about both the use of UAVs and cyber warfare, so to an extent this chapter will inevitably revisit some of the discussion of how the law applies to their (potential) use. However, we will attempt to do so in a way that tries to bring together both some of the operational considerations relating to their use alongside the legal challenges and combine both perspectives to some extent. This may help the discussion along and contribute to placing the use of these techniques of remote warfare into a clearer perspective.

This chapter is structured as follows. In the following section, some of the principal operational uses and possibilities of the use of UAVs and cyber warfare will be presented. Their main characteristics, uses and limitations will be set out. In the third section, some of the main legal challenges relating first to the use of UAVs and then in relation to cyber warfare will be discussed in two separate subsections. The chapter will be rounded off with a number of conclusions bringing together both the operational and the legal perspectives. It should be borne in mind that this chapter is not intended to provide a comprehensive analysis of all of the operational and legal questions relating to these two types of remote warfare, nor still less, to explore the concept of remote warfare as a whole in depth or provide conclusive answers to the questions discussed. However, to the extent it provides a useful overview of some of the main

---

[3] See Section 2 below for a more detailed discussion of the (potential) uses of UAVs and cyber warfare in contemporary operations.

questions relating to the interaction of operational and legal factors in their (potential) employment, it will have succeeded in its purpose.

2. OPERATIONAL PERSPECTIVES IN REMOTE WARFARE

2.1 Some Operational Characteristics and Capabilities of UAVs

Over the last decade and a half, there has been an exponential growth in the military use of UAVs, both for intelligence, surveillance and reconnaissance (ISR) purposes[4] as well as for conducting attack missions.[5] Though drones come in all sizes and shapes, those that are capable of acting as a strike platform are usually able to operate at considerable altitudes and ranges and are dependent on highly sophisticated communications equipment, a category of drones often referred to as medium-altitude long-endurance (MALE) UAVs.[6] Military powers like the United States, the United Kingdom and China are known to possess armed drone technologies, but it is less well-known that countries such as Pakistan, Iraq, Iran and South Africa also possess them.[7] Most recently, it was revealed that Nigeria too operates armed drones after it released a video showing a Nigerian Air Force drone bombing a Boko Haram logistics base in the northeast of the country.[8] Dozens of other nations are either openly or discreetly trying to obtain similar technologies. The US *Predator* and *Reaper* drones are perhaps the most commonly known types, yet due to strict US export regulations, Israeli and Chinese-made aircraft are becoming more popular. The wide proliferation of armed UAVs has caused considerable controversy regarding legal and ethical

---

[4] In fact, the vast majority of UAVs in use are solely dedicated to ISR tasks. UAVs are also referred to as 'remotely piloted aircraft' (RPA), but because the term UAV is more widely known, it will be used here.

[5] Kelley Sayler, *A World of Proliferated Drones: A Technology Primer* (Center for a New American Security 2015).

[6] See for instance House of Commons Defence Committee, Remote Control: Remotely Piloted Air Systems – current and future UK use (Tenth Report of Session 2013–14, vol I) 52.

[7] Clay Dillow, 'All of These Countries Now Have Armed Drones' Fortune online (12 February 2016), accessed 4 May 2017 at http://fortune.com/2016/02/12/these-countries-have-armed-drones/.

[8] Kelsey Atherton, 'Watch Nigeria's First Confirmed Drone Strike – Against Boko Haram' (3 February 2016), accessed 4 May 2017 at http://www.popsci.com/watch-nigerias-first-confirmed-drone-strike.

issues, yet despite the critiques, countries continue pursuing their use, particularly due to their (perceived) operational and strategic value.[9] There is, however, some debate as to the novelty of the use of armed drones. One view is that they do not add a significantly new capability compared to other weapons. Armed drones are, in many respects, very similar to manned aircraft, performing similar functions and missions, with the obvious exception that the operator of a UAV is physically removed, sometimes by thousands of kilometres, from the aircraft. Its main characteristics, such as covertness, precision, or the ability to engage targets without putting an operator at risk, are also provided by assets such as long-range artillery, cruise missiles or special forces. While this is undoubtedly the case, the alternative view is that the *combination* of characteristics of armed UAVs sets it apart from more traditional capabilities, bringing together valuable features otherwise only separately found in other types of weapons.

Most of the armed drones currently in existence were developed to be used in irregular type conflicts that have been predominant over the last couple of decades, such as those in Afghanistan (from 2001) and Iraq (from 2003), resulting in both the technical specifications of the aircraft and the weapons they carry to have been tailored to the requirements of this type of operation. It must be mentioned that, due to some of those characteristics, current armed UAVs would only have limited value in large-scale, interstate conflict against a technologically sophisticated adversary. The necessary data link between the aircraft and a ground station, for instance, could be susceptible to jamming, spoofing or hacking, as was recently shown when media released information showing that American and British intelligence services had tapped into Israeli and Iranian drone footage.[10] Likewise, current types of UAVs are relatively slow, making them vulnerable to ground-based air defense systems, and commonly have no means of defending themselves against

---

[9] See section 3 for some of the legal issues. See, for possible drivers of drone proliferation, Michael Horowitz and Matthew Fuhrmann, 'Droning One: Explaining the Proliferation of Unmanned Aerial Vehicles' (October 2015), accessed 4 May 2017 at http://papers.ssrn.com/sol3/papers.cfm?abstract_id=2514339.

[10] Cora Currier and Henrik Moltke, 'Spies in the Sky: Israeli Drone Feeds Hacked by British and American Intelligence', The Intercept (29 January 2016), accessed 4 May 2017 at https://theintercept.com/2016/01/28/israeli-drone-feeds-hacked-by-british-and-american-intelligence/; however Israeli officials deny the reports, see http://www.timesofisrael.com/us-uk-didnt-crack-israeli-drone-encryption-officials-say/ (accessed 4 May 2017).

air-to-air threats, reducing their ability to operate in high-threat environments where air supremacy or air superiority are not a given factor. And when looking at the number of weapons armed UAVs can carry, this is generally quite limited, mostly only a couple of bombs or missiles, rarely more than six, having limited capacity against large formations or area targets. That is not to say that armed drones will not play a role in interstate conflict in the future. Several states are known, or are believed, to be developing next generation drones capable of carrying increased payloads on long-range, stealth (sometimes partly autonomous) operations.[11] Such developments are most likely to continue, giving armed drones an increasing range of usages in the future.

Notwithstanding some of the previously mentioned drawbacks, the armed drones currently in use possess a certain combination of characteristics that makes them highly valuable to conduct operations against non-state groups of fighters—be it terrorist, insurgent or other types—whose modus operandi is to conceal their actions by blending in with the civilian population. Three of those characteristics will be discussed: the possibility of operating them without putting personnel at risk; the ability to provide high-quality ISR; and the opportunity to respond immediately and precisely if a target is identified.

The first—and defining—feature of UAVs is the fact that the human operator is physically remote from the aircraft, providing a number of advantages over manned aircraft. Operations may be conducted without putting a pilot at risk, which in general terms prevents or minimizes casualties on the side that is using the technology. This is especially relevant given the fact that most of these types of conflicts are regarded as so-called 'wars of choice' rather than ones of necessity, for which constituencies are highly reluctant to accept body bags returning home. At the same time, it makes feasible operations, or provides access to areas, that would otherwise be considered too dangerous or politically sensitive. Having an unmanned aircraft shot down or crash will generally have fewer political and diplomatic costs and does not evoke difficult questions of launching risky search and rescue operations. Given the fact that irregular forces often operate in inhospitable areas or across international borders, this makes UAVs particularly suitable compared to manned aircraft or ground forces in these types of conflicts.

The second feature of UAVs is their ability to provide high-quality ISR. They can send real-time battlefield information either to command centres or to ground forces directly, thus providing situational awareness

---

[11] Sayler (n 5) 24–7.

and enhancing decision-making.[12] Their use of the air (third dimension) provides additional information compared to ground sources, makes them flexible and responsive, and gives access to otherwise inhospitable areas whether because of terrain or battlefield conditions. Since flight endurance is only limited by technical features of the aircraft and not by human endurance—multiple remote operators may work in shifts—most MALE UAVs are capable of sustaining missions between 12 and 24 hours, offering persistent observation over extended periods of time and vast areas, even considering their generally narrow fields of view. Normal operating altitudes are between 5,000 and 30,000 feet (above ground level), which, combined with their limited sound profile, also makes them suitable to conduct missions covertly. This combination of persistent and covert observation is particularly well suited for gaining information on irregular forces as it takes advantage of those who erroneously deem themselves unobserved and may thus persist in activities they would otherwise have discontinued or tried to conceal. In this respect, UAVs provide more and better information on the identity or characteristics of potential targets, on their status as lawful military objectives, and on the general pattern of life in their direct vicinity, so as to make better informed judgments on whether and how to strike and how to limit undesired effects.

Once a legitimate target has been identified, immediate response may be necessary as the opportunity to attack might be lost, for instance when a person moves into a densely populated civilian area or into a building. If an unarmed UAV had been used, other assets would have to be tasked to strike the target,[13] with the obvious disadvantages of requiring availability of those assets, requiring close coordination between relevant players, and most likely having lengthened response times. UAVs that offer a strike capability have the opportunity to respond immediately, diminishing the so-called sensor-to-shooter time, which prevents missing the window of opportunity and could generally support a high tempo of operations. Weapons that are carried are commonly precision-guided munitions, also capable of hitting moving targets. Although precision is

---

[12] It also provides the possibility to have high-level commanders, intelligence analysts or military lawyers join in observing the video images and participate in the decision-making.

[13] When other assets are used to attack the target, UAVs could still be valuable during the strike phase, for instance by providing a mensurated coordinate for the use of a GPS-guided cruise missile, by 'painting' the target with its laser designator, or by using its streaming video to call for and adjust artillery or mortar fire.

no silver bullet in itself—attacking the wrong person is not solved by doing so precisely (it may only be solved by better intelligence and analysis)—it may prevent or minimize undesired consequences. Having one's weapons detonate at the intended location prevents, or at least limits, death, injury, or damage to surrounding persons or objects, while at the same time allowing the use of smaller warheads, thereby exponentially resulting in fewer undesired effects compared to less precise weapons aimed at the same location. Nevertheless, even precision weapons may cause unexpected civilian casualties, since, just as any other type of weapon, drones are not exempted from the impact of human error or other untoward occurrences. In any event, the type of weapons commonly used by drones are among the most precise available[14] and since most are laser-guided, they offer the opportunity to steer away an already launched missile or bomb into an open area when changing circumstances so require, a procedure that is referred to as 'shift cold'.

When combining the above-mentioned features, armed UAVs provide persistent, real-time, and covert ISR, combined with immediate precision strike capability, while not putting personnel at risk, a mix that is unique in the history of warfare. They add value in situations where persistent ISR is required to identify potential targets that offer small windows of opportunity, requiring immediate action. The targets they can be most effective against are relatively small, perhaps moving targets, for which precision weapons are necessary. This makes them highly suitable to conduct so-called hunter-killer missions against enemy personnel. On some occasions persons have been pre-identified as legitimate targets and need to be located and their identities confirmed (possibly with the aid of other intelligence sources such as geolocation and voice recognition from cell phones) before being engaged. These are attacks that are often referred to as personality strikes. While on other occasions persons are targeted who, during persistent observation, reveal behaviour that qualify them as legitimate targets. Such attacks are often referred to as signature strikes. Due to their characteristics and the tactical/operational advantages they offer, it is unsurprising that armed drones are extensively used in counter-insurgency and counter-terrorism operations.[15] There remains, however, a wide debate as to their effectiveness in the longer term and

---

[14] Often used weapons are the Hellfire and Brimstone missiles and the GBU-12 laser-guided and GBU-38 GPS-guided bombs. They have a circular error probable (CEP) of between 3 and 30 feet, meaning that 50% of munitions will land within that distance of the aim point.

[15] See for an overview of the relative impact of drones on different types of situations Michael Horowitz et al, 'The Consequences of Drone Proliferation:

their strategic value. Are there just a lot of nails for which this is a powerful hammer, or is this a hammer that makes every problem look like a nail? It is often claimed that drones are counter-productive by having the effect of hurdling opponents to pick up arms and join the fight. Without joining sides in this debate, it must be mentioned that what stands out is that this debate is often highly speculative. Answering the question whether or not drones are an effective counter-terrorism tool relies partly on answering the question on what the alternatives would have brought—including doing nothing. As solving conflicts would most likely always require a combination of political, military, economic and other tools, perhaps posing such a question puts an unrealistically high burden on just a single piece of military equipment.

2.2 Some Operational Considerations and Characteristics of Cyber Warfare

"A domain aimed at the free, unhindered flow of data is quickly becoming the proving ground for states and non-state actors. The Internet has been dubbed a new war fighting domain, a battle space that is proving to be the host of most of 21st century conflict. These conflicts will revolve around influencing others through information, about creating, collecting, controlling, denying, disrupting or destroying that information. A cyber prefix flood has securitized every aspect of information technology; White Papers have earmarked cyber systems, cyber citizens, cyber threats and many more as part of the cyber environment. William Gibson's pristine cyberspace—coined in his 1984 book *Neuromancer*—has become militarized, spawning a new breed of military action dichotomized as cyber warfare."[16]

A wide variety of actors are active in the domain dubbed 'cyberspace' by governmental decision-makers in various publications and doctrines.[17] Many actors vie for control over this domain to guarantee 'information

Separating Fact from Fiction' (January 2016), accessed 4 May 2017 at http://papers.ssrn.com/sol3/papers.cfm?abstract_id=2722311.

[16] Jelle van Haaster, Rickey Gevers and Martijn Sprengers, *Cyber Guerilla* (Boston: Syngress 2016) p. xiii.

[17] See for instance, The Joint Chiefs of Staff, *Joint Publication 3-12 (R): Cyberspace Operations* (The Joint Chiefs of Staff 2013b); The White House, 'Securing America's Cyberspace, National Plan for Information Systems Protection: An Invitation to a Dialogue' (The White House 2000); United States Army, *Cyberspace Operations Concept Capability Plan 2016–2028* (The United States Training and Doctrine Command 2010).

superiority'; these range from non-state (criminal, activist, terrorist and 'regular' individuals and organizations) to state actors (law enforcement, intelligence agencies, armed forces). The subsections will discuss how cyber capacities are used and highlight some of their potential advantages and disadvantages.

2.2.1 Using cyber capacities in the military context

Cyber capacities can be used 'stand-alone' or in support of traditional operations (land, maritime, air and space). They may be utilized for various purposes, namely: intelligence, surveillance and reconnaissance (ISR); defense or offense.[18] The intelligence and defensive uses have a supporting role to both offensive activities and traditional operations. Without a defensive baseline guaranteeing freedom of movement to some extent, an actor cannot effectively engage in intelligence and offensive activities or effectively conduct traditional operations.

The cyber activities aimed at intelligence are used as a resource in the 'all-source intelligence process', which is used to support decision-making at the strategic, operational and tactical levels of war for both defensive and offensive cyber activities and traditional operations.[19] The cyber activities aimed at intelligence, surveillance and reconnaissance seem to be most prevalent. Most of the malware found in recent years had the ability to collect and extract information regarding the target system, network and users.[20] The intelligence derived from the cyber ISR

---

[18] The Joint Chiefs of Staff, *Memorandum for Chiefs of the Military Services, Commanders of the Combatant Commands, Directors of the Joint Staff Directorates: Joint Terminology for Cyberspace Operations* (The Joint Chiefs of Staff 2010) 5–6; The Joint Chiefs of Staff, 'Joint Publication 3-12 (R): Cyberspace Operations' (The Joint Chiefs of Staff 2013b) II-2–II-5.

[19] The Joint Chiefs of Staff, *Joint Publication 2-0: Joint Intelligence* (The Joint Chiefs of Staff 2013a) I-23–I-25.

[20] F-Secure, *The Duke: 7 Years of Russian Cyberespionage* (F-Secure F-Secure Labs Threat Intelligence Whitepaper, 2015); GReAT, 'Equation: The Death Star of Malware Galaxy' (2015), securelist.com/blog/research/68750/equation-the-death-star-of-malware-galaxy/; GReAT, 'Poseidon Group: A Targeted Attack Boutique Specializing in Global Cyber-Espionage' (2016), securelist.com/blog/research/73673/poseidon-group-a-targeted-attack-boutique-specializing-in-global-cyber-espionage/; Kaspersky, 'Adwind: Malware-as-a-Service Platform that Hit more than 400.000 Users and Organizations Globally' (2016), kaspersky.com/about/news/virus/2016/Adwind; Serge Malenkovich, 'Indestructible malware by Equation cyberspies is out there – but don't panic (yet)' (2015), blog.kaspersky.com/equation-hdd-malware/7623/; Michael Mimoso, 'Inside NLS_933W.DLL, the Equation APT Persistence Module' (2015),

activities can be used as a basis for offensive and defensive cyber activities but also for traditional operations.

Having established a defensive baseline and detailed target knowledge via ISR activities, offensive cyber activities can be used to project (military) power in or through cyberspace. These activities can be used stand-alone or in support of traditional operations (land, maritime, air and space). When utilized in support of traditional warfare activities, cyber capacities are used to shape preferential circumstances for traditional operations by affecting elements in or contributing to cyberspace. An often-used example of the use of cyber capabilities in support of traditional operations is *Operation Orchard* whereby the Israeli Defence Forces reportedly used cyber capacities to affect the Syrian air defense system. Although there is no conclusive evidence that cyber capacities definitely were used in that operation, there are numerous reports that cyber capacities were used to manipulate the software of the Syrian air defense system in order to temporarily render it inoperable, thereby creating safe passage for an airstrike on the facility.[21] Irrespective of whether cyber capacities were actually used, this example serves to illustrate how cyber capacities may be used to support traditional operations.

The variety of objectives and targets that can be affected by offensive cyber activities increases on a daily basis as information technologies permeate and proliferate throughout societies.[22] Offensive activities can target most connected systems within society, even when these are air-gapped, that is, not connected.[23] Offensive cyber capacities have targeted a wide range of these systems, from critical industries (for

---

threatpost.com/inside-nls_933w-dll-the-equation-apt-persistence-module/111128; Pierluigi Paganini, 'Duqu 2.0: The Most Sophisticated Malware Ever Seen' (2015), resources.infosecinstitute.com/duqu-2-0-the-most-sophisticated-malware-ever-seen/ (all accessed 4 May 2017).

[21] Galrahn, 'Electronic War in IAF Strike in Syria' (2007), information dissemination.net/2007/10/electronic-war-in-iaf-strike-in-syria.html; Charles Recknagel, 'Five Things you Should Know About Syria and Russia's S-300 Missile System' (2013), rferl.org/content/explainer-russia-syria-s-300-missile-system-/25003647.html; Sharon Weinberger, 'How Israel Spoofed Syria's Air Defense System' (2007), wired.com/2007/10/how-israel-spoo/ (all accessed 4 May 2017).

[22] See for instance: 'IoT List: Discover the Internet of Things' (2016), iotlist.co/ (accessed 4 May 2017).

[23] Dan Goodin, 'Meet "badBIOS," the Mysterious Mac and PC Malware That Jumps Airgaps' (2013), accessed 4 May 2017 at arstechnica.com/security/2013/10/meet-badbios-the-mysterious-mac-and-pc-malware-that-jumps-airgaps/;

example, water, power and petrol) to hospitals;[24] social-media accounts to governmental databases;[25] and individual mail accounts to military networks.[26]

2.2.2 Advantages

Cyber capacities have various advantageous characteristics that make them valuable for employment in offensive and ISR activities, namely: (1) rapid strike ability; (2) asymmetry; (3) anonymity; (4) data availability; and (5) non-destructive/lethality. These characteristics will be briefly discussed below.

One of the defining characteristics of cyber capacities is their ability to strike any place in the world instantly—at least so it appears. When using cyber capacities, the relevance of classical factors in traditional operations such as (geographical) space and time is diminished. Although the planning time necessary for employing cyber capacities may resemble or exceed that of traditional operations, the time required to create an effect

---

Michael Hanspach and Michael Goetz, 'On Covert Acoustical Mesh Networks in Air' (2013) 8 Journal of Communications 758.

[24] Alex Dobuzinskis and Jim Finkle. 'California Hospital Makes Rare Admission of Hack' (2016), reuters.com/article/us-california-hospital-cyberattack-idUSKCN0VS05M; Paul Roberts, 'Homeland Security Warns SCADA Operators of Internet-Facing Systems' (2011), threatpost.com/homeland-security-warns-scada-operators-internet-facing-systems-121211/75990/; Jeff Stone, 'U.S. Confirms BlackEnergy Malware Used in Ukrainian Power Plan Hack' (2016), ibtimes.com/us-confirms-blackenergy-malware-used-ukrainian-power-plant-hack-2263008; Dmitry Tarakanov, 'Shamoon The Wiper: Further Details (Part II)' (2012), securelist.com/blog/incidents/57784/shamoon-the-wiper-further-details-part-ii/ (all accessed 4 May 2017).

[25] Brian Fung and Andrea Peterson, 'The Centcom "hack" That Wasn't' *Washington Post* (12 January 2015), accessed 4 May 2017 at washingtonpost.com/news/the-switch/wp/2015/01/12/the-centcom-hack-that-wasnt/; 'US Government Hack Stole Fingerprints of 5.6 Million Federal Employees' *The Guardian* (23 September 2015), accessed 4 May 2017 at theguardian.com/technology/2015/sep/23/us-government-hack-stole-fingerprints.

[26] Scott Shane and Michael S Schmidt, 'Hillary Clinton Emails Take Long Path to Controversy' *New York Times* (8 August 2015), accessed 4 May 2017 at nytimes.com/2015/08/09/us/hillary-clinton-emails-take-long-path-to-controversy.html; 'US Military's Joint Staff Hacked as Official Point the Finger at Russia' *The Guardian* (7 August 2015), accessed 4 May 2017 at theguardian.com/technology/2015/aug/06/us-military-joint-chiefs-hacked-officials-blame-russia.

may be nearly instantaneous. Most of the time this does not require physical presence in proximity of a target.[27]

Another defining characteristic is asymmetry; the effort and resources needed to defend against a threat far exceed the effort needed to launch a cyber action. Orchestrating a distributed denial of service (DDoS) can cost as little as five dollars for a couple of minutes to 500 dollars a month (as a service).[28] This small investment is dwarfed by the financial damage caused and costs to mitigate such an attack. Even in the case of more advanced malware, the costs to create them are asymmetrical to the costs to mitigate the effects of the malware. Contrary to traditional operations where attackers classically need a preferable ratio to overcome defensive positions, in the realm of cyber activities the circumstances almost always favor the attacker and the defender has to commit considerably more resources than an attacker.

A third defining characteristic is anonymity, or alternatively plausible deniability. As attribution is difficult on the Internet, primarily due to the many ways of obfuscating tracks and denying control over a system used in an attack, an attacker has little to fear from retaliatory activities.[29] As such, cyber capacities are an attractive option for high-profile covert action where compromising the source of the action would damage the attacking actor. Even in less high-profile activities cyber is an attractive course of action as there currently is little to no consequence for engaging in offensive, intelligence or criminal activities as the target has difficulties in determining and responding to the attack source.[30]

---

[27] Sometimes being near a target offers additional possibilities for reconnaissance and entry.

[28] Brian Donohue, 'How Much Does a Botnet Cost?' (2013), threatpost.com/how-much-does-botnet-cost-022813/77573/; Brian Krebs, 'Six Nabbed for Using LizardSquad Attack Tool' (2015) krebsonsecurity.com/tag/lizard-stresser/; William Turton, 'Lizard Squad's Xbox Live, PSN Attacks Were a "Marketing Scheme" for New DDoS Service' (2014) dailydot.com/crime/lizard-squad-lizard-stresser-ddos-service-psn-xbox-live-sony-microsoft/ (all accessed 4 May 2017).

[29] Some of the obfuscation techniques that can be used are mac address changers, virtual private networks, proxy servers and the Onion Ring network (TOR).

[30] See for instance: Roger A Grimes, 'Why Internet Crime Goes Unpunished' (2012) infoworld.com/article/2618598/cyber-crime/why-internet-crime-goes-unpunished.html accessed 25 February 2016; Asheeta Regidi, 'Internet Immunity?' (2015), firstpost.com/india/internet-immunity-why-does-india-have-an-abysmal-0-7-conviction-rate-for-cyber-crimes-2566380.html (both accessed 4 May 2017).

A fourth advantageous characteristic is the ability to gain detailed knowledge regarding potential target persons and systems, in other words, the potential for comprehensive intelligence. As we increasingly extend our social and professional activities onto the Internet and integrate devices into our daily lives, we create vast amounts of data. These data include insights on, location, relationship status, financial status, sentiment at a given time, social network(s), employment history, education, political preferences, sexual preferences, shopping habits, devices used to browse the Internet, Internet protocol (IP) address, media access control (MAC) address and many more. Gathering this data and fusing it with other sources may result in comprehensive knowledge regarding a target.

Closely related to the comprehensive intelligence is a fifth characteristic, namely additional methods for non-destructive and non-lethal targeting. Although cyber capacities may cause a destructive effect, they are particularly suited for non-destructive effects. These capacities can be used to temporarily deny a system to an actor and reverse this effect after it has no further use to the attacker. This reversibility may limit the long-term harmful consequences of military and non-military activities to a target environment. The non-lethal character of most cyber operations may also create more options for non-lethal targeting of groups or individuals, for instance by dissuading them from engaging in harmful activities via information campaigns via social-media and other online platforms.

2.2.3 Disadvantages

Although there are certain advantageous characteristics to cyber capacities, they also have certain disadvantages or issues that should be taken into account. These characteristics, amongst other, include (1) intertwinement; and (2) conclusive attribution.

From its origins to its contemporary omnipresence, the Internet has shifted from a military network to a primarily civilian network to being one of the largest so-called dual-use objects in existence. As armed forces and belligerent non-state actors (armed groups, terrorists and so forth) rely on the same physical and logical infrastructure as non-belligerent civilian actors, the danger arises of civilians being unintentionally targeted by cyber activities. As many of these activities are non-lethal and non-destructive, the real-world damage may be limited. It may, however, blur the concept of protecting civilians from the harmful consequences of conflict. Even when dedicated to discriminating between military and non-military targets, it is doubtful whether actors are technologically capable of distinguishing between military and non-military targets. The

Internet by its nature is an amorphous intertwined whole and becoming increasingly complex due to the emergence of more interrelated information technologies.

A second issue is the inconclusiveness and difficulty of attribution on the Internet. Attribution in cyber activities often is very inconclusive, although many ascribe it with prophetical characteristics as being able to pinpoint individuals on the basis of device identifiers; this often is a(n) (educated) best guess. It is paramount to realize that information gathered via cyber activities should be enriched with other information sources to become actionable intelligence. For instance an Internet Protocol (IP) address without access to an Internet Service Provider's (ISP) records listing IP addresses with home addresses or other sources will only point conclusively to the country and region. All other factors (city, latitude and longitude, address, and so on) are best guesses or need inference from other sources. A complicating factor is that all identifying information in information technologies can be changed, spoofed or otherwise obfuscated.

Another issue with attribution is that device identifiers used in networking identify devices instead of users. Devices and credentials can be exchanged between users; hence operating on this information alone may lead to faulty insights and decisions. When this type of information is used in lethal targeting decisions, for instance basing a strike on a location attached to a social-media post or a device identifier,[31] positive visual identification remains the only sure way of matching a person to a device.[32]

Although these are some of the complicating factors when conducting offensive and intelligence cyber activities, there are currently more advantages than disadvantages. Cyber capacities produce a high-yield (asymmetry), an extended spectrum of policy and military options (non-destructive/(non) lethal), comprehensive knowledge and situational understanding (data availability) and do so at little risk (anonymity). These advantageous characteristics result in many actors planning on creating or increasing the ability to project power in or through cyberspace.

---

[31] Douglas Ernst, 'Terrorist "Moron" Reveals ISIS HQ in Online Selfie' *Washington Times* (4 June 2015), accessed 4 May 2017 at washingtontimes.com/news/2015/jun/4/air-force-bombs-islamic-state-hq-building-after-te/.

[32] Sean Gallagher, 'Opposite of OPSEC: Russian Soldier Posts Selfies—From Inside Ukraine' (2014), accessed 4 May 2017 at arstechnica.com/tech-policy/2014/08/opposite-of-opsec-russian-soldier-posts-selfies-from-inside-ukraine/.

3. SOME LEGAL CONSIDERATIONS AND CHALLENGES POSED BY REMOTE WARFARE

3.1 Unmanned Aerial Vehicles and The International Law of Military Operations

The use of UAVs in contemporary remote warfare has received considerable attention and a degree of criticism from a legal perspective. Let us start with a definition and a number of propositions to lay the ground for further discussion. The term 'International Law of Military Operations' refers to those areas of international law which interact and regulate military operations of all types taking place outside a state's territory. It includes a variety of different areas or sub-disciplines of public international law such as the law relating to the use of force, the law of armed conflict (also known as international humanitarian law) and international human rights law, all of which are crucial in determining whether, where and how a particular target may be engaged. In addition to these key areas of international law relating to the existence or lack of a legal basis to resort to force and the attendant legal regimes governing the application of force inside and outside the context of an armed conflict, there are other areas of international law which can be relevant to the conduct of operations including, for example, air law, the law of the sea and rules of general international law relating to state sovereignty and non-intervention alongside general principles and techniques of legal methodology to interpret how the law should be applied and to reconcile conflicts of obligation between different rules and legal regimes when they arise.[33]

The first proposition is that any use of force across international borders must have a recognized legal basis for it to be in conformity with international law. The contemporary international legal order is (still) largely based on the twin pillars of state sovereignty and the prohibition of the use of force in international relations. Consequently, the use of UAVs in the airspace of another state must have either the lawful consent of the host state or a credible legal basis which would preclude the wrongfulness of its presence on another state's territory and if force is employed, which would provide a sufficient legal basis for the use of

---

[33] Terry Gill and Dieter Fleck, 'Concept and Sources of the International Law of Military Operations' in Terry Gill and Dieter Fleck (eds), *The Handbook of the International Law of Military Operations* (2nd edn, Oxford University Press 2015) 3ff.

such force. Absent either of these, the use of UAVs to conduct strikes against any target, even one which unequivocally constituted a military objective, under the law of armed conflict would be illegal. A legal basis for the use of armed force will determine whether force may be used, but the question of how and against whom or what force may be employed is primarily determined by the application of the relevant legal regime.[34] There are two such regimes that can be applicable; the law of armed conflict and international human rights law. Each of these has its own sphere of application and the question of whether a particular strike is lawful will largely depend upon which of these are applicable and whether the rules pertaining to the application of force arising from whichever one of them is applicable have been adhered to, taking into account the relevant factual situation pertaining at the time the strike was conducted.

The second proposition is that the law of armed conflict (also referred to as international humanitarian law and by the acronyms LOAC and IHL) only applies when the material conditions for the existence of an armed conflict have been fulfilled. This may be tantamount to stating the obvious, but in view of some of the uncertainties surrounding the use of armed UAVs in particular strikes, it needs emphasizing. Likewise, if it does apply, it may be subject to geographical and temporal considerations relating to its scope of application. These could arise from the law of armed conflict itself, or from other bodies of law, such as the rules relating to respect for a third state's sovereignty or the principles of necessity and proportionality, which are an integral part of the law governing the exercise of self-defense.[35] Both of these will receive further attention below.

The third proposition is perhaps more controversial than the preceding two, but will nevertheless be posited as a guiding consideration which is related to the two preceding propositions. If the law of armed conflict is

---

[34] *Legality of the Threat or Use of Nuclear Weapons*, Advisory Opinion, International Court of Justice, ICJ Reports, 1996, 226 at 245, para 42. See also Shima Keene, *Lethal and Legal? The Ethics of Drone Strikes*, US Army War College Strategic Studies Institute (December 2015) 8.

[35] Christopher Greenwood, 'The Relationship of the *Ius ad Bellum* and the *Ius in Bello*' (1983) 9(4) Rev Int Leg St 221–34; Greenwood, 'Self-Defence and the Conduct of International Armed Conflict' in Yoram Dinstein (ed), *International Law at a Time of Perplexity: Essays in Honour of Shabtai Rosenne* (Martinus Nijhoff Publishers 1989) 273–88; Terry Gill, 'Some Considerations Concerning the Role of the *Ius ad Bellum* in Targeting' in Paul Ducheine, Michael Schmitt and Frans Osinga (eds), *Targeting: The Challenges of Modern Warfare* (TMC Asser Press/Springer Academic Publishers 2016) 101–20.

not applicable to a particular situation, either because the material conditions for its applicability have not been met, or because the state which conducts a particular strike is not party to a particular ongoing conflict with the consequence that its conduct would not be governed by the law of armed conflict, or because of geographical or other considerations which could stand in the way of its applicability to a particular strike, then some other legal regime *must* apply. There can be no situation in which strikes involving the application of lethal force are conducted without being subject to legal constraints. This would be tantamount to saying that in the absence of a de jure applicable regime, a state is free to target specific individuals and objects at will on the basis of its own determination of what the applicable standards are. If the law of armed conflict does not apply for any of the above-mentioned reasons, the application of lethal force must be subject to some other legal regime which determines how and under which conditions such force must be applied (the existence or lack of a legal basis does not govern how force must be employed). In short, a legal vacuum is an impossibility as a matter of law, as well as for compelling policy and ethical considerations. The obvious 'other' candidate for an applicable legal regime when the law of armed conflict is inapplicable is international human rights law (IHRL). It applies, in principle, to any force used outside the conduct of hostilities in armed conflict and to situations not amounting to armed conflict whenever the state in question has either effective control over persons or territory.[36] There are to be sure potential obstacles that arise in relation to the application of human rights conventions in an extra-territorial context, especially when the state concerned does not exercise physical control over either the territory where the prospective target is located, or the individual(s) who are to be targeted.[37] However, these obstacles are arguably not insurmountable, at least in relation to the applicability of specific human rights. One compelling reason for assuming the applicability of at least the prohibition of arbitrary deprivation of life, is that this particular human right is generally acknowledged as having a peremptory character, albeit subject to certain exceptions arising either from the law of armed conflict, or from international human rights law itself. If the regime of the former body of law (LOAC) is not applicable, it follows that the exceptions under the latter (IHRL) must be

---

[36] Gill and Fleck (n 33) 10.

[37] See Jann Kleffner, 'Human Rights and International Humanitarian Law: General Issues' in Gill and Fleck (n 33) 35, 53–7, where the different standards of human rights bodies relating to when human rights apply extraterritorially and what constitutes effective control for the purposes of applicability.

determinative to the question under which conditions lethal force may be applied. Armed conflict and its more permissive scope for the application of lethal force are exceptional situations. If they do not apply as a matter of law, then the less permissive scope for the deprivation of life under international human rights law is the default legal regime that must be adhered to.[38] Any other conclusion leads to unacceptable consequences for any number of considerations, the most compelling of which is, simply stated, that one cannot kill outside the law.

Assuming for the purposes of our further analysis the above-mentioned three propositions to be true, let us now proceed to a general discussion of a few of the main considerations and controversies surrounding the use of armed UAVs for the conduct of strikes against particular individuals or groups of individuals.

To start with, if a strike by a UAV (or for that matter any other permissible weapon or weapons system) is conducted in the context of an armed conflict to which the state in question is party and is carried out in conformity with the rules governing an 'attack'; namely is directed against a lawful military objective, whereby all feasible precautions have been taken to avoid or in any event minimize harm to civilians and civilian objects; and the prohibition of attacks which would result in excessive collateral injury and damage to civilians and civilian objects is adhered to, then such an attack would be lawful. A UAV is simply a remotely piloted aircraft and is subject to the exact same rules governing attacks as any other aircraft (or platform carrying weapons). It is stating the obvious that if the law governing attacks is adhered to, there is no legal reason why an attack conducted by a UAV should be presumed to be illegal. It is not an inherently indiscriminate weapons system, nor is it one which inherently causes unnecessary suffering and superfluous injury to persons subject to attack and is also not a prohibited weapon as such. It is capable of adhering to the rules governing attack under LOAC, and to the extent it does where that body of law is applicable, such an attack is lawful under the law of armed conflict.[39] Its characteristics make it capable of identifying targets with a very reasonable degree of accuracy and conducting an attack in conformity with the above-mentioned rules governing attack. Moreover, UAVs are regularly used for many purposes

---

[38] Nils Melzer and Gloria Gaggioli, 'Conceptual Distinction and Overlaps between Law Enforcement and the Conduct of Hostilities' in Gill and Fleck (n 33) 63, 70–79.

[39] Advisory Committee [to the Netherlands Government and Parliament] on Matters of Public International Law (CAVV) Advisory Report on Armed Drones, Report no. 23, July 2013, 4.

that do not involve them engaging targets on their own. These include the location and fixing of a target which is engaged by conventional manned aircraft or ground forces and long range reconnaissance and surveillance, none of which are controversial in themselves in relation to the question of the ability of a UAV to adhere to LOAC. In any case, it should be emphasized that assuming LOAC is applicable in a material sense (the threshold of either international or non-international armed conflict has been met), is applicable in the geographic space where a UAV strike is conducted, and that the rules and principles of LOAC in conducting attacks are adhered to, the strike will be lawful in principle. There are sometimes questions relating to whether specific individuals are subject to attack as in the case of so-called 'signature strikes' against persons based on age, gender and location, which are not in themselves determinative criteria as to whether someone is a lawful target;[40] or whether a particular strike causes excessive civilian casualties in relation to the anticipated military advantage. But these questions can also arise in relation to an attack by a manned aircraft or other weapon and can only be answered on a case-by-case basis, taking all relevant facts into consideration. In any case, the rules of LOAC are the same with regard to UAV strikes as with manned aircraft.

Likewise, the fact that its operator(s) may be located at a (very) considerable distance from the target or area in which hostilities are being conducted does not, in itself, make it less legal (or illegal if the above-mentioned rules governing attack are *not* adhered to). Simply stated, the distance of the operator(s) from the target; their 'remoteness' from the area of operations is, in itself, irrelevant in legal terms. While 'remoteness' from a battlefield and the virtual impossibility of the targeted individual being able to harm the operator(s) may seem at variance with traditional notions of warfare, this is not unique to UAVs or even to what is defined here as 'remote warfare'. The pilot of a modern fighter aircraft operating at an altitude well above the range of the type of weapons in use by most armed groups and delivering weapons from a distance of many miles from the target is not essentially different in terms of 'remoteness' from the target or being subject to any appreciable risk of being hit, outside of an untoward mechanical failure of the aircraft. For that matter, the artillery battery firing at a range of perhaps dozens of miles from a targeted group of infantry or vehicles is virtually impervious to counter-fire, unless the opposing side also has similar weapons in its inventory that are capable of engaging in effective counter

[40] Ibid.

battery fire. Nor, it should be stressed, is warfare a joust or tournament in which opposing forces must have parity for it to be legal. Chivalry may still have a role in contemporary warfare, but it does not, nor has it ever precluded, the use of weapons that give an attacker an inherent advantage against an opponent.[41]

If a UAV can be used in conformity with LOAC, what are the main dilemmas and sources of criticism of their use? While there are a number of criticisms raised, there are three main points of controversy that will receive attention here (other controversies of a legal or ethical nature exist, but those will not be discussed here as they are essentially subsumed in the main critiques and because they would expand this discussion beyond what is feasible). They are separate, but nevertheless to a significant degree are interrelated.

First, there is controversy concerning the existence of a legal basis in some cases in which UAVs have been and are employed. This controversy concerns in some cases whether the state where the strikes are conducted has consented to such operations being carried out on its territory. In the absence of consent, there must be some other legal basis and the one potential other basis and the one put forward most often is self-defense. Much has been written on whether self-defense applies to attacks conducted by non-state entities, on whether the 'unwilling or unable test' allows a state which has been, is or is likely to be, threatened by attack by a non-state armed group to exercise self-defense on another state's territory in situations in which the state where the armed group is (partially) located and operates from is not capable of preventing attacks by the armed group against other states, or simply fails to prevent such attacks occurring.[42] These controversies are wider than simply whether

---

[41] Ibid. See also Terry Gill, 'Chivalry a Principle of the Law of Armed Conflict?' in Marielle Mathee et al (eds), *Armed Conflict and International Law: In Search of the Human Face* (TMC Asser Press/Springer Academic 2013) 33, 47.

[42] The literature on these controversies is far too extensive to cite here in detail. The International Court of Justice inferred that the right of self-defense was only applicable in relation to armed attacks by states in its *Advisory Opinion on the Legality of a Wall in Occupied Palestinian Territory*, ICJ Reports (2004) 194, paras 138–139. It repeated this inference in its decision in the *Armed Activities in the Democratic Republic of the Congo* case (*Democratic Republic of the Congo v Uganda*), ICJ Reports (2005) 223, para 146. In both cases this inference was the cause of vigorous dissent on the part of several members of the Bench. See, for example, the Declaration of Judge Kooijmans in the *Armed Activities* decision at 313–14, paras 25–29; Separate Opinion of Judge Simma in the same case, 336–7, paras 7–11 and Judge Higgins in the *Wall* Advisory

certain UAV strikes are lawful, but since UAVs have often been used in these contexts, the controversies surrounding these issues inevitably are at the forefront of the debate concerning their employment to conduct strikes in these circumstances.

The second controversy is related to the first, although it has a different frame of reference. It concerns the question of the geographical scope of application of the law of armed conflict, as well as the question of how the targeting of individuals may or may not be affected by considerations from other bodies of international law that are potentially relevant in determining where a party to an armed conflict may operate.[43] Related to this is the broader question of the scope of applicability of LOAC to begin with. Is LOAC applicable simply because a state considers itself to be 'at war' with a particular armed group, and if it is, does this apply everywhere members of such an armed group are located, or to groups which may share a common ideology and modus operandi, but which may not have a common command structure, objectives or even the same opponents? What if the group splits and pursues related but separate objectives in another country or even several countries, does it still constitute the party with which one considered itself to be 'at war' with in the first place? Finally, how long does a resort to self-defense, which may contribute to meeting the threshold conditions for the applicability of the law of armed conflict, continue to confer a right to conduct

Opinion, 215, para 33. The UN Security Council in contrast has deemed attacks by armed groups which act autonomously as armed attacks in inter alia Resolutions 1368 and 1373 (2001) in relation to the '9/11' attacks on the Pentagon and World Trade Center. This controversy carries over until the present with the states supporting Iraq by conducting airstrikes against ISIS in Syria, basing their action on the right of collective self-defense—a position which is contested by inter alia the Syrian government and Russia.

[43] See the sources cited in n 35 above in support of the position that other legal considerations arising from neutrality law, the law relating to the use of force and rules of general international law relating to respect for territorial sovereignty and non-intervention can impact upon the geographical scope of targeting in international and non-international armed conflict alongside LOAC. For the contrary position see inter alia Geoffrey Corn, 'Geography of Armed Conflict: Why it is a Mistake to Fish for the Red Herring' (2013) 89 International Law Studies 77; Michael Lewis, 'Drones and the Boundaries of the Battlefield' (2012) 47 Texas Intl L J 293. For a general discussion of the issue see, for example, Jessica Dorsey and Christophe Paulussen, 'The Boundaries of the Battlefield: A Critical Look at the Legal Paradigms and Rules in Countering Terrorism' ICCT Research Paper (2013), accessed 4 May 2017 at http://www.icct.nl/download/file/ICCT-Dorsey-Paulussen-Boundaries-of-the-Battlefield-Report-April-2013_docx.pdf.

operations within the scope of LOAC?[44] In short, what are the limitations, if any, to the material, geographical and temporal scope of applicability of the law of armed conflict?[45]

The third controversy is whether certain UAV strikes are not in fact subject to a wholly different legal regime in which the employment of UAV strikes against particular individuals would rarely, if ever, be lawful. In other words, if the scope of applicability of LOAC to strikes conducted far from a traditional area of operations is (at least in some cases) questionable, the applicability of human rights law to such strikes enters the picture and makes the use of a UAV firing an air to ground missile, which can and often does cause significant collateral effects, difficult, if not impossible to justify under human rights law. Under IHRL deadly force is reserved for exceptional situations and the way in which precautions and proportionality apply within it differ radically from how they apply within LOAC. From this perspective 'personality strikes' conducted by UAVs far from any traditional battlefield look very similar to 'extra judicial execution', without recourse to any arbiter other than the party conducting the strike.[46]

Much has been written on all these controversies, including some views on all these questions by one of the authors of this chapter. They remain, however, controversial and do not look like they will be resolved in the near future. This chapter is not intended to help resolve them, simply because they are not capable of being resolved in any one contribution to this debate by any one person or group of persons and there are many views in circulation which come to sometimes similar, sometimes widely diverging, conclusions on all these issues. Nevertheless, it has hopefully been useful to briefly outline where the main fault lines in the debate lie. Of course, there are more (for example, do some

---

44 The applicability of LOAC is governed by the material conditions within LOAC for meeting the threshold of an armed conflict of either an international or non-international character. If self-defense is invoked, it will usually contribute to meeting those conditions, including the applicability of LOAC rules on targeting. In such cases, the question of its duration can be relevant. See in this respect, Terry Gill, 'When Does Self-Defence End?' in Marc Weller (ed), *The Oxford Handbook of the Use of Force in International Law* (Oxford University Press 2015) 737ff.

45 See sources cited in n 43 and n 35 above. See also Keene (n 34) 13–14; Kleffner (n 37) 49–51.

46 See inter alia Jens David Ohlin, 'Acting as a Sovereign versus Acting as a Belligerent' in J D Ohlin (ed), *Theoretical Boundaries of Armed Conflict and Human Rights* (Cambridge University Press 2016) 118; Melzer and Gaggioli (n 38) 77–8.

or all UAV strikes cause inordinate collateral damage, do they result in a 'play-station' mentality with regard to using deadly force, do they make the conducting of controversial strikes easier because of their lower visibility in comparison to strikes by conventional manned aircraft, is there adequate oversight and accountability in conducting certain types of strikes and so on), but the answers to these are essentially contained in the main controversies set out above. Whatever one's views on this form of 'remote warfare' may be, it is useful to know what the main points of controversy are.

3.2 Cyber Warfare and the International Law of Military Operations

'Cyber warfare' is a term used in a variety of different ways and in different contexts. In the *Tallinn Manual on the International Law Applicable to Cyber Warfare* (hereinafter referred to as the *Tallinn Manual*), it is defined as cyber operations constituting a use of force and/or within the context of an armed conflict, which are governed by the international law governing the use of force (jus ad bellum) and the law of armed conflict (LOAC/IHL/jus in bello) respectively.[47] It is in that sense that it will be used here to demarcate it from the broader notion of 'cyber security' in which activities not related to 'warfare', remote or otherwise, such as cyber peacetime espionage, cyber criminal activity and cyber intellectual property theft, all of which take place on a systematic and large scale but are not 'warfare' in the cyber domain any more than they are in the physical world.[48] 'Cyber warfare' as used here will therefore denote situations in which an armed conflict is presumed to be in progress. It does include operations not having a physical effect in the sense of causing casualties or physical destruction, such as intelligence gathering and support to traditional forms of warfare, so long as these are conducted in the context of an armed conflict.

Cyber warfare in this sense is a new and still emerging phenomenon and has not yet been conducted on a large scale as an independent method of warfare, or as an adjunct to more traditional forms of warfare in recent and ongoing armed conflicts until now.[49] Nevertheless, it has

---

[47] Michael Schmitt (ed), *Tallinn Manual on the International Law Applicable to Cyber Warfare* (Cambridge University Press 2013) 4.

[48] Ibid.

[49] Computer data systems have become an indispensable part of contemporary warfare. Starting already in the 1990s with the Gulf War (Iraq/Kuwait) they have been and increasingly are used for everything from controlling various

received considerable attention and this is undoubtedly due to its potential use, which is likely to play an increasingly important role in military operations and armed conflicts. A number of states have adopted cyber strategies and incorporated them into military planning and doctrine and if history is any guide, whenever a means or method of warfare is developed, it is likely to be used at some point in the future. In the preceding section, a number of operational applications of cyber warfare were identified and in this subsection a number of comments will be made concerning some of the legal certainties and challenges this form of remote warfare poses.

One of the main controversies which figured prominently in the early debates and discussions relating to the phenomenon of cyber warfare seems to have largely been settled. When discussion and consideration of cyber warfare first started in the 1990s and continued while states were considering and formulating new cyber strategies and during the drafting process of the *Tallinn Manual*, one of the main and most fundamental questions was whether traditional international law applied or could be applied to cyber activity, including cyber warfare in the narrow sense used here.[50] 'Cyber' seemed to some to be incapable of being regulated by rules and principles of international law that were developed to regulate physical activities within a defined geographical area. By now, notwithstanding certain challenges relating to how the law should best be applied in relation to a phenomenon which is only partly perceptible

---

weapons systems through planning and calibrating targeting to logistical support of operations (see previous Section). However, to date, they have not been used on a large scale to conduct attacks on opposing forces or in direct support of kinetic attacks in the sense of being employed as a 'weapon'. The distributed denial of service (DDOS) attacks on Estonia in 2007 did not constitute a use of force, nor did they occur within the context of an armed conflict. The use of computers in the 2008 war between Russia and Georgia for propaganda purposes was likewise neither a use of force, nor in direct support of kinetic operations. The 2010 Stuxnet attacks on the Iranian nuclear programme were also not conducted within the context of an armed conflict. The only known example of cyber warfare in direct support of kinetic operations to date was the reported use by Israel of cyber warfare akin to electronic jamming against the Syrian air defense system in direct support of a conventional kinetic airstrike on a nuclear facility at Al Kibar in northern Syria in 2007. See Andrew Garwood-Gowers, 'Israel's Airstrike on Syria's Al Kibar facility; a Test Case for the Doctrine of Pre-Emptive Self-Defence?' (2011) 16 J of Conflict & Security L 263; David Fulghum and Douglas Barrie, 'Israel Used Electronic Attack in Airstrike Against Syrian Mystery Target' *Aviation Week* (8 October 2007) 28. See also n 21 above.

[50] *Tallinn Manual* (n 47) 3.

within the physical domain and takes place across spatial frontiers, it seems to be widely accepted that international law not only applies, but is capable of being applied, albeit with some adaptations and the liberal use of analogous interpretation, to cyber activities across a wide spectrum, including cyber warfare. More and more states have adopted policy positions and expressed views reflecting the position that international legal regulation of cyber activities is not only possible, but even imperative.[51] Likewise, the growing amount of legal discussion in the literature has followed this trend and accepts that existing international law can and does apply to cyber activity including to cyber warfare. While there may be differences of opinion on how it applies and whether there are gaps in the law which need to be addressed, most authorities now take for granted that international law is applicable to cyber activities, including warfare. That is the position taken here as well and our contribution will be directed at identifying where some of the main areas of controversy may lie. We will not, however, attempt to resolve these questions, or still less to prescribe how the law should best be applied, as this would be both impossible in a single short contribution and would ignore the fact that this debate is ongoing. It will ultimately be settled as state practice develops in this area; the future application of the law will be shaped and interpreted by states and other actors and by authorities from a wide range of states, and with different perspectives.

There are three areas of controversy in relation to cyber warfare that will receive attention here. They are, first, how the fundamental principle of distinction within LOAC applies to a form of warfare which is both remote and to a significant extent anonymous and can be conducted by individuals located in different countries without necessarily having the normal degree of organization and hierarchy associated with military

---

[51] Examples of cyber strategies which have been adopted under the premise that international law is applicable to cyber activity, including military activity, include those of the United States, The White House *National Security Strategy* (2010) 27; The White House, *International Strategy for Cyberspace: Prosperity, Security and Openness in a Networked World* (2011) 9; The United Kingdom, HM Government, *A Strong Britain in an Age of Uncertainty: The National Security Strategy* (2010); *The UK Cyber Security Strategy: Protecting and Promoting the UK in a Digitized World* (2011) 22; Canada, Government of Canada, *Canada's Cyber Security Strategy* (2011) 7ff and the Netherlands, *National Cyber Security Strategy* 2 (2014) 21. The United Nations 'Group of Governmental Experts on Developments in the Field of Information and Telecommunications in the Context of International Security' issued a report acknowledging the full applicability of international law to cyber activity in June 2013. See UNGA A/68/98 (24 June 2013).

operations. Second, the question of whether 'data' is an object which would place it within the framework of the rules of LOAC which govern attacks against persons and objects will be given some brief consideration. Third, the question of how collateral damage may be assessed in the context of conducting cyber operations amounting to an 'attack' in the sense this term is used in LOAC will be briefly addressed. Each of these points has received attention and consideration elsewhere, but the purpose of the consideration given here is to identify where the main points of contention lie and this may contribute to resolving them as the law develops further.

Before entering into a discussion of these controversies, a number of considerations relating to how cyber activities may be used in an armed conflict and how the law relates to cyber activities inside and outside the context of an armed conflict are important to bear in mind. First, armed conflict rarely involves the use of only one particular means of warfare or weapons system. Consequently, while it is generally accepted that an armed conflict could theoretically take place wholly in the digital domain, this is unlikely at best.[52] If the threshold of armed conflict, either international or non-international has been met, cyber techniques of warfare may well have figured in reaching that threshold or, more likely, will be employed in the context of an ongoing armed conflict, both on their own alongside kinetic operations and directly in support of kinetic operations, but the likelihood of an armed conflict which is purely 'cyber on cyber' is remote. On the other hand, once an armed conflict is in progress cyber operations undertaken in the context of the conduct of hostilities, either kinetic or cyber, will be governed by LOAC, as will a number of other situations which are governed by LOAC, such as certain actions directed against protected objects or persons under LOAC. Hence, cyber activity which, for example, impeded humanitarian relief operations to the civilian population, or with the functioning of hospitals engaged in treating protected persons, either military or civilian, would be governed by the relevant provisions of LOAC.[53] Likewise, cyber operations undertaken outside the context of armed conflict, or by persons with no nexus to the armed conflict (for example, law enforcement activities in relation to criminal activity with no connection to the parties to the conflict) are not governed by LOAC and are not 'warfare' in any legal sense. Finally, while LOAC does apply to certain cyber

---

52 See e.g. Thomas Rid, 'Cyber War Will not Take Place' (2012) 35 J of Strategic Stud 5–32.

53 See *Tallinn Manual* (n 47) rule 20 and accompanying commentary.

activities which do not constitute the conduct of hostilities as in the previously mentioned examples, the rules of LOAC relating to the conduct of hostilities, such as the taking of precautions in attack and the rule of proportionality in LOAC, are only relevant when the cyber action constitutes an 'attack', either on its own or in conjunction with kinetic operations amounting to an attack in the sense of LOAC.[54] Consequently, cyber operations such as interception of communications, gathering intelligence in a general sense or conducting information operations and psychological warfare do not constitute acts of violence directed against the adversary and are not governed by the LOAC provisions relating to 'attack', although they may in some cases be governed by other rules of LOAC.[55]

We will now turn to the first of the controversies identified above, the question of how the principle of distinction applies in the cyber domain. The principle of distinction is one of the fundamental principles of the law of armed conflict and requires parties to an armed conflict to distinguish at all times between persons and objects subject to attack and civilians and civilian objects and to direct their operations (meaning essentially attacks) exclusively against military objectives. This fundamental principle underlies all the rules governing the conduct of hostilities and applies in any armed conflict, whether international or non-international in character. It should not be confused with the requirement for combatants to distinguish themselves from the civilian population (at least during the conducting of an attack), but that requirement is nevertheless related to the principle at least to some extent. However, that particular requirement will not be considered here.

There are a number of ways in which the application of this principle poses significant challenges in the cyber context. We will consider two of them here. One of these is the fact that aside from discrete purely military computer systems, such as those presumably used to control particular weapons systems, the bulk of military use of the cyber domain

---

[54] Ibid rule 30 and accompanying commentary. See also Noam Lubell, 'Lawful Targets in Cyber Operations: Does the Principle of Distinction Apply?' (2013) 89 Intl L Stud 252, 261 ff; Eric Talbot Jensen, 'Cyber Attacks: Proportionality and Precautions in Attack' (2013) 89 Intl L Stud 198, 200–201; Terry Gill, 'International Humanitarian Law Applied to Cyber Warfare: Precautions, Proportionality and the Notion of "Attack" under the Humanitarian Law of Armed Conflict' in Nicholas Tsagourias and Russell Buchan (eds), *Research Handbook on International Law and Cyberspace* (Edward Elgar 2015) 366 ff.

[55] One example would be the use of cyber to instill terror among the civilian population. See *Tallinn Manual* (n 47) rule 36, 122–4.

takes place on the Internet (and related civilian cyber infrastructure), which is by its nature a civilian object. To the extent a party to an armed conflict relies on the Internet for various purposes during an armed conflict, it will become a 'dual-use' object and hence a military objective by virtue of use or purpose. However, the Internet is not just 'any' dual-use object, but one that is so essential to contemporary life and the functioning of civilian society that undertaking operations that cannot be directed exclusively against military objectives can be particularly challenging. This poses a problem that arguably goes beyond the usual type of calculation of proportionality that weighs the definite military advantage anticipated from an attack on a specific military objective against foreseeable collateral injury and damage (which will be considered separately below). In view of the 'interconnectedness' of the Internet and the digital domain as a whole and the way data is transferred, it will require a party to an armed conflict to take particular care in refraining from operations, the effects of which cannot be directed against a specific military objective, or that cannot be controlled and are likely to strike civilian as well as military uses of the digital domain without distinction. Such attacks are indiscriminate (disproportionate) by nature and are prohibited.[56] It is not always possible to ascertain in advance what the military uses of the Internet are or are likely to be and to reasonably predict the effects of an attack on a particular segment of the Internet, much less on the Internet as a whole. The example of the Stuxnet virus is an example which shows that an attack conducted by cyber means is capable of being directed against a particular objective and that its effects can be limited in a manner required by LOAC. Although it did not take place within the context of an armed conflict, it does serve to illustrate that attacks can be both directed and limited to striking a military objective in principle.[57] However, it required considerable research and development and computer resources to achieve its objective in conformity with these requirements and it is by no means a foregone conclusion that all parties to an armed conflict could or even would attempt to limit the effects of their attacks to a comparable extent. To the extent they cannot or will not, such attacks would clearly be illegal, but it would be naive to assume this would necessarily prevent their occurrence.

Another challenge to the principle of respecting the principle of distinction arises from the problem of determining whether a particular

---

[56] Article 51(4) Additional Protocol I 1977 (hereinafter API) reflecting customary international law; *Tallinn Manual* (n 47) rule 49 with accompanying commentary, 156 ff.

[57] Jensen (n 54) 203.

individual or group of individuals is subject to attack. While purely military computer systems and their operators may be assumed to constitute lawful military objectives by nature, the possibility of civilians directly participating in hostilities through the use of cyber means, sometimes without any visible organizational structure or direct connection to each other, is one which is likely to occur in future armed conflicts. While individual civilians who directly participate in hostilities lose their protection for the duration of their direct participation, they also regain protection, in principle, after their direct participation ceases.[58] In the cyber context this could be an effective bar to actually being able to target them in conformity with this requirement, at least in many cases. Likewise, while members of an organized armed group (with a continuous combat function) may lose their protection against attack in a non-international armed conflict, the level of organization of members of a group of 'patriotic hackers' might in many cases be either difficult to ascertain or fall short of what is considered to constitute an 'armed group' in the physical domain.[59] If their operations caused significant damage to a party, it would be faced with the dilemma of either stretching the limits of the law in order to attack them anyway, or having no effective recourse. In the physical domain, irregular warfare often poses a significant challenge to adhering to the law; in the cyber context, this could well be even more so. This is by no means an exhaustive summary of the problems posed in cyber warfare to the respect of the principle of distinction, but it serves to illustrate that while the law is capable of being applied, there are nevertheless particular challenges posed to adhering to the law in the cyber context.

Another question that has received some attention is whether data is or should be considered an object for the purposes of determining whether the rules and principles governing attack in LOAC apply to it. In the *Tallinn Manual*, the majority of the experts concluded that there was an arguable case for determining that disruption of data which caused large scale adverse consequences, but without causing or intending to cause physical harm to persons or damage or destruction of physical objects, might be considered an attack. Nevertheless, they also concluded that the law of armed conflict did not, at least at present, include such immaterial

---

58 ICRC, *Interpretive Guidance on the Notion of Direct Participation in Hostilities* (2009) 70.

59 This problem was discussed in the drafting of the *Tallinn Manual*. The experts struggled with the obstacles posed by virtual organization in the cyber domain in relation to the condition of organization in non-international armed conflict. See *Tallinn Manual* (n 47) rule 23, commentary at 88–90.

effects within the notion of 'attack'. The members of the 'Group of Experts' found that whenever physical harm or damage resulted as a secondary effect of destruction or disruption of data this constituted an attack. Likewise, a majority found that if the functionality of an object had been disrupted to the extent that it required repair of its physical components, this too would constitute an attack. However 'data' in itself was not considered a physical object and consequently if it were adversely affected without causing any physical consequences, this would fall short of an attack.[60] This means that cyber operations which do not have effects in the physical domain are not governed by the LOAC rules pertaining to attacks, including much of the scope of the aforementioned principle of distinction, the requirement to take all feasible precautions prior to and during an attack to avoid or minimize harm to civilians and civilian objects and the principle of proportionality which prohibits attacks which would result in excessive death, injury to civilians or damage to or destruction of civilian objects. This position taken in the *Tallinn Manual* has been criticized on the grounds that even in the absence of direct physical harm, disruption of data systems with significant harmful consequences, such as the blocking of digital communications or GPS navigation, should be considered as causing a violent effect and hence an attack.[61] This question is presently unsettled in state practice, but it is likely to figure prominently in the future and quite possibly result in a wider interpretation of when the rules pertaining to attack would apply in the cyber context.

A third and final question that may pose specific challenges in the application of LOAC is the degree to which 'cascading effects' should be factored into the taking of precautions in attack. If a computer attack aimed at, say, the guidance system of a weapons system or the navigational system of military aircraft had unforeseen knock on effects resulting in damage or injury not calculated into the original proportionality assessment, would this result in the attack being disproportionate? The short answer is that only effects that are reasonably foreseeable at

---

[60] *Tallinn Manual* (n 47) rule 30 and commentary at 108–9.

[61] Kubo Macak, 'Military Objectives 2.0; The Case for Interpreting Computer Data as Objects under International Humanitarian Law' (2015) 48 Israel L Rev 55–80; Heather Harrison Dinniss, *Cyber Warfare and the Laws of War* (Cambridge University Press 2012) 179ff. Marco Roscini suggests that the concept of 'violence' as used in Article 49 API should be reinterpreted to include not only damage to physical objects, but also incapacitation of infrastructures without physical destruction. See Roscini, *Cyber Operations and the Use of Force* (Oxford University Press 2014) 181.

the time the attack is being planned and executed are relevant to the assessment of what constitutes a disproportionate attack. This is true as a matter of *lex lata* within the cyber domain as well as in the physical one. However, because of the interconnected nature of the digital domain and the difficulty of predicting with certainty what the effects of a particular attack might be in terms of longer term or unforeseen consequences the problem is arguably greater in the cyber context than in the physical domain.[62] Essentially it boils down to what is reasonably foreseeable at the time the attack is contemplated and executed and taking relevant factual circumstances into account. 'Feasible precautions' are those that are reasonable under the attendant circumstances and no commander can be expected to factor in every eventuality. However, in the cyber context, it may be necessary to refrain from, cancel or suspend an attack if there is at least a distinct possibility that harm to civilians or civilian objects *might* occur, particularly if there is at least an appreciable chance that such collateral effects could potentially be substantial. Take, for example, a cyber attack on the Global Positioning System. As it is used for military purposes, it would constitute a military objective in an armed conflict. However, a cyber attack which had the *potential* effect of spreading throughout the system and effectively making normal air and sea navigation impossible, even *potentially* resulting in the widespread loss of life due to crashes of civil aircraft (or interference with navigational systems of protected medical aircraft) would, from this perspective, almost invariably be disproportionate (in the sense of failure to take adequate precautions in attack) even if these consequences were less than definite or likely, so long as there was an appreciable possibility of their occurrence. This is probably an extension of what is strictly required under Article 57 API (and related customary law), but arguably at least an interpretation that may be compelling in the cyber context, where collateral effects are often less predictable than in the physical domain. It remains to be seen how the law will be applied and may develop in this context.

These three challenges to the proper interpretation and application of the law are not the only ones posed by cyber warfare, but they are probably the most significant ones. Neither has this discussion more than outlined the issues that have been mentioned. All of them have received attention elsewhere, so no claim of originality is being made here either. However, this brief *exposé* has been hopefully useful in focusing on some

62 Jensen (n 54) 207–10.

of the main controversies and identifying where some of the principal fault lines lie.

4. SUMMARY AND SOME TENTATIVE CONCLUSIONS

The foregoing discussion of some of the most pertinent operational characteristics and legal challenges posed by these two types of remote warfare has illustrated a number of possibilities and leads to a number of conclusions, some of which are tentative in nature as the debate concerning their use is ongoing and some of the operational modalities of their application are still developing, making in some cases definite conclusions as to how the law will be applied premature.

First, both UAVs and cyber warfare have distinct operational applications that make them in some respects quite different from traditional modes of warfare. Each has potential and real uses that have to an extent, and increasingly will, influence how military operations are conducted and how wars may be fought. UAVs add strike and ISR capabilities that combine a certain degree of covertness, range and endurance and make the conducting of strikes in remote and hostile areas significantly more feasible than through the use of more traditional methods and systems. Cyber capabilities vastly increase the ability of an attacker to carry out operations anonymously (or at least with a high degree of plausible deniability) and create effects and influence the adversary's actions at a low cost and risk. It has exponentially expanded the capability of an actor to gain valuable insight and information regarding the adversary's capabilities, strengths and vulnerabilities and can be used for a variety of purposes ranging from intelligence gathering to direct (support of) kinetic and non-kinetic attacks with varying consequences.

Nevertheless, despite the new and to an extent revolutionary impact of these modes of remote warfare, it is clear they are and are, in principle, capable of being governed by the framework of international law, including the law relating to the use of force and the legal regimes which govern how and against whom or what force may and must be applied, in particular the humanitarian law of armed conflict and international human rights law where these are applicable. While each of these modes of remote warfare poses particular challenges to how the law should be applied and there are ongoing debates on whether particular rules need adjustment, there is by now a general consensus that existing international law applies to these new modes of warfare as it has applied to similar developments in the past, such as aerial warfare, submarine warfare and nuclear warfare to name a few examples. All of those forms of warfare raised questions as to how the law could and should apply and most, if not necessarily all of them, have been resolved.

Some of the main challenges to how the law applies and should be applied to the two modes of warfare covered here have been briefly discussed. While opinions may differ on particular points, it seems to be accepted that these controversies are not incapable of being resolved. Both UAVs and cyber warfare can be used and conducted in conformity with or in violation of the law. This is no less true for them than for any other means or method of warfare. Since they are here to stay and are likely to become more and more pervasive in the future, it is in the common interest to identify where the potential problems lie and point in the direction of the way controversies and challenges can be resolved. There are no easy answers to these questions, which is why we have not attempted to provide them. It will require more study and further effort and practice on how to conduct operations utilizing these new modes in conformity with the law. However, if one assumes that the law applies and shares the conviction that it must be adhered to, it is not a challenge that cannot be met. To the extent this chapter has shed some light on these controversies, it has fulfilled its purpose.

PART III

REMOTENESS THROUGH AUTONOMOUS WEAPONS

11. Remote and autonomous warfare systems: precautions in attack and individual accountability

Ian S Henderson, Patrick Keane and Josh Liddy[*]

1. INTRODUCTION

The long and tragic history of human warfare manifests an endless quest for more effective ways to conduct attacks and defeat adversaries. This has in turn driven innovation in means and methods of defence against attacks. Remote warfare exemplifies both streams of development. The ability to conduct effective attacks at great distance from your own forces is a significant advantage in attack. Similarly, the ability to distance military personnel from the effective range of enemy weapons is a significant advantage in defence. Military forces have sought these advantages since time immemorial.

In medieval times, heavily armoured knights were a powerful force on the battlefield. With their speed, mobility and heavy armour they could wreak havoc among the foot soldiers with relative impunity. However, developments in projectile weapons, and in particular the development and proliferation of gunpowder, contributed to ending the superiority of the mounted knight.[1] It allowed combatants with far less training and equipment to apply force remotely, at a distance where the weapons of the knights were ineffective.

The development of weapons to apply force at ever greater distances has continued down the centuries. From the early firearms that ended the

---

* The authors are legal officers in the Royal Australian Air Force. This chapter was written in their personal capacities and does not necessarily represent the views of the Australian Department of Defence or the Australian Defence Force.

[1] David Schwope, 'The Death of the Knight: Changes in Military Weaponry during the Tudor Period' (*Henderson State University*, 2003–04) 133–5, accessed 5 May 2017 at http://www.hsu.edu/academicforum/2003-2004/2003-4AFThe Deathof%20the%20Knight.pdf.

reign of armoured knights through to intercontinental ballistic missiles, humankind has employed great ingenuity in developing means of warfare to attack adversaries remotely. Modern remote weapon systems, including vehicles that can be operated remotely, are but the latest element in this continuum of development.

For a variety of reasons, the means of remote warfare that involve armed remotely operated vehicles have evoked significant social, political and legal reactions. This is despite the fact that, from an effects perspective, there is little difference between a warhead delivered by other means, such as a remotely operated vehicle and a similar warhead delivered by a long-range artillery or missile system.[2] There is no point of legal distinction, in terms of the precautions that must be taken in attack, between a weapon system that is operated by a human inside it compared to one that is operated by a human remotely.[3]

The reactions to remotely operated vehicles are at least partially driven by the way in which these systems might be employed. The ability to gain pervasive and persistent intelligence through remotely operated aircraft enhances the capability to conduct dynamic targeting, particularly in areas where the attacker does not have ground forces. It can be argued that because there are no personnel at risk in a remotely operated vehicle, states are more likely to carry out an attack than if they had to risk a manned platform or ground forces. It is also suggested that the lack of risk to military personnel may encourage states to more readily violate the sovereignty of another state in order to conduct an attack (for example an attack on non-state actors within the territory of another state). However, these are not issues that affect precautions in attack under the law of armed conflict, which must be applied regardless of whether a system is remotely operated or, for that matter, autonomous.

The potential development of remotely operated autonomous weapons systems (AWS) is contentious at the present time, with many states, non-government organisations and commentators raising concerns from

---

[2] For example, Lockheed Martin, 'High Mobility Artillery Rocket System', accessed 5 May 2017 at www.lockheedmartin.com.au/us/products/himars.html. This system has a short time of flight and high accuracy and can deliver similar tactical effects to attacks by remotely operated aircraft. The issue is the accuracy and timeliness of target information rather than the means or method of delivery of the effect.

[3] Note, though, that while the legal tests are the same, the factors to be weighed when applying those tests may vary. However, that is no different from any other targeting engagement, as each case must be assessed individually and on its merits.

legal and ethical perspectives. Efforts to develop remote AWS are a logical extension of the development of remotely operated systems. From a military perspective, AWS would solve some of the existing limitations on remotely operated vehicles. AWS would be less susceptible to being neutralised by the adversary through interruption of communications. They could potentially work under water, inside buildings and in caves and tunnels under the ground where remotely operated systems may not be able to reach. They would also be less manpower intensive to operate; while lacking a pilot on board, some current remotely operated systems require a large footprint of personnel in order to function effectively.[4]

However, the concept of autonomous decision-making by machines to conduct attacks on human beings arouses visceral trepidation for many and has given rise to calls for bans before the technology can be developed.[5] The potential development of remote AWS is an important topic for legal and ethical discussion by governments, non-government organisations and scholars. This chapter will focus on the legal dimensions of remote AWS, analysing how the precautions in attack in Article 57 of Additional Protocol I (API)[6] might apply to such systems and how the persons creating or using such systems may be held accountable under international humanitarian law (IHL) for outcomes of their use.

---

[4] It can take as many as 170 persons to launch, fly, and maintain such aircraft as well as to process and disseminate its intelligence, surveillance and reconnaissance products. See Jeremiah Gertler, 'US Unmanned Aerial Systems' (*Congressional Research Service*, CRS Report for Congress, 3 January 2012) 26, accessed 5 May 2017 at https://www.fas.org/sgp/crs/natsec/R42136.pdf.

[5] See Frank Sauer, 'Banning Lethal Autonomous Weapon Systems (LAWS): The Way Forward' (*International Committee For Robot Arms Control*, 13 June 2014), accessed 5 May 2017 at http://icrac.net/2014/06/banning-lethal-autonomous-weapon-systems-laws-the-way-forward/; Human Rights Watch, 'Losing Humanity: The Case Against Killer Robots' (*Human Rights Watch,* November 2012), accessed 5 May 2017 at https://www.hrw.org/report/2012/11/19/losing-humanity/case-against-killer-robots; Human Rights Watch, 'Making the Case: The Dangers of Killer Robots and the Need for a Preemptive Ban' (*Human Rights Watch*, December 2016), accessed 10 May 2017 at https://www.hrw.org/report/2016/12/09/making-case/dangers-killer-robots-and-need-preemptive-ban; Will Ockenden, 'Australian AI Expert Toby Walsh takes fight to ban 'Killer Robots' to United Nations after thousands sign petition' (*ABC News Online*, 20 October 2015), accessed 5 May 2017 at http://www.abc.net.au/news/2015-10-20/australian-ai-expert-calls-for-ban-on-killer-robots-at-un/6868834.

[6] Protocol Additional to the Geneva Conventions of 12 August 1949, and relating to the Protection of Victims of International Armed Conflicts (adopted 8 June 1977, entered into force 7 December 1978) 1125 UNTS 3 ('API').

2. DEFINING REMOTE WARFARE AND AWS

As a term, 'remote warfare' does not have any accepted legal definition. Rather, the term is a *concept* that covers a spectrum of different means and methods of warfare. In one sense it immediately brings to mind machines that are operated by remote control. In another sense it may be applied to any weapon that can have an effect at great distance from the person that operates it. It may also be applied to any system that operates in a way such that it gives a degree of safety from rapid counter-attack by the adversary.

These aspects of the concept of 'remoteness' are not exclusive and defining remote warfare is problematic. Combatants might operate a small remote controlled vehicle but be well within the range of the weapons of the adversary. Ballistic and cruise missile systems operate at great distances from the persons operating them but if the adversary has similar capabilities then the operators may face a similar threat. Manned military aircraft may be carrying out attacks with relative safety in an unopposed operating environment where the adversary lacks weapons that are effective against the aircraft. Should any of these examples be regarded as remote warfare?

It would seem that the concept of 'remoteness' is relative. At the least remote end, the term could be used to describe any form of weapon system where the attacker is not in appreciably increased physical danger from enemy attack while using that weapon system. Using that as the definition, the use of some weapons is almost never going to be considered remote warfare—for example, a knife. At the most remote end, some weapon systems will almost always amount to remote warfare—for example, an unmanned combat aerial vehicle that can be operated by combatants who are themselves located thousands of kilometres away from the intended target.

It also seems impossible to separate the concept of remoteness from the military capabilities of the adversary. Attacking an enemy with artillery where that enemy had only small arms would be a form of remote warfare; however, it would not be remote warfare if the enemy had its own long-range ground attack capability (for example, artillery, surface to surface missiles, or ground attack aircraft).

It may be that nothing of legal significance turns on a definition of remote warfare, except perhaps to illustrate that the term may be applied to a wide variety of means and methods of warfare and that the aspect of remoteness does not, by itself, create any distinct legal issues under the law of armed conflict.

Similarly, there is no internationally agreed definition of an AWS.[7] For the purposes of this chapter, we have used the version put forward by the US Department of Defense in its Directive on Autonomy in Weapon Systems, namely: 'A weapon system that, once activated, can select and engage targets without further intervention by a human operator'.[8] This is very similar to the definition used by Human Rights Watch, which is '[r]obots that are capable of selecting targets and delivering force without any human input or interaction';[9] and by the Geneva Academy, which is 'weapon systems that can select and engage targets without a human override'.[10]

It is important to note that not all AWS will necessarily fall within the spectrum of remote warfare. There are existing weapons that may satisfy the definition of AWS but which are operated in relatively close proximity to combatants. For example, certain surface-to-air weapons are programmed to operate autonomously when a target meeting pre-set criteria is identified. Surface-to-surface counter-battery weapons[11] can be programmed in the same way.

3. PRECAUTIONS IN ATTACK

The principal treaty provision on conducting attacks is API, Article 57 *Precautions in attack*.[12] In this section, each relevant sub-article of

---

7 As at May 2016.

8 United States Department of Defense, *Directive No 3000.09 Autonomy in Weapon Systems* (21 November 2012) Glossary.

9 Human Rights Watch, 'Losing Humanity: The Case Against Killer Robots' (n 5).

10 Geneva Academy, *Autonomous Weapon Systems under International Law* (Academy Briefing no. 8, November 2014) 6.

11 Counter battery fire is fire delivered for the purpose of destroying or neutralising the enemy's fire support system. See North Atlantic Treaty Organisation, 'NATO*Term*', accessed 5 May 2017 at https://nso.nato.in/natoterm/Web.mvc. While a little dated, a good description of counter-battery fire and the associated technology can be found at GlobalSecurity.org, 'Counter-Rocket, Artillery, Mortar (C-RAM)', accessed 5 May 2017 at http://www.globalsecurity.org/military/library/budget/fy2006/dot-e/army/2006cram.pdf.

12 For a list of non-binding sources discussing the customary international law status of IHL/LOAC rules, see Michael Schmitt and Jeffrey Thurnher, '"Out of the loop": Autonomous Weapon systems and the Law of Armed Conflict' (2013) 4 Harvard National Security Journal Features 231 fn 9.

Article 57 is discussed in relation to remote warfare in general, and remote AWS in particular.[13]

Article 57(1)

> In the conduct of military operations, constant care shall be taken to spare the civilian population, civilians and civilian objects.

The relevance of this sub-article overlaps with the discussion below on sub-article 57(2)(a)(ii). There is nothing inherent in this sub-article that legally preferences remote or non-remote warfare. Rather, to the extent that either remote or non-remote warfare is more or less likely to spare the civilian population, civilians and civilian objects, there should be a preference for choosing that form of warfare. However, the effect on the civilian population, civilians and civilian objects is only one factor and it is not an absolute requirement. To give but one example, it is permissible, within limits, to cause incidental loss of civilian life, injury to civilians and damage to civilian objects.[14]

The legal issue contained in this sub-article is cited by some commentators as a factor in favour of AWS, or at least a reason for continued research and development. To summarise the argument, as explained by Sassòli: 'if autonomous systems are better than human beings, such as in taking precautions, and a State and a commander have them in their arsenal and don't need to reserve their use for other militarily more important tasks or tasks involving higher risks for civilians, they must use them.'[15] As a general premise, that is a sound enough legal argument. However, a legal point that should not be overlooked is that sub-article 57(1) is a general requirement and as such would give way to the specific. For example, a prohibited means or method of warfare could not be used; notwithstanding that it might be more discriminate or cause less collateral damage compared with the next most feasible non-prohibited means or method—an attacker could not use a chemical weapon to clear a building in preference to explosives.

If a separate legal prohibition on a method of remote warfare, and in particular an AWS, existed, then sub-article 57(1) (or the rest of article 57

---

13 Note, '[t]here is universal consensus that the law of armed conflict applies to [AWS].' (ibid 243).

14 API (n 6) Article 57(2)(a)(iii).

15 Marco Sassòli, 'Autonomous Weapons and international Humanitarian Law: Advantages, Open Technical Questions and Legal Issues to be Clarified' (2014) 91 International Law Studies 308, 320.

for that matter) would not authorise its use. However, the existence of sub-article 57(1) reminds us that care should be taken in considering the overall implications for the civilian population, civilians and civilian objects before creating a new prohibition on remote and autonomous warfare.

Article 57(2)

> With respect to attacks, the following precautions shall be taken: (a) those who plan or decide upon an attack shall: (i) do everything feasible to verify that the objectives to be attacked are neither civilians nor civilian objects and are not subject to special protection but are military objectives within the meaning of paragraph 2 of Article 52 and that it is not prohibited by the provisions of this Protocol to attack them;

The legal issue contained in this sub-article, namely the requirement to do everything feasible to verify that a target is a lawful target,[16] is cited by both proponents and opponents of remote warfare, specifically in relation to remotely operated weapon systems and AWS.

An essentially technological argument put forward by opponents of AWS is that the limited technological capability of sensors and software means that a human is better able to distinguish lawful targets.[17] Conversely, it is suggested that an anticipated advantage stemming from the use of AWS is its processing capability. Computers today can already process more data, and more quickly, than a human mind and further advances in computer technology, such as quantum computing, are likely. In the future, it is entirely foreseeable that an AWS will be able to receive information from a wide selection of sensors and sources in order to create an accurate picture of a potential target. In comparison to a human combatant, who is limited to information discernible by the five senses at the time of an attack and the human mind's ability to process that information, an AWS could receive information from its own sensors,

---

[16] It is generally agreed that in this context, *feasible* 'means that which is practicable or practically possible, taking into account all circumstances prevailing at the time, including humanitarian and military considerations.' International Humanitarian Law Research Initiative, HPCR Manual on International Law Applicable to Air and Missile Warfare (2009), online: Program on Humanitarian Policy and Conflict Research at Harvard University ('HPCR Manual') rule 1.q, accessed 5 May 2017 at http://ihlresearch.org/amw/HPCR%20Manual.pdf.

[17] The other main technological argument made against AWS, namely that they could not comply with the principle of proportionality, is dealt with below.

including from spectrums otherwise invisible to human perception without technological assistance, as well as data from other surveillance platforms. The amount of data an AWS could receive and process would overwhelm the processing capability of a human combatant in the same situation.

The use of AWS may also remove some of the aspects of human decision-making that can cloud the verification of a target as a military objective. Fear, hate, prejudice, unconscious bias, groupthink, fatigue, the desire for success—these are elements of the human condition that may contribute to the misidentification of a target as a military objective. As Sassòli notes: 'human beings often kill others to avoid being killed themselves. The robot can delay the use of force until the last, most appropriate moment, when it has been established that the target and the attack are legitimate.'[18] In short, and assuming proper distinction can be made between a civilian and a military objective, it is foreseeable that an AWS could be more readily able to comply with the requirements of sub-article 2 than a human combatant. The obstacle that needs to be overcome, and some commentators have suggested that it is an insurmountable one,[19] is the ability of the AWS to distinguish between legitimate objects of attack and protected objects.

Ultimately, an aspect of this argument is really an engineering issue (whether an AWS can perform a certain function) and there is little lawyers can meaningfully comment on. It is possible that AWS will never develop the programs or sensors necessary to be able to satisfy the requirements of distinction, let alone making appropriate proportionality calculations, when launching an attack. However, there is a valuable role for lawyers in setting out the correct interpretation of the legal tests that should be applied.

The first point to note is the 'standard' against which AWS should be judged. As Sassòli notes, '[t]here is widespread agreement that the ability to use autonomous weapons in compliance with IHL should not be evaluated against a hypothetical ideal, but instead the comparison should be to human beings'.[20] Nonetheless: 'Determining the data-match acceptance criterion is uncontestably a sensitive issue. Should it be 100 per

---

[18] Sassòli (n 15) 310; See also Schmitt and Thurnher (n 12) 264.

[19] Human Rights Watch, *Mind the Gap: The Lack of Accountability for Killer Robots* (April 2015) 8.

[20] Sassòli (n 15) 319 (footnote omitted). See also Schmitt and Thurnher (n 12) 247; Alexander Bolt, 'The use of autonomous weapons and the role of the legal advisor' in Dan Saxon (ed), *International Humanitarian Law and the Changing Technology of War* (Martinus Nijhoff 2013) 123, 133–4.

cent? Would it be sufficient if the system could identify the correct target 80 per cent of the time?'[21] The authors are unaware of any comprehensive study on the reliability of human assessment of whether potential targets are in fact lawful targets. It is a safe assumption that human combatants do make mistakes but there is no data on how frequently this occurs. Accordingly, while not disputing that the standard against which AWS should be judged is human beings, there would be great difficulty in translating that into an identification standard that could be coded into a computer program.[22]

When Sassòli writes that a 'robot must be able to sense all the necessary information in order to distinguish between targets in the same manner as a person',[23] it is unclear whether he means AWS must make the decision on distinction based on the same inputs that a human would, or merely to no lesser degree of reliability. Humans are limited to our five primary senses, while the methods that an AWS might employ continue to be developed. For example:

> Significant work is underway to produce integrated systems where cross-cueing of intelligence, surveillance, and reconnaissance sensors allows for improved detection rates, increased resolution, and ultimately better discrimination. Multi-sensor integration can achieve up to 10 times better identification and up to 100 times better geolocation accuracy compared with single sensors.[24]

It is suggested that the better view is that AWS must perform to at least the same degree of reliability, but may use different inputs. As noted above, one of the potential advantages of an AWS is the possible ability for an AWS to receive information from its sensors in spectrums otherwise invisible to human perception (without technological assistance through an interface) and cross reference that data with intelligence databases to achieve greater reliability in target identification.

---

[21] Vincent Boulanin, 'Implementing Article 36 Weapon Reviews in the Light of Increasing Autonomy in Weapon Systems' (*SIPRI Insights on Peace and Security*, November 2015) 15–16, accessed 5 May 2017 at http://books.sipri.org/files/insight/SIPRIInsight1501.pdf.

[22] Alan Backstrom and Ian Henderson, 'New capabilities in warfare: an overview of contemporary technological developments and the associated legal and engineering issues in Article 36 weapons reviews' (Summer 2012) 94 No 886 International Review of the Red Cross 483, 495 (footnotes omitted).

[23] Sassòli (n 15) 327.

[24] Backstrom and Henderson (n 22) 489.

As the sub-article requires that 'everything feasible' is done to verify that a proposed target is a lawful target, proponents of remote warfare argue that if in a given situation a remote means of warfare would be more discriminate than a non-remote means, then subject to use of that remote means also being feasible in the circumstances, there is a positive legal obligation to employ the remote means.[25]

A separate legal issue arises specifically in relation to AWS. Some commentators argue that the decision whether a proposed target is a lawful target requires human judgement.[26] For example, Akerson states that AWS 'are inherently illegal under IHL for three reasons. First, the fundamental rules of IHL—including distinction and proportionality—require the application of judgment and discretion. These terms necessarily refer to human judgment and discretion …'.[27]

The authors are not persuaded by the inherently unlawful arguments when based on the principle of distinction. To see why not, it is helpful to consider targeting objects and targeting people separately. But before doing so, it is worth addressing a purported distinction between offensive and defensive AWS.

Akerson distinguishes between offensive and defensive AWS and says only the former are inherently illegal. He is not the only commentator to draw the distinction. For example, the 'open letter' calling for a ban on AWS concludes with: 'Starting a military [artificial intelligence] arms race is a bad idea, and should be prevented by a ban on *offensive* autonomous weapons beyond meaningful human control.'[28]

While there may be moral, ethical or political reasons for drawing a distinction between offensive and defensive AWS, there is no proper legal basis. Article 49 of API provides that: '"Attacks" means acts of violence against the adversary, whether in offence or in defence.'[29] As such, the rules concerning targeting in API, and in particular the precautions in attack in Article 57, apply equally to both attackers and defenders.

---

[25] Sassòli (n 15) 320.

[26] Sassòli, who does not agree in general, provides a good overview of the arguments. See Sassòli (n 15) 331–5.

[27] David Akerson, 'The illegality of offensive lethal autonomy' in Dan Saxon (ed), *International Humanitarian Law and the Changing Technology of War* (Martinus Nijhoff, 2013) 65, 69–70. Note, Akerson limits this inherent illegality to 'offensive' AWS, in distinction to 'weapons that are defensive in design or character.' (Akerson, 73).

[28] 'Autonomous Weapons: An Open Letter from AI & Robotics Researchers' (*Future of Life Institute*, 2015), accessed 5 May 2017 at http://futureoflife.org/open-letter-autonomous-weapons/ (emphasis added).

[29] API (n 6) 49.

Accordingly, as a strict matter of law, Akerson is not correct when he writes 'when a rocket is fired at a naval vessel with an Aegis system, the principle of distinction is obviated, and the principle of proportionality is scrutinized more leniently'.[30] With respect to distinction, it seems Akerson assumes that the Aegis system will be 100 per cent reliable in only responding when a threat actually exists and when doing so only target the threat. With respect to proportionality, he explains that 'the application of proportionality is treated more forgivingly than an attacking force because most precautions are unfeasible due to the exigencies of the attack'.[31] While that is partially correct concerning precautions like choice of weapon and choice of target, it is not legally true with respect to the final proportionality decision on expected collateral damage not being excessive in relation to anticipated military advantage. For example, what if the threat posed to the naval vessel was from another vessel with civilians on board? The proportionality decision applies with just as much force when defending your ship from attack as it does when deciding to target an enemy ship—albeit the military advantage is not necessarily, and indeed almost certainly will not be, equivalent.

With that preliminary point dealt with, starting with objects as potential targets, the main argument for why an AWS would not be able to lawfully select and engage objects is that the test for whether an object is a military objective is not reducible to a mathematical formula, but rather turns on whether a commander has a reasonable belief that it is a lawful target. While there is some element of truth to this for complex targets like dual-use objects,[32] the argument loses force with respect to military equipment with little or no civilian counterpart.

In 2006, Canning suggested a concept of operation for AWS to limit them to being employed against other weapon systems.[33] As explained by Backstrom and Henderson:

> [I]f a commander was prepared to forgo some theoretical capability, it is possible in a particular armed conflict to produce a subset of objects that are at any given time targetable. As long as the list is maintained and reviewed, at

---

[30] Akerson (n 27) 74.

[31] Ibid.

[32] See Markus Wagner, 'Autonomy in the battlespace' in Dan Saxon (ed), *International Humanitarian Law and the Changing Technology of War* (Martinus Nijhoff 2013) 99, 111–12.

[33] John S Canning, 'A Concept of Operations for Armed Autonomous Systems', 3rd Annual Disruptive Technology Conference (Washington, DC 2006), accessed 5 May 2017 at http://www.dtic.mil/ndia/2006disruptive_tech/canning.pdf.

> any particular moment in an armed conflict it is certainly possible to decide that military vehicles, radar sites, etcetera are targetable. In other words, a commander could choose to confine the list of targets that are subject to automatic target recognition to a narrow list of objects that are clearly military objectives by their nature—albeit thereby forgoing automatic target recognition of other objects that require more nuanced judgement to determine status as military objectives through their location, purpose, or use.[34]

Boothby notes that: '[t]here is, at the time of writing, established technology which enables sensors to detect and recognise pre-determined categories of military equipment, such as artillery pieces, tanks, armoured personnel carriers, anti-aircraft batteries and so on.'[35] For some time this is likely to be coupled with limiting the area of operations in which an AWS is set to function.[36]

Moving from objects to people, for at least some human targets, the test can be reduced to a relatively simple test: a member of the enemy armed forces, who is not a non-combatant, is lawfully targetable unless *hors de combat.* If a (partial) list of known enemy combatants was used, the main legal issue would be determining the required degree of confidence and failure rate for an AWS that could positively identify whether a person matched one of the names on the list. The other legal issue would be to resolve how an AWS could identify whether a combatant was *hors de combat.*

The more legally difficult issue would be for an AWS to determine whether a civilian was targetable due to taking a direct part in hostilities.[37] This is a more legally complex test than whether a person is a combatant and because of the temporal element of 'for such time', whether a person is targetable as a civilian taking a direct part in hostilities is much less susceptible to being resolved via the list method identified above for combatants. This does not mean that the problems are insurmountable. At present, a human decision-maker may determine that an individual is taking a direct part in hostilities based on whether they are armed, the proximity of the fighting and the direction and

---

[34] Backstrom and Henderson (n 22) 492.

[35] William Boothby, 'How far will the law allow unmanned targeting to go?' in Dan Saxon (ed), *International Humanitarian Law and the Changing Technology of War* (Martinus Nijhoff 2013) 45, 55. See also Wagner (n 32) 113; Geneva Academy (n 10) 14.

[36] Schmitt and Thurnher (n 12) 241.

[37] See API (n 6) Article 51(3); Sassòli (n 15) 328–30.

manner in which an individual is moving.[38] It may become possible to program these variables into an AWS.

However, while the application of the legal test might be more difficult, the operational solution is straightforward. Unless and until an AWS can be designed that can, to the required degree of confidence, determine whether a person is taking a direct part in hostilities, the simple operational solution is to not employ an AWS in that role. For an AWS to be lawful, it need not be able to do everything that a combatant could theoretically do. Rather, the AWS must be able to perform any assigned role in a lawful manner.

There is one other subsidiary technological argument that is made, namely that even if an AWS could distinguish a combatant from a civilian, it would have significant, if not insurmountable, difficulty in determining whether that combatant was *hors de combat*.[39] Article 41 of API, which reflects customary international law,[40] provides:

> 1. A person who is recognized or who, in the circumstances, should be recognized to be *hors de combat* shall not be made the object of attack. 2. A person is *hors de combat* if: a) he is in the power of an adverse Party; b) he clearly expresses an intention to surrender; or c) he has been rendered unconscious or is otherwise incapacitated by wounds or sickness, and therefore is incapable of defending himself; provided that in any of these cases he abstains from any hostile act and does not attempt to escape.[41]

The issue is whether an AWS could recognise that a combatant was expressing an intention to surrender, or was unconscious or incapacitated by wounds or sickness.[42] As well as technological and legal issues, there also appear to be ethical issues that need to be considered.[43]

The discussion about recognising *hors de combat* illustrates an interesting issue. The starting point for most discussions about the ability of an AWS to comply with the LOAC is whether an AWS could perform as well as a human. While from an ethical perspective Sparrow posits that

---

38 See the cases and reports discussed in Ian Henderson and Bryan Cavanagh, 'Unmanned Aerial Vehicles: Do They Pose Legal Challenges?' in Hitoshi Nasu and Robert McLaughlin (eds), *New Technologies and the Law of Armed Conflict* (Springer 2014) 193, 205.

39 See Human Rights Watch, 'Mind the Gap' (n 19).

40 ICRC, *Customary International Humanitarian Law* (2005) rule 47.

41 Sub-article (3) omitted.

42 Sassòli (n 15) 328; Boothby (n 35) 59.

43 See Robert Sparrow, 'Twenty Seconds to Comply: Autonomous Weapon Systems and the Recognition of Surrender' (2015) 91 International Law Studies 699.

'we might well expect more of robots than this',[44] nonetheless the legal position is that 'as a matter of law, more may not be asked of [AWS] than of human-operated systems'.[45] However, it is arguable that there is a key legal difference between target recognition and *hors de combat*. Article 57(2)(a)(i) of API requires those who plan or conduct attacks to do 'everything feasible to verify that' targets are lawful targets. Whereas Article 41(1) of API requires that a 'person who is recognized or who, *in the circumstances, should be recognized* to be *hors de combat* shall not be made the object of attack'.[46] The legal interplay between these two articles is not clear.

On the one hand, it could be argued that as Article 57(2)(a)(i) of API refers to 'do everything feasible to verify that ... it is not prohibited by the provisions of this Protocol to attack them', those who plan and conduct attacks need to adopt all feasible means to check whether or not the subject of a planned attack is *hors de combat*. Alternatively, it could be argued that Article 41(1) sets the standard for assessing *hors de combat* and there is no *legal* obligation to adopt means and methods of warfare that facilitate recognition of surrender. For example, existing methods of warfare such as rocket attacks, bombing, strafing and artillery barrages are far from likely to facilitate attackers being able to determine whether a combatant is or is not *hors de combat*, especially after a lengthy attack has commenced on a large number of combatants.[47] The rule is explained as follows:

> Combatants ... must communicate clearly their intention to surrender before becoming immune from attack. If a combatant ... does not indicate an intention to surrender in a way that the enemy can perceive and understand, this person is still liable to be attacked. For example, the crew of an attacking aircraft conducting a beyond-visual-range attack may be unaware that the forces they are attacking wish to surrender. As long as the lack of knowledge is reasonable in the circumstances, the attack may lawfully be conducted because the desire to surrender has not been effectively communicated to the aircrews (or other forces which could pass that information to the crew in adequate time).[48]

---

[44] Ibid 710.

[45] Schmitt and Thurnher (n 12) 246.

[46] Emphasis added.

[47] See Boothby (n 35) 59.

[48] International Humanitarian Law Research Initiative, HPCR commentary on the HPCR manual on international law applicable to air and missile warfare (2010), online: Program on Humanitarian Policy and Conflict Research at Harvard University ('HPCR Commentary'), accessed 5 May 2017

The issue of effective communication may go some way towards dealing with any technological issues associated with an AWS recognising surrender. Whilst an AWS may not be able to identify whether a combatant is *hors de combat*, neither can an artillery shell or missile, both long accepted as means of warfare. To effectively communicate surrender to forces conducting attacks remotely, the onus is on the surrendering forces to communicate as best it can with the forces of the adversary. With respect to an AWS it may be necessary, just like other forms of remote warfare, to communicate surrender to other enemy forces who can arrange for the cessation of the attack by the AWS. Of note, in air-to-air combat there are not even 'widely-agreed upon mechanisms to allow aircraft to surrender'.[49]

In light of existing state practice, the better view is that Article 41(1) sets the standard for determining *hors de combat*. Therefore, the legal issue is not whether a different means or method of warfare would have enabled better recognition of whether a person was *hors de combat*, but rather, based on the actual means or method employed, should a person have been recognised as being *hors de combat*.[50]

Article 57(2)(a)(ii)

> take all feasible precautions in the choice of means and methods of attack with a view to avoiding, and in any event to minimizing, incidental loss of civilian life, injury to civilians and damage to civilian objects;

This sub-article is similar to the opening sub-article, but is more precise in its wording. The requirement to take 'all feasible precautions' is identified by proponents of remote warfare as a factor that, in the appropriate circumstances, points in favour of remote warfare.[51]

First, it is argued that for human-operated weapon systems, being remote from harm reduces the personal anxiety and fear of the combatant, thereby allowing a more measured judgement concerning who is a lawful target and whether civilians are in the vicinity. In other words, it is argued that given

---

at http://ihlresearch.org/amw/Commentary%20on%20the%20HPCR%20Manual.pdf 95, [5]. See generally HPCR Commentary, Rule 128 and accompanying commentary, 267–8.

[49] Sparrow (n 43) 717.

[50] But see Boothby (n 35) 60.

[51] See generally Schmitt and Thurnher (n 12) 262; Kenneth Anderson, Daniel Reisner and Matthew Waxman, 'Adapting the Law of Armed Conflict to Autonomous Weapon Systems' (2014) 90 International Law Studies 386, 393; Geneva Academy (n 10) 16.

the choice between a combatant who is personally at risk and an operator of a weapon system who is personally remote from danger, the remote weapon system operator is more likely to make a conservative decision.

Second, it is argued that the nature of remote warfare means that decision-makers have access to more information and intelligence, and an ability to consult experts, than a combatant who is not remote. This is a situational argument. Technology varies from military to military, and from operating environment to operating environment. However, as long as it is phrased as a possible factor to be considered in a given case, and not as an absolute truth, the point is well made.

Third, concerning AWS in particular, in certain circumstances the AWS may be better able than a human decision-maker to distinguish lawful from non-lawful targets. This is similar to, but slightly different from, the issue discussed concerning sub-article 57(2)(a)(i). Take the example of a person standing at a window. Under sub-article 57(2)(a)(i), the issue is whether that person is a lawful target. Whereas under sub-article 57(2)(a)(ii), the issue is whether there are also other people in the room (or the building depending upon the likely means of attack); and if so, whether or not they are also lawful targets.

Article 57(2)(a)(iii)

> refrain from deciding to launch any attack which may be expected to cause incidental loss of civilian life, injury to civilians, damage to civilian objects, or a combination thereof, which would be excessive in relation to the concrete and direct military advantage anticipated

This sub-article is one expression in rule form of the principle of proportionality.[52] Like sub-article 57(2)(a)(1), there are both technological and legal issues associated with AWS and assessing proportionality. Along with the principle of distinction, the capacity to assess proportionality is one of the most contentious aspects of AWS. '[N]o existing weapon systems have a sufficient level of situational awareness to autonomously evaluate military advantage and balance it against expected collateral damage',[53] and Sparrow goes so far as to write that 'whether robots will ever be capable of making the required proportionality calculations remains highly controversial'.[54]

---

52 See also sub-article 51(5)(b) of API.
53 Boulanin (n 21) 10.
54 Sparrow (n 43) 702.

Two particular legal issues arise with this sub-article. First, and the simplest to deal with from a legal perspective, is how the 'expected' collateral damage would be assessed. The second issue is, for an AWS, how would the calculation be conducted? It is generally recognised that the comparison between collateral damage and military advantage is a subjective assessment, albeit there are differing views as to what is meant by 'subjective' in this context.[55] It is clear that subjective must mean more than just personal preference (for example, whether or not the reader likes wine), but equally the calculation cannot be reduced to an objective formula. Nonetheless, there are a number of ways in which military personnel employing AWS could conduct a 'subjective' proportionality assessment.

First, and perhaps the technologically most straightforward, is for AWS to conduct attacks in ways that pose negligible risk of civilian casualties and civilian property damage against targets the destruction of which represents a significant military advantage. For example, suppose an AWS was tasked with locating and attacking enemy military fighter aircraft. Among the many possibilities, the attack could be conducted while the fighter aircraft were on the ground by way of a small AWS flying into the jet engine intake and detonating a very low yield explosive.

A second and only slightly more complex approach would be for the AWS to conduct attacks in a geographic area where the likelihood of civilian casualties and civilian property damage can be reduced to practically zero.[56] This could be achieved, for example, by limiting an AWS's operating parameters to areas in which it has been assessed that any people present are lawful targets by virtue of either being combatants (or the equivalent in a non-international armed conflict) or civilians taking a direct part in hostilities. For example, the UK Brimstone, a fire-and-forget, anti-armour missile, 'can be programmed not to search for targets until they reach a given point … or only to accept targets in a designated box area, thus avoiding collateral damage'.[57] Of course, 'thus avoiding collateral damage' is an assertion. For the purposes of this chapter, the point is that militaries already have weapons that are designed to mitigate collateral damage risk by operating in limited geographical areas.

---

[55] Sassòli, who actually advocates for introducing greater objectivity into the test, provides a good overview of the arguments. See Sassòli (n 15) 331–5.

[56] See Boothby (n 35) 57; Anderson, Weisner and Waxman (n 51) 402.

[57] Royal Air Force, Aircraft & Weapons 87 (*Royal Air Force,* August 2003), accessed 5 May 2017 at http://www.raf.mod.uk/rafcms/mediafiles/0186cc2a_1143_ec82_2ef2bffff37857da.pdf. Interestingly, the second edition of this publication makes no mention of this capability.

A similar but alternative approach is an AWS that is tasked with locating and attacking enemy targets that may be in unknown geographic areas but nonetheless are extremely likely to be free of civilians. For example, airborne enemy military fighter aircraft. If an AWS were programmed to conduct attacks only against airborne enemy aircraft, there would be a low probability of civilian casualties. This probability could be reduced still further by limiting the airspace in which an attack could occur to airspace over large expanses of water or other very low density population areas (for example, deserts or large forested areas).

A third option would be for a commander to pre-determine the upper limit for civilian casualties and various types of property damage prior to authorising the AWS mission.[58] Staying with our anti-fighter aircraft AWS, at any given point in the conflict a commander could assign an upper limit for civilian casualties. This limit might vary depending upon the location of the enemy fighter. Simplistically, at a given time a commander might determine that the death of no more than three civilians would not be excessive when compared to the military advantage anticipated from destroying an enemy fighter aircraft at an enemy military air base. In such a scenario, the AWS would have to be able to assess whether or not humans were present, but in the simplest case need not have to have the capability to distinguish civilians from lawful targets. The AWS could be programmed to treat any human as a civilian when calculating proportionality.

For these three options, the proportionality assessment has effectively been made in advance by the personnel who deployed the system based on the known capabilities of the AWS and the military situation. The anticipated military advantage of destroying an enemy fighter aircraft has been weighed against the expected loss of civilian life, injury to civilians, and damage to civilian objects, and a decision has been made to deploy the AWS. Limiting the operating parameters of the AWS allows assumptions to be made about both the nature of the anticipated military advantage and the expected damage.

A fourth and currently quite technologically daunting option would be for an AWS to 'learn' how to apply the principle of proportionality.[59]

Put simply, an AWS could be presented with a series of scenarios and told whether in each scenario the 'correct answer' was to attack or not attack. In this context, the 'correct answer' means the answer that the

---

[58] See Boothby (n 35) 57; Schmitt and Thurnher (n 12) 256.

[59] Backstrom and Henderson (n 24) 493–4. Sassòli queries whether the ability of AWS to 'learn' could not be applied to other circumstances—Sassòli (n 15) 313.

instructor had assessed as being legally correct. In some respects, this is no different from what might occur in a classroom or field exercise for junior commanders.[60] The difference would be that the AWS could be presented with vastly more scenarios than the average human student.

A necessary step for developing this fourth option would be to 'test' the AWS by presenting it with training scenarios and seeing if it assesses the expected collateral damage as proportional or not. Assuming at some point that the AWS gets the required percentage of answers correct, then functionally speaking it would be acting like any other combatant that had also 'passed' this test.[61]

Regardless of the level of autonomy AWS are generally able to achieve, human beings will still have some involvement in the use of AWS on the battlefield. This could be the commander deciding to employ an AWS or the operator involved in the AWS's mission. The policy of the United States stipulates that there must be a person in the decision-making loop for autonomous weapons.[62] That policy identifies three methods of human involvement in the decision-making for an AWS to use lethal force, particularly in an offensive role: 'in-the-loop', 'on-the-loop' and 'outside-the-loop'.[63] While many other states are yet to develop policies on this issue (or declare them if they already done so), it would seem logical that comparable concepts and policies for human involvement in AWS operations will be adopted.

A human operator 'in-the-loop' means that the human operator of the AWS holds the authority for when the system utilises lethal force. While the AWS will likely handle routine tasks without human intervention, such as driving/flying/sailing the platform and navigation, the AWS cannot use force unless specifically authorised by a human operator. This concept is not dissimilar to current weapon systems already in use, such as unmanned aerial vehicles. In this case, accountability for the failings of the AWS can be easily attributed to the human operator who controls the final decision to use force.[64]

---

[60] See generally Anderson, Reisner and Waxman (n 51) 403.

[61] See Jens David Ohlin, 'The Combatant's Stance: Autonomous Weapons on the Battlefield' (2016) 92 International Law Studies 1, 16–19 for a discussion of when an AWS might be functionally indistinguishable from a human combatant despite the external observer having no knowledge of the AWS's programming.

[62] US Department of Defense (n 8).

[63] Ibid.

[64] Michael N Schmitt, 'Autonomous Weapon Systems and International Humanitarian Law: A Reply to the Critics' [2013] Harvard National Security Journal Features 1.

A human operator 'on-the-loop' means that while the AWS makes the decision on whether lethal force will be used, a human operator has the ability to override this process or decision as required, effectively exercising the power of veto.[65] This system would be designed to mitigate the risk of non-compliance with a state's legal obligations and also potentially makes the human operator accountable for failings of the AWS.

A human operator 'outside-the-loop' is a system whereby there is no involvement by a human operator in the AWS's decision to use lethal force. A human operator will no doubt be supervising the AWS, just as a human commander would supervise human subordinates, but that person is not required to approve the AWS's decision to use lethal force. Instead, the system will rely upon the parameters assigned to it to determine when force will be used. Such systems already exist, albeit they are used in limited defensive situations. Essentially, there is no, or very limited, opportunity for a human operator to intervene in the decision-making process of the AWS before force is used. Furthermore, the system is not designed to require human input. Current use of such systems is generally limited to defensive systems, such as counter battery artillery systems[66] and anti-missile systems.[67]

The outside-the-loop option, and potentially the on-the-loop option, attract the argument that they would be illegal because IHL assessments and decision-making must involve human judgement and discretion.[68] While views vary as to whether this is a legal requirement, assume for the moment that this is a correct statement of the law. For the reasons discussed above, it can be argued that human judgement and discretion is being applied by human decision-makers electing to employ AWS at a certain location or time, or against certain kinds of targets, with knowledge of how the system will operate in that environment.

---

[65] P W Singer, *Wired for War: The Robotics Revolution and Conflict in the 21st Century* (The Penguin Press 2009) 126.

[66] See *NATOTerm* (n 11).

[67] United States Navy, 'Fact File: MK 15 – Phalanx Close-In Weapons System (CIWS)' (*United States Navy*), accessed 5 May 2017 at http://www.navy.mil/navydata/fact_print.asp?cid=2100&tid=487&ct=2&page=1.

[68] Akerson (n 25); but see Duncan Hollis, 'Setting the Stage: Autonomous Legal Reasoning in International Humanitarian Law' (5 January 2016) Temple International & Comparative Law Journal, Forthcoming, accessed 5 May 2017 at http://ssrn.com/abstract=2711304.

Article 57(2)(b)

> an attack shall be cancelled or suspended if it becomes apparent that the objective is not a military one or is subject to special protection or that the attack may be expected to cause incidental loss of civilian life, injury to civilians, damage to civilian objects, or a combination thereof, which would be excessive in relation to the concrete and direct military advantage anticipated

The application of this rule to remote and autonomous warfare is problematic. Much turns on the understanding of 'becomes apparent'. Sassòli writes:

> First, artillery and missiles are not—in the same manner as a sniper—able to cancel or suspend an attack at the last moment based on changing circumstances. Nevertheless, no one claims that such weapons are inherently unlawful. What counts is that either the system itself through technical means, or the human beings using it, are able to acquire information indicating that the attack must be interrupted and either the machine or its human operators are able to react to such information.[69]

With respect, we do not fully agree with this statement. As explained by Henderson elsewhere,[70] this sub-article is only enlivened if it does become *apparent* that circumstances have changed. What it does not do is impose a legal obligation to adopt a method of warfare that facilitates last minute observation of the target. If there were such a legal obligation, then 'fire and forget' weapons, such as GPS guided bombs, would be unlawful unless targets were kept under constant observation and there was an ability to guide-off the bomb or in some other way neutralise the attack. That is neither the law nor state practice.

Presumably, in order to meet the obligation under Article 57(2)(b), an AWS should be programmed to cancel or suspend an attack that had commenced if the target no longer met the parameters of a valid target under its programming.

Article 57(3)

> When a choice is possible between several military objectives for obtaining a similar military advantage, the objective to be selected shall be that the attack on which may be expected to cause the least danger to civilian lives and to civilian objects.

---

69 Sassòli (n 15) 320 (footnotes omitted).

70 Ian Henderson, *The Contemporary Law of Targeting: Military Objectives, Proportionality and Precautions in Attack under Additional Protocol I* (Martinus Nijhoff 2009) 182–5.

This is an interesting sub-article in the context of remote and autonomous warfare as it has the possibility of affecting the choice of means or method of attack. Suppose Target Alpha and Target Bravo both offer a similar military advantage if attacked by way of bombing. Further, suppose the expected danger to civilian lives and civilian objects is higher for Target Alpha compared to Target Bravo.[71] But also assume that Target Bravo is in a location with significant air-defence. Without remote warfare as an option, an attacker would be legally permitted to favour the attack on Target Alpha. However, if the attacker has the ability to conduct a remote attack with the same likelihood (or higher) of mission success, and if the remote warfare means or method does not increase the expected danger to civilian lives and civilian objects then an attacker may be required to preference attacking Target Bravo by the remote means or method. In other words, risk to own forces is an element of assessing competing military advantage when a choice is possible between several military objectives.

Despite the use of 'shall', it is important to note that the attacker is only obliged to preference attacking Target Bravo and not absolutely obliged to attack Target Bravo in preference to Target Alpha. For instance, the attacker may have a very limited number of remote warfare weapons available and there may be other targets that can be successfully attacked only by remote means.

Article 57(4)

> In the conduct of military operations at sea or in the air, each Party to the conflict shall, in conformity with its rights and duties under the rules of international law applicable in armed conflict, take all reasonable precautions to avoid losses of civilian lives and damage to civilian objects.

The relevance of this sub-article to remote warfare is similar to that discussed at Article 57(2)(a)(ii). If a means or method of remote warfare would be less likely to endanger the civilian population, then it should be preferred, but not necessarily absolutely so, over a non-remote means or method.

---

[71] For completeness, also assume that an attack on Target Alpha would nonetheless be proportional when looked at in isolation (i.e., the anticipated military advantage is not excessive compared to the expected collateral damage).

4. INDIVIDUAL ACCOUNTABILITY

The issue of individual accountability for remote warfare involving a human operator is relatively straightforward. It is a matter of identifying the individual responsible for carrying out an attack and what he or she knew or should have known at the time of relevant actions or decisions. However, for AWS, accountability for when things go wrong is one of the more contentious issues. Many commentators are concerned with the morality and ethical issues associated with a machine making the decision whether to kill a human being,[72] and some argue that the issues with attributing accountability for war crimes committed by an AWS are insurmountable. This is raised as another reason for seeking a pre-emptive ban on the development of AWS.[73]

When only a theoretical consideration of how an AWS might operate is conducted, questions of legal accountability for breaches of IHL prove complicated, if not impossible, to resolve. To explore the topic fully it is necessary to consider the practical aspects of how AWS are likely to develop and how they might be used on the battlefield in the context of the current framework of IHL.

The starting point in considering accountability for attacks conducted by an AWS on the battlefield is the capability of the AWS itself and whether it reaches the standard of target discrimination required by law. Article 57 of API sets the standard required of a combatant launching an attack. In order for the AWS to be lawfully used in the first place, the system must be able to satisfy the requirements laid out in Article 57. If the AWS cannot reach that standard, then the question of accountability is a moot point, because the AWS cannot comply with IHL. Its use would be illegal and the individuals employing that system may be held accountable. The AWS need not be better than a human combatant in satisfying the requirements of Article 57, but, as discussed above, the manner in which the AWS complies with that standard will likely be different to how a human combatant would do so. The different approach does not make AWS inherently illegal, or the methodology suspect. Ultimately, AWS will be reacting to the data it receives in accordance with its programming, including any mission specific parameters applied by human operators.

---

72 See generally (n 5).

73 Human Rights Watch, 'Losing Humanity: The Case Against Killer Robots' (n 5).

Assuming that the AWS is capable of discrimination to the standard required by law, the position of the human operator relative to the AWS decision loop to use force will be an important factor in considering individual accountability. However, while the three system models outlined in the current US policy[74] articulate where the human operator is situated with respect to the AWS decision loop, it does not provide the complete answer on where accountability could lie for war crimes. Obvious candidates for individual accountability with respect to AWS extend beyond the human operator to commanders, programmers and manufacturers.

It is possible that weapon systems will never be able to develop to the point of being able to participate in armed conflict without human involvement in its decision-making process.[75] Certainly some commentators would suggest that it will be impossible for AWS to ever be able to distinguish between combatants and civilians, or even between friendly and enemy forces.[76] In other words, AWS may only develop to the level of remotely operated systems with some automated features but with a human operator making the key decision on use of force.

However, assuming that a true AWS could be developed, an analysis of how accountability would work when an AWS attack allegedly constitutes a war crime can be considered using a relatively simple example in the context of Article 57. Assume that an AWS has targeted a civilian[77] and killed that person. The killing could arise in three ways: a deliberate act with the civilian's death an intended result; an accidental death where the attack itself was not unlawful, but the civilian's death was an unintended consequence; and the reckless attack, where there has been a failure to adequately consider the likelihood that the attack could or would result in the civilian's death.

For the case of the intentional attack, a wilful attack on the civilian population or individual civilians, or an indiscriminate attack that affects the civilian population or civilian objects, when the attacker has knowledge that the military advantage from the attack will be outweighed by excessive loss of life or injury to civilians, or damage to civilian objects,

---

74 Ohlin (n 61).

75 Chantal Grut, 'The Challenge of Autonomous Lethal Robotics to International Humanitarian Law' (2013) 18 Journal of Conflict & Security Law 5, 13.

76 Ockenden (n 5).

77 For the purposes of this example, there is no question of the civilian taking an active or direct part in hostilities or otherwise forfeiting their protection under IHL.

is a grave breach of API.[78] Article 85(3) of API addresses failure to adhere to the requirements of Article 57 (emphasis added):

> In addition to the grave breaches defined in Article 11, the following acts shall be regarded as grave breaches of this Protocol, when committed *willfully*, in violation of the relevant provisions of this Protocol, and causing death or serious injury to body or health:
>
> (a) making the civilian population or individual civilians the object of attack;
> (b) launching an indiscriminate attack affecting the civilian population or civilian objects in the *knowledge* that such attack will cause excessive loss of life, injury to civilians or damage to civilian objects, as defined in Article 57, paragraph 2 (a) (iii)
>
> ...

The operative words in the context of an act that breaches Article 57 are the terms 'willfully' and 'knowledge'. The Commentary on API tells us that 'willfully' means that:

> the accused must have acted consciously and with intent, i.e., with his mind on the act and its consequences, and willing ('criminal intent' or 'malice aforethought'); this encompasses the concepts of 'wrongful intent' or 'recklessness', viz., the attitude of an agent who, without being certain of a particular result, accepts the possibility of it happening; on the other hand, ordinary negligence or lack of foresight is not covered, i.e., when a man acts without having his mind on the act or its consequences (although failing to take the necessary precautions, particularly failing to seek precise information constitutes culpable negligence punishable by disciplinary sanctions).[79]

The parameters that caused the AWS to target the civilian with the intention of causing that person's death have been provided to the AWS by a human being, either through its programming or by direction given to it by its human operators. The mental elements of intention and knowledge would be present in either the programmer or manufacturer of the AWS or the human operator who employed the AWS knowing it would act in this fashion. The material elements and mental element necessary to establish a war crime would be present; the AWS is merely the 'innocent instrument' of the criminal act. For this example, Ohlin is correct when describing the AWS as merely the cog in the machine that

---

[78] API (n 6) Article 85(83)(b).

[79] Yves Sandoz, Christophe Swinarski and Bruno Zimmerman (eds), *Commentary on the Additional Protocols of 8 June 1977 to the Geneva Conventions of 12 August 1949* (Martinus Nijhoff 1987) 994.

commits the war crime, harking back to the Nuremberg Trials.[80] As Ohlin points out, this concept is not a new one in IHL, particularly in relation to the prosecution of war crimes. In the Ponzano Case during the Nuremberg Trials a British Military Court noted that:

> [a] person can be concerned in the commission of a criminal offence, who, without being present at the place where the offence was committed, took such a part in the preparation for this offence as to further its object; in other words, he must be the cog in the wheel of events leading up to the result which in fact occurred.[81]

If we consider that the AWS is simply a cog in a larger machine that uses force, then it is far easier to attribute liability for wrongdoing caused by the AWS—and that liability will inevitably lie at the feet of the human operator that put that machine in motion with the intention of killing a civilian or with the knowledge that it would cause excessive civilian loss of life, injury or damage. An AWS would operate in accordance with its programming. A deliberate breach of IHL by an AWS would come about because of the parameters or programming given to the AWS. In short, the AWS is instructed to use force when it receives data that satisfies a pre-programmed level of certainty that the target can be engaged. In this case, it is immaterial where the human operator sits in the AWS's decision loop. The wilful breach of IHL was committed by the human operator and he or she is responsible for an intentional breach of IHL perpetrated by an AWS.

A commander, knowing that the AWS will behave in manner that would be inconsistent with IHL, such as knowing that the AWS's attack would indiscriminately affect civilians, would be liable for the AWS's 'breach' by virtue of the doctrine of command responsibility. This is no different to a breach of IHL committed by a subordinate human combatant. In the armed forces of most states, individual combatants are given orders for the use of force. If such orders are in breach of IHL, then the commander could be held responsible for the issuing of such orders; particularly if the commander is aware of the illegality of the orders, or that the conduct of the combatants under his or her command are likely to be in breach of IHL because of the orders or parameters provided to them. The distinction with AWS is that the human combatant following

---

80 Ohlin (n 61) 2–3.

81 'Feurstein and Others (Ponzano Case): British Military Court sitting at Hamburg, Germany Judgment of 24 August 1948' 5 Journal of International Criminal Justice 238, 239.

unlawful orders in breach of IHL, where that person knows, or should have known, the orders were unlawful, can also be held criminally liable for actions contrary to IHL concurrently with the commander.

Accountability differs when the death is accidental. This is no different to the current state of the law with regard to human combatants. A mistake that inadvertently causes the death of a civilian, be it by a combatant or an AWS, is unlikely to be a war crime unless the conduct meets the test for criminal negligence that applies within an applicable jurisdiction. Weapon systems will, from time to time, malfunction and fail to operate as intended. It is possible those weapon malfunctions will cause unintended death, injury or damage and sometimes a protected person or object will suffer as a result. A weapon malfunction is unlikely to be considered a war crime as it is highly unlikely to have been a wilful act.[82] It is no mistake that the threshold for liability outlined in Article 85 is a high one.

Weapons and human operators are not perfect now and the law does not impose a higher standard upon an AWS simply because of the absence of a human operator. Errors in programming that cause accidental death and injury should be identified and resolved. If the error cannot be resolved, or if the error is known but ignored, then further use of the AWS may breach Article 57.

If, regardless of the flaw in the AWS, the AWS continues to be used and more deaths ensue, the issue of accountability must logically shift. The utilization of an error-prone AWS is no different to any other malfunctioning weapon system. If a combatant employing a weapon system is aware that the weapon is so error prone as to be indiscriminate, or is so likely to malfunction that the anticipated military advantage to be gained from an attack will be outweighed by expected collateral damage, then that combatant would be in breach of IHL. Decisions to launch attacks in those circumstances could hardly be considered to be taking 'constant care' to spare the civilian population from the effects of an attack.

Another issue with attribution in cases of accident, for systems where there is a human in the loop, is the extent to which the human operator may be held responsible for failing to properly supervise the automated functions of the system. The human operator may demonstrate a bias against questioning the assessment or decision of the AWS about targets engaged—in effect negating the purpose of being 'in-the-loop' and providing proper oversight. This was demonstrated during the 2003 Gulf War when a US patriot missile battery was involved in a friendly fire

---

82 API (n 6) Article 85(3).

incident, shooting down a British GR-4 Tornado when the aircraft was misidentified by the weapon's targeting computer as an enemy anti-radiation missile.[83] While the report into the incident highlighted some training deficiencies in the missile battery crew, what the report also discovered was that there was an inappropriate over reliance upon the automation of the system. This created a bias against questioning the assessment made by the weapon's targeting computer that designated the aircraft an enemy missile.[84]

The issue of accidental deaths caused by AWS with human operators 'in-the-loop' and 'on-the-loop' may raise complex questions about determining accountability between the human operators and the commanders who made decisions to deploy the systems. While these modes of operation appear to provide the military advantages associated with using AWS with the benefit of a 'failsafe' of human override, the human oversight could in reality be nothing more than a placebo. If AWS continue to develop, it is natural to expect that the speed at which the systems identify a target, assess as valid and engage it will become faster and faster. The potential speed of AWS engagements will likely present a new challenge in assessing accountability as the human operator on-the-loop may have no realistic ability to intervene when the AWS is acting contrary to expectations or requirements. Attributing accountability to the human operator becomes problematic when the cycle of identification, assessment and engagement becomes so quick, and so fluid, that a human operator 'on-the-loop' cannot react quickly enough to be able to provide any realistic oversight of the AWS's response to a situation.

If the proper supervision of the AWS is in fact beyond the capacity of the human operator, then accountability would likely revert to commanders who made the decision to deploy the system with knowledge of the likely consequences. This leads to the question of whether the death should be regarded as an accident or whether it was caused by recklessness.

The reckless attack causing the civilian's death is the most difficult in terms of determining accountability. However, as Ohlin notes, this is no different under the current state of the law with human combatants.[85] Recklessness in this context is when an AWS engages the civilian despite there being a risk that the target is not a legitimate one and, despite the

---

[83] John K Hawley, 'Not by widgets alone' (*Armed Forces Journal*, 1 February 2011), accessed 5 May 2017 at http://www.armedforcesjournal.com/not-by-widgets-alone/.

[84] Ibid.

[85] Ohlin (n 61) 21.

risk, continues with the attack anyway. The awareness of that risk by human operators could come about for any number of reasons, but essentially it comes down to the probability that the target is indeed a civilian. It differs from the situation of intentionally killing a civilian as described earlier because of knowledge. Ohlin correctly points out that international law, particularly international criminal law, is ill equipped to deal with the reckless breach of IHL, especially in the situation of the example that results in the death of a civilian.[86]

The concept of recklessness is an aspect of individual accountability against which the operating parameters and capabilities of the AWS, and the decision to deploy the AWS, could be tested. Let us assume that, based on empirical testing, the state fielding the AWS is satisfied that it meets the requirements of Article 57(2) in terms of capability to identify military objectives. A commander deploys the AWS in a particular conflict and while largely successful it misidentifies civilians and civilian objects as military objectives at a rate of about 5 per cent.

It is arguable that through continued use of the AWS with knowledge of the rate of misidentification, the commander is wilfully making the civilian population the subject of attack through the concept of recklessness; that is, without being certain of a particular result, the commander adverts to the probability of it happening. The counter-arguments to this proposition would be that most weapons have an error rate where they do not function as expected and that the AWS is sufficiently accurate in its target identification and attack that it should not be regarded as unlawfully indiscriminate.

This would lead to difficult questions about what rate of error should cause an AWS to be regarded as unlawfully indiscriminate for its use to be regarded as causing the wilful death of civilians through recklessness. However, this is a question that requires resolution under the current law of armed conflict, as it is a consideration for the legality of any new weapon system. From a humanitarian perspective, there is little difference between an AWS that misidentifies a target and strikes it accurately causing harm to civilians and a human combatant who aims to target a military objective but, due to inherent failure rates in the weapon, misses the target and causes harm to civilians. Despite being an extant issue, it is one that nonetheless has not been well explored and settled in either the decisions of courts and tribunals or in learned commentary.

The other human in the equation of accountability of an AWS wrong doing is the programmer or manufacturer of the AWS. Schmitt argues

---

86 Ibid.

that at some point in the operation of an AWS, a human being has made a decision to determine how the AWS will act and that that individual would be accountable for war crimes committed by the AWS.[87] While Schmitt argues that 'self-evidently' the programmer would be accountable for the war crime caused by the programming,[88] the reality may not be so straightforward. Human involvement in the performance of an AWS actually encompasses a broad period of time, from its initial programming up to the individual commander's decision to employ the AWS. To hold the commander accountable, one would have to consider whether the AWS was utilised in a manner consistent with its intended purpose and programming or given a task beyond its capability so that it was destined to perpetrate a grave breach. It is also highly unlikely that an AWS programmed to breach IHL will ever be employed without the knowledge of someone within the chain of command of the armed forces employing it. Such functionality will have already been detected during the development stage of the AWS. It is improbable that a manufacturer of AWS will be able to cause an IHL breaching AWS to be deployed onto a battlefield without the state's tacit approval. If a manufacturer today produced a weapon that was inherently unlawful, it seems unlikely that the manufacturer of that weapon will be held accountable if a combatant uses that weapon during an armed conflict. While domestic law might prohibit the manufacture of certain weapons, IHL (and international criminal law) deals with the use of weapons.

For those AWS already in use today, the criteria for their employment is quite limited such that it is unlikely that a civilian will meet engagement criteria. From the outset, the commander employing such systems, aware of the parameters in which the AWS operates, has already made the assessment that civilians and civilian objects are unlikely to be targeted by the AWS. Obviously, if the situation changes, the decision to employ the AWS will need to reassessed. For example, the Phalanx ship defence system could be 'safely' employed on the high seas during an armed conflict. Its targeting parameters are limited to automatically attacking objects heading towards the ship travelling at a certain speed and trajectory: the obvious conclusion from that information is that it could be nothing other than a weapon that poses a threat to the ship.[89] The chances of the object being anything other than a valid military target are remote. However, if the ship were to go into port, the

---

[87] Schmitt (n 64) 33.
[88] Ibid.
[89] United States Navy (n 67); Grut (n 75) 12.

commander would need to consider whether the system should remain switched on: could it inadvertently target a civilian person or object? These systems, when operational, are effectively using the human 'on-the-loop' control system. Liability for the systems failings would logically fall to the commander who decided to turn on the system in circumstances when there was a chance that IHL would be breached.

The discussion of AWS control systems demonstrates that, in reality, these systems may operate in a manner that is no different to command and control arrangements within many modern armed forces. Depending upon a number of factors, pre-determined by commanders, combatants may be required to seek approval from higher up the chain of command before initiating an attack. This is particularly relevant for questions of determining whether the collateral damage anticipated from attack will outweigh the expected military advantage to be gained. Oftentimes, senior commanders will hold the authority to make such determinations and will put in place appropriate command and control systems. There is no reason to suppose that a similar constraint would not be imposed upon an AWS. To suggest that an AWS will be allowed to override these decision-making processes is contrary to current practice in modern armed forces. There may be cases, and such situations exist today, whereby the AWS can engage a target without relying upon human intervention, when the parameters of the attack are such that the commander of the AWS can be satisfied that the AWS will comply with the state's requirements under IHL. These situations will have been considered ahead of time, and the commander making the decision to deploy the system will have taken into account his or her responsibilities under Article 57 in doing so.

This discussion does highlight the high threshold for attributing responsibility for an AWS engagement that prima facie amounts to a war crime to a combatant commanding or operating the system. An AWS will be an incredibly complex system, both in terms of hardware and software. The legal fault for an AWS engagement that is contrary to IHL could lie in a single line of code and how it interacts with another line of code. If different people are responsible for the conflicting lines of code it becomes increasingly difficult to attribute responsibility to a person, or persons, as a result. Again, this is not dissimilar to a human combatant operating a complex piece of military equipment: he or she will have been trained on how to properly operate the equipment, but not necessarily on the technicalities of how the equipment performs its functions. This brings the unintended result into the realm of accident.

5. CONCLUSION

Due to the background and expertise of the authors, this chapter has been limited to a discussion of the law of armed conflict and individual accountability as it applies to remote warfare and AWS. It has not addressed, among other things, arguments concerning human rights law, peace and security (for example, non-proliferation) or military strategy. These are all valid issues; however, there is a benefit to look at each area separately before bringing the various discussions together. To that end, we set out to address law of armed conflict and individual accountability issues discretely.

To consider these issues properly from a legal standpoint it is necessary to separate issues of law from issues of ethics and philosophy. It is also necessary to analyse the concepts of remote warfare and autonomy separately for they are logically distinct and merit individual consideration. There must also be a separate consideration between the means and the methods of warfare; too often questions are aimed at the former that should properly be directed towards the latter.

The authors suggest that this analysis of the law has demonstrated neither remote warfare systems nor autonomous weapon systems are means of warfare that are currently proscribed by the law of armed conflict. Like all means of warfare, there are methods of use of these systems that may contravene the law of armed conflict.

Outside consideration of the current state of the law, philosophical and ethical considerations may contribute to the development of positions on what the law should be in relation to technological developments in warfare. The issues have been canvassed well by other authors[90] including the arguments that such systems will increase the likelihood of armed conflict and the argument that the use of force must be confined by human decision-making.

It may be argued that the use of remote and autonomous systems where there is no risk to human operators contributes to the willingness of parties to utilise armed force and to use force in ways that may be antithetical to norms of sovereignty. Therefore, it could be argued, these technologies may contribute to the prevalence of armed conflict and to the violation of the sovereign rights of states that lack the means to effectively prevent their use within sovereign territory. The potential issues, perceived and actual, relating to accountability for war crimes involving remote weapon systems and AWS, adds to these concerns.

[90] See Sassòli (n 15) 313–8.

The logical underpinning of some of these arguments assumes a societal value set that places the safety of the human operators above the monetary cost of systems and the risk to capability should that system be lost or compromised. It assumes that entities will be more likely to take an action when only money and machines are at risk compared with placing human agents at risk of death, harm or capture. Different societies and governments will approach this calculation from different perspectives of history, culture, resources and perception of threat when making decisions about the use of armed force. In other words, some states may ill afford the risk to the limited supply of a complex remote operating capability when compared with risking the lives of personnel to achieve an outcome.

Of course the key factors are the fidelity of the intelligence and the capability to deliver the effect remotely and precisely within a limited time window. Consider, for example, if advances in satellite and missile technology provided sufficiently high fidelity targeting intelligence to allow the use of a small warhead hypersonic missile to conduct an attack remotely, within a limited time window and with limited risk of civilian casualties. Such a capability with remote sensing and high-speed weapons may deliver the same effect that gives rise to arguments about impunity in relation to remote and autonomous systems.

Accordingly the argument that remote systems such as remotely operated vehicles may contribute to the prevalence of armed conflict and violation of sovereign rights has limited logical force when focussed on the system. It is the effect that may give rise to the perception of impunity (that is, the capability to conduct attacks remotely, precisely and in real time) rather than the exact means.

As for arguments that decisions on killing another human being must be a human decision, if those arguments are correct, nonetheless there are a range of human decisions underlying the use of the AWS and its programming. For example, it would be a human decision to deploy the AWS to operate in a particular area, with knowledge of the way that the AWS operates. It would be a human decision that the machine is sufficiently accurate in identifying military objectives such that it should be utilised.

These are relevant considerations for states in charting the development of remote and autonomous systems and whether there is a need to develop new legal norms for those systems. In considering the ethical and philosophical issues relating to remote and autonomous systems, it is probably also worth considering how these systems might compare to the violations of IHL perpetuated by human actors. To quote Sassòli:

> It is perhaps because I have been confronted in actual armed conflicts with so many violations committed by human beings, but inevitably never with atrocities by robots (although admittedly, they did not exist in the armed conflicts I witnessed), that my first feeling is not skepticism, but hope for better respect of IHL.[91]

The deployment of offensive AWS on a battlefield, able to use force independent of a human operator, is an event many years into the future, if indeed that event ever arrives. Fully autonomous weapon systems will not just suddenly appear on the battlefield. While some doubt whether AWS can even be developed, the majority view is that such systems will become operational over time.[92] It can be assumed that there will be a gradual development of such systems over time, likely over many years. Eventually, such development may reach the stage when a state contemplates using an AWS that truly is fully autonomous in combat.

AWS will develop incrementally over time and it is likely that this will include the development of degrees of autonomy in various subsystem components. This is already occurring in some remotely operated systems, which have semi-autonomous functions. For example, for some remotely operated aircraft, the operator does not 'fly' the aircraft but rather tells it where to go. Most of the flying functionality is carried out by the system itself.

Assessments will be made of the system components as they are implemented, and the gradual refinement of the autonomous aspects of such systems means that the process which states undertake to assess the introduction of a new weapon into service will enable states to assess its compliance with IHL.

For example, the current US directive on the development of AWS makes it a requirement for AWS to undergo 'rigorous hardware and software verification and validation ... and realistic system developmental and operational test and evaluation'.[93] Like pieces of a puzzle, the states using and developing these weapons will be required to ensure that each new advance accords with IHL, such that the system as a whole, when 'pieced together', operates in accordance with the state's international law obligations. The end result should be that if a truly

---

[91] Sassòli (n 15) 310 (footnotes omitted).

[92] Robin Geiß, 'The International-Law Dimension of Autonomous Weapons Systems' (*Friedrich Ebert Stiftung*, October 2015) 3–4, accessed 5 May 2017 at http://library.fes.de/pdf-files/id/ipa/11673.pdf.

[93] US Department of Defense (n 8) 2.

autonomous weapon system is developed, it will have developed over time, and in such a manner, that its use will comply with IHL.

The alternative is that development may only reach a stage of partial autonomy. If that 'next step' in evolution of an AWS does not comply with IHL, and this means the weapon system as a whole could no longer comply with the state's obligations under IHL, then the previous compliant step in development will be the 'high water mark' of a system's autonomy.

At this early stage of development of the AWS, an opportunity exists to address the issues associated with liability arising out of the use of such systems. It could be determined that all AWS must be developed with a human operator either 'in-the-loop' or 'on-the-loop', with the ability to override the AWS's operation and intervene.[94] A 'kill switch' would give some guidance to determine liability issues arising out of the use of AWS on the battlefield, but that would require use of AWS in an environment where there is reliable communication—which cannot be assumed.

Ultimately, it will most likely be commanders that will be held accountable for the actions of an AWS, for both their triumphs and their failings. If commanders are satisfied that a particular AWS can meet the requirements of IHL, particularly the requirements laid out in Article 57, then under the current state of the law, the use of AWS will be lawful. AWS, like any other system, will at times not otherwise operate as intended. Unfortunately, this may result in civilian casualties. Provided such casualties are unintentional, then there is no reason to suppose that commanders will be held criminally liable for the unintended actions of the AWS. It will, however, be necessary for commanders, just like any other weapon system or even subordinate human combatants, to continue to ensure that IHL obligations are adhered to. If the commander fails to take appropriate action following a 'breach', then that commander could be held criminally responsible for later actions—it would mean that the commander has failed to take 'all feasible precautions to ensure in the choice of means or methods of attack'[95] so as to avoid civilian casualties or cause damage to civilian objects.

There is no doubt that continued developments in remote and automated systems will inspire further study of the law as it applies to such systems and further advocacy for what the law ought to be. It also seems

---

[94] Anderson, Reisner and Waxman (n 51) 407.
[95] API (n 6) Article 57(52)(a)(ii).

that the developments in remote and autonomous systems may be destined to change the role played by combatants; but it appears unlikely that this will reduce the role played by lawyers.

12. Autonomous weapons systems: a paradigm shift for the law of armed conflict?

Robin Geiß and Henning Lahmann*

1. INTRODUCTION

The development of autonomous weapons systems (AWS) is widely considered a genuine revolution in weapons technology and of military affairs.[1] In the future, humans might only decide on the deployment of AWS in general, whereas the system itself controls all decisions necessary for the mission in question, including the decision to employ lethal force.[2]

To date, truly autonomous weapons systems do not yet exist. Some robotics engineers even doubt that such systems could ever be developed.[3] However, the majority of experts believe that it is only a matter of time until such systems will be ready for deployment. The US Department of Defense has officially announced its intention to develop and put into service increasingly autonomous weapons systems before the year

* This chapter builds on a study written for the Friedrich Ebert Foundation, R. Geiß, 'The International-Law Dimension of Autonomous Weapons Systems', October 2015, accessed 5 May 2017 at http://library.fes.de/pdf-files/id/ipa/11673.pdf and on an intervention delivered to the third CCW meeting of experts on lethal autonomous weapons systems (LAWS) Geneva, 11–15 April 2016, see R Geiß, Autonomous Weapons Systems: Risk Management and State Responsibility, accessed 5 May 2017 at http://www.unog.ch/80256EDD006B8954/(httpAssets)/00C95F16D6FC38E4C1257F9D0039B84D/$file/Geiss-CCW-Website.pdf.

1 See e.g. Peter Singer, *Wired for War* (Penguin 2009) 179 et seq.

2 UN General Assembly, Report of the Special Rapporteur on extrajudicial, summary or arbitral executions, Christof Heyns, A/HRC/23/47, 9 April 2013, at 28.

3 See Mary Ellen O'Connell, 'Banning Autonomous Killing: The Legal and Ethical Requirement That Humans Make Near-Time Lethal Decisions' in Matthew Evangelista and Henry Shue (ed), *The American Way of Bombing. Changing Ethical and Legal Norms, from B-17s to Drones* (Ithaca 2014) 224, 226.

2038.[4] At the same time, the International Committee of the Red Cross (ICRC) has rightly pointed out that, already today, different critical functions within existing weapons systems are carried out autonomously, that is, without human intervention.[5]

Taking into account such announcements by the United States and other nations, some pundits have expressed concerns regarding the threat of a newly revived arms race.[6] It is expected that modern-day military necessity considerations alone will push states' armed forces towards increasing the automation and autonomization of military technology. Considering today's realities on the battlefield, it is increasingly problematic for the armed forces to keep up with the sheer amount of critical information to be processed and the demands for speedy decision-making and real-time reactions. Therefore, advancing autonomous technologies is perceived as almost inevitable in order to manage the dynamics and complexities of the modern battlefield, which, in turn, entails a virtually inevitable acceleration of the international arms race.[7]

The announcements of some states, most prominently the United States, to further promote autonomous military technologies in the future, have triggered an international debate on the ethical and legal implications of such systems in recent years. So far, the discussion has focused on so-called 'lethal autonomous robots' (LARS) or 'lethal autonomous weapons systems' (LAWS). However, in the long run, we can expect autonomous technologies to be gradually expanded to all levels of military and strategic decision-making.

Proponents of AWS see a number of advantages. For instance, AWS are assumed to be far more capable of collecting and processing new information than humans. They could act with more precision, speed and flexibility concerning their decisions as well as the execution of attacks. Such systems would replace humans on the battlefield and would thus

---

[4] US Department of Defense, Unmanned Systems Integrated Roadmap FY2013-2038 (2013), accessed 5 May 2017 at http://archive.defense.gov/pubs/DOD-USRM-2013.pdf.

[5] United Nations, General Assembly, 69th session, First Committee, statement by the ICRC, New York, 14 October 2014, accessed 5 May 2017 at https://www.icrc.org/en/document/weapons-icrc-statement-united-nations-2014.

[6] See e.g. Wendell Wallach, 'Terminating the Terminator: What to Do About Autonomous Weapons' science progress, 29 January 2013, accessed 5 May 2017 at http://scienceprogress.org/2013/01/terminating-the-terminator-what-to-do-about-autonomous-weapons/.

[7] Hans-Arthur Marsiske, 'Können Roboter Kriege humanisieren?' Telepolis, 13 April 2014, accessed 5 May 2017 at http://www.heise.de/tp/artikel/41/41439/1.html.

directly reduce the threat of human loss in armed conflict. Furthermore, due to their lack of emotions or physical exhaustion, they would be much more capable of carrying out both dull routine jobs and very dangerous missions. And finally, as robots do not experience fear, anger or hate, the danger of excessive behavior in emotionally stressful situations is nil.

Critics of the new technology, however, warn that human life would be devalued fundamentally if the decision about life and death were to be left to a machine to make. It is precisely the absence of emotions which is perceived as problematic in this respect: algorithmically controlled, autonomous weapons systems would operate clinically, without ever resorting to actions involving mercy or empathy. Furthermore, the so-called 'video game mentality' that has been observed in soldiers remotely controlling drone missions might be further augmented by the employment of AWS.[8] It also remains an unsettled question whether such systems could ever be programmed in such a way as to sufficiently preclude the danger of serious malfunctioning. Some experts moreover argue that a class of weapons which completely eradicates any risks for the employing party would be inherently unethical, due to the significant asymmetry that such a situation would necessarily entail.

Non-governmental organizations concerned with human rights and international humanitarian law have become particularly vocal critics of the emerging technology. As early as 2009, the 'International Committee for Robot Arms Control' was initiated.[9] In October 2012, a number of NGOs launched the 'Campaign to Stop Killer Robots'[10] in order to advance the public debate on such weapons on the international level. As of today, even the European Parliament has publicly announced its opposition to the development, production and employment of completely autonomous weapons systems.[11]

The scientific and political discussion concerning the advantages and risks of autonomous weapons is part of a broader societal discourse on the implications of tendencies of increasing automation in various

---

[8] Geneva Academy of International Humanitarian Law and Human Rights, Academy Briefing No. 8: Autonomous Weapons Systems Under International Law, November 2014, 5, accessed 5 May 2017 at http://www.geneva-academy.ch/academy-publications/academy-briefings/1190-briefing-no-8-autonomous-weapons-systems-under-international-law.

[9] http://icrac.net/ (accessed 5 May 2017).

[10] http://stopkillerrobots.org/ (accessed 5 May 2017).

[11] European Parliament resolution of 27 February 2014 on the use of armed drones (2014/2567(RSP), accessed 5 May 2017 at http://www.europarl.europa.eu/sides/getDoc.do?type=TA&language=EN&reference=P7-TA-2014-0172.

spheres of daily life. The military aspect of this debate is only the tip of the iceberg. On a fundamental level, it needs to be asked how much 'de-humanizing' of societal mechanisms humankind can, or is willing to, afford to tolerate, before the social costs outweigh the benefits.[12] To name but one example, the problems that might surface in connection with the increasing automation of civil air traffic have been well documented in recent years.[13] Aside from merely technical questions concerning the reliability and security of such systems, the ethical side of the development has more and more become the focus of public debate. If computer-controlled machines autonomously assume tasks in ever more spheres, then society as a whole needs to assess how algorithms are supposed to act and react in moral borderline cases. For instance, how should a self-driving car react if a child suddenly crosses the street? Should it drive into oncoming traffic in order to avoid collision with the child, thereby compromising the occupant's safety? Do we want a robot to decide on its own when to administer strong pain medication, without oversight by a human doctor? In view of such issues, it hardly comes as a surprise that the possibility of machines making decisions regarding the use of lethal force are even more disconcerting for large parts of society. Often, the technological development appears inevitable, which leads to a widespread feeling of discomfort.

Against this backdrop, a larger debate about the implications and consequences of the employment of AWS—taking into account ethical and political as well as (international) legal perspectives—is indicated. In this sense, it is no news that the politics of international law face a recurring problem: more often than not, states and other stakeholders on the international level tend to start dealing with a new technology only after it has already been deployed. Indeed, one might even go so far as to suggest that international law is frequently 'one war too late'. However, as we are dealing with a technological paradigm shift when it comes to the development of AWS, and not just some other new weapon, this dilemma entails potentially fatal ramifications. Some commentators contend that the debate is already overdue, perhaps even coming too late given the current state of technology. Various systems with offensive

---

[12] Paul Ford, 'Our Fear of Artificial Intelligence' MIT Technology Review, 11 February 2015, accessed 5 May 2017 at https://www.technologyreview.com/s/534871/our-fear-of-artificial-intelligence/.

[13] See e.g. William Langewiesche, 'The Human Factor' Vanity Fair, October 2014, accessed 5 May 2017 at http://www.vanityfair.com/news/business/2014/10/air-france-flight-447-crash; Nicholas Carr, *The Glass Cage: How Our Computers Are Changing Us* (W W Norton 2014).

military capabilities are already largely automated, with the consequence that human soldiers struggle with the resulting information overflow, leading to the inability to make well-informed, autonomous decisions.[14] In other words, the de-humanization of warfare is already well underway. There is only a small step left towards fully autonomous weapons.

Taking the above deliberations into consideration, this chapter attempts to examine AWS within the context of current international law. After outlining the legal issues connected with the employment of AWS more generally, the chapter will focus on the questions of accountability and responsibility as regards the conduct of autonomously acting weapons. If a machine's actions amount to war crimes or other breaches of a norm of international law, who can be held legally responsible, and according to which regime?

2. AUTONOMOUS WEAPONS SYSTEMS WITHIN THE CURRENT LEGAL FRAMEWORK

Although so far there is no established, generally accepted definition of what counts as an autonomous weapons system, most scholars and decision-makers now agree that the crucial factor is the machine's ability to make decisions on the basis of algorithms without any further human intervention. 'Autonomy' in this sense is indeed not to be confused with the concept found in moral philosophy: the free will of the human individual, the ability to make moral decisions. An autonomous robot is invariably merely able to act within the boundaries set by the programmers who wrote its code. This shows that, at some point, all actions by AWS involve human decision-making. Those who have suggested that this fact alone allows for the conclusion that there should exist no actually autonomous systems, however, seem to misunderstand the problem.[15] To be sure, human influence on AWS will never cease in its entirety—it will be human beings who decide whether an AWS will be deployed on a mission, under which circumstances, and within which

---

14 Niklas Schörnig, 'Automatisierte Kriegsführung – Wie viel Entscheidungsraum bleibt dem Menschen?', Aus Politik und Zeitgeschichte 35-37/2014, 18 August 2014, accessed 5 May 2017 at http://www.bpb.de/apuz/190115/automatisierte-kriegsfuehrung-wie-viel-entscheidungsraum-bleibt-dem-menschen?p=all.

15 See e.g. Michael N Schmitt and Jeffrey S Thurnher, '"Out of the Loop": Autonomous Weapons and the Law of Armed Conflict' (2013) 4 Harv Natl Sec J 231, 280.

framework. Moreover, at least for the time being, it will be humans who activate and deactivate the machines. However, if the actual moment of deciding whether to use lethal force is beyond human control, it is justified to speak of an 'autonomous' weapon. Within the context of the modern-day theater of war, with its highly complex, asymmetrical constellations of conflict, it is in no way predictable with which kind of scenario the AWS will be confronted during its mission.

As opposed to autonomous systems in this sense, automatic systems merely execute pre-programmed commands. They are incapable of reacting to unforeseen events independently. Therefore, such systems will typically be employed to perform certain precisely defined tasks within a clearly bounded and limited geographical area and timeframe—for instance the defense against anti-ship missiles in a five-mile radius around the vessel. In this way, a mine might be considered a typical example for an automated weapons system, albeit in its most primitive form. However, it is difficult to find a clear-cut line between automated systems on the one hand and autonomous on the other. It should indeed rather be regarded as a sliding scale. The more complex and broader an automatic machine's tasks become, or the bigger or more diverse its area of operation, the more it would seem justified to speak of autonomy. In any case, it does not seem necessary to actually make that distinction unambiguously in order to deal with both the ethical and the legal issues raised by AWS. According to the ICRC, for example, it is crucial whether the system's critical functions, most importantly independent decisions over life and death, are beyond human control.[16] This approach is convincing. Aside from technical definitions, the focus of the debate must be to find a workable distinction between such decisions that may be delegated to the machine on one side and those decisions that must remain in the hands of human operators on the other. At the same time, the ICRC's method convincingly implies that not all autonomous systems are necessarily problematic in and of themselves. On the contrary, an autonomous submarine designed to sweep mines does not raise the same set of legal and ethical questions that a lethally armed robot deployed into an urban warfare setting would raise.

The following section will nevertheless briefly outline the models and various attempts to define the concept of AWS that are commonly used in contemporary debate.

---

[16] United Nations, General Assembly, 69th session, First Committee, statement by the ICRC, New York, 14 October 2014, accessed 5 May 2017 at https://www.icrc.org/en/document/weapons-icrc-statement-united-nations-2014.

The most widely cited working definition circumscribes AWS as robots which collect information about their surroundings by means of integrated sensors, and process that information, in order to make a decision that will finally be carried out by means of their installed components such as weapons.[17] The US Department of Defense considers weapons autonomous if, after being activated, they are able to identify and attack targets independently, that is, without further human intervention. The fact that humans retain the possibility to intervene in order to stop the machine or alter its behavior does not speak against autonomy.[18] As already mentioned, the ICRC focuses mainly on the system's critical components and their ability to choose, approach and attack the target autonomously.[19]

A commonly cited approach was developed by Human Rights Watch. It is a three-step model that principally takes into account the degree of human interaction with the weapons system.[20] On this basis, HRW distinguishes between systems that leave a human 'in the loop', 'on the loop', or 'out of the loop'. 'In the loop' means that the machine is incapable of acting without human decision-making. One or more crucial steps need to be controlled by the person in charge. Such systems are not autonomous. One example of such systems are the drones that the United States currently employs in the airspace over Pakistan, Afghanistan and Yemen, which are being steered by soldiers in control centers on US soil. So-called semi-autonomous weapons, which carry out an attack on their own after a human soldier has selected a target, also belong to this category. In this case, human involvement is sufficiently critical in order to not deem such systems autonomous. 'On the loop' means that, in principle, the machine is capable of executing all operations on its own, without human influence. However, a human operator monitors the machine's decisions and retains the ability to intervene if necessary, being able to shut down the system's actions. This variant may be called

---

[17] Heyns (n 2) 38.

[18] US Department of Defense, Autonomy in Weapons Systems, Directive No. 3000.09, 21 November 2009, 13 et seq., accessed 5 May 2017 at http://www.dtic.mil/whs/directives/corres/pdf/300009p.pdf.

[19] Report of the Expert Meeting on Autonomous Weapons Systems: Technical, Military, Legal and Humanitarian Aspects, 26–28 March 2014, Geneva, 9 May 2014, 1, accessed 5 May 2017 at https://www.icrc.org/eng/assets/files/2014/expert-meeting-autonomous-weapons-icrc-report-2014-05-09.pdf.

[20] See William Marra and Sonia McNeill, 'Understanding "The Loop". Regulating the Next Generation of War Machines' (2013) 36 Harv J of L & Pub Poly 1139.

'monitored autonomy'. Still, in this context, UN Special Rapporteur Heyns rightly points out that the possibility to actually intervene is inherently limited if the machine is capable of making decisions within milliseconds. In such cases, actual control over the system's actions may be no more than an illusion.[21] Finally, if human beings are entirely 'out of the loop', a system acts entirely on its own terms. There are no possibilities for a human to directly intervene. The system is autonomous in regard to all steps that are necessary for an attack. Even more so, in the future, experts expect machines to be able to learn from experience and autonomously adapt to new situations and environments.[22]

To be sure, the three steps proposed by Human Rights Watch can only schematically represent reality, where the transitions from one step to the next happen, rather, on a sliding scale. At times it will be difficult to make a clear-cut decision as to which step a particular system is to be attributed to.[23] Scenarios, in which human soldiers will formally remain 'in the loop' but essential aspects of the decision-making process will be delegated to a machine, will become increasingly likely in the near future. Even if crucial steps, such as the final decision whether to resort to lethal force, remain with a human actor, conditions such as stress or time pressure—which will typically prevail in combat situations—will likely lead the human to rely on the machine's target selection. Such behavior—the tendency to trust an artificial intelligence system even when clear signs exist that it is unreliable or even erring in a certain situation—is commonly referred to as the 'automation bias'.[24] Though technically possible, in such cases, true human control over the machine is but an illusion.

3. NEW WEAPONS TECHNOLOGIES AND INTERNATIONAL LAW

During the course of its history, international law has consistently had to deal with the emergence of new weapons technologies: developments that were continuous and at times swift.[25] However, while technological

---

[21] Heyns (n 2) 41.

[22] Schmitt & Turnher (n 15) 240.

[23] Schörnig (n 14).

[24] Peter M Asaro, 'Modelling the Moral User' (2009) 28(1) IEEE Technology and Society Magazine 20–4, 2009, 22.

[25] Robin Geiß, 'The Law of Weaponry from 1914 to 2014: Is the Law Keeping Pace with Technological Evolution in the Military Domain?' in Jose

innovation must be considered the norm, international law has persistently struggled to keep up, due to its static and, therefore, lengthy, tedious law-making mechanisms. Indeed, new weapons technologies invariably pose a challenge. Naturally, this challenge increases drastically if a developing technology is not merely to be regarded as some new weapon but moreover likely to alter the way armed conflicts are fought more generally, that is, if the technology changes, or even revolutionizes, military conduct altogether. Autonomous weapons systems, it is submitted, are such a technology.

To be sure, not all legal or moral questions surrounding the introduction of AWS to the battlefield are entirely new, or unheard of.

On the one hand, autonomous robots have the principal aim of sparing human soldiers, whose presence on the battlefield can be reduced through the deployment of autonomous systems. Therefore, the employing side of the conflict minimizes the risk of harm to its own forces. In this sense, AWS are only the latest invention in a long line of long-distance weapons that had been introduced with the same goal in mind, such as the bow and arrow, crossbow, gunpowder, artillery and air force, all the way to drones. And each single time, the new technology was met with criticism that called it unethical, being contrary, for instance, to some moral code of knighthood.[26] Viewed from this perspective, AWS are nothing fundamentally new, and they are, in many ways, linked to the debate surrounding the use of (remotely controlled) drones. However, beyond this aspect, AWS moreover serve the function of better perceiving and processing the increasing amount of information and data in modern-day situations of armed conflict. Autonomous systems are built in order to better manage and optimize military decision-making. By doing so, human decision-makers are taken out of the equation concerning numerous critical decisions and replaced by algorithms. It is this aspect in particular, not so much the reduction of risk for its own soldiers, which raises the most crucial ethical and legal questions.

Traditionally, international law has responded to the introduction of new weapons systems on two distinct levels. On the one hand, the use of certain categories of weapons was prohibited outright. The relevant treaties concluded to this end are based on considerations of international

---

Delbrück et al (ed), *Aus Kiel in die Welt: Kiel's Contribution to International Law* (Duncker & Humblot 2014) 229, 237.

[26] Herfried Münkler, 'Neue Kampfsysteme und die Ethik des Krieges' [New combat systems and the ethics of war], speech at the Heinrich-Böll- Stiftung, 21 June 2013, accessed 5 May 2017 at http://www.boell.de/de/node/277436.

humanitarian law. According to the fundamental principles of humanitarian law, certain kinds of weapons are inherently unethical, for example because they cause unnecessary suffering, or because one of their main features is the fact that they kill without distinguishing between protected civilians and combatants. On the other hand, there are disarmament treaties, which emerged in greater numbers after World War II, during the Cold War. Such agreements, instead of being based on jus in bello considerations aim to reinforce the jus contra bellum, which is principally enshrined in the Charter of the United Nations. Reducing the amount of weapons in the states' arsenals is supposed to reduce the threat of armed conflicts. Recent treaties have begun to merge both aspects. Ensuing from merely restricting a certain weapon's use—such as, for instance, the Geneva Protocol for the Prohibition of the Use in War of Asphyxiating, Poisonous or other Gases, and of Bacteriological Methods of Warfare[27]—many modern-day agreements moreover contain comprehensive clauses regarding research, development, storing, or distribution of such systems. One example of such a conjunction of elements of disarmament and humanitarian considerations is the 1993 Chemical Weapons Convention. Instead of prohibiting specific weapons, it outlaws an entire class of weapons.[28]

As already mentioned, however, history shows that most treaties prohibiting weapons were usually concluded 'one war too late'—that is, after the devastating weapon had already been used against humans. For example, states could only agree on the Geneva Protocol after chemical weapons had been employed during World War I, with disastrous consequences. This is not a problem of the far past: more recently, anti-personnel mines or cluster bombs could only be restricted by way of treaty after decade-long political struggles, mainly carried out by human rights NGOs. The only class of weapons that could ever be prohibited prior to its initial use in a conflict setting was that of blinding laser weapons. Due to its effect of permanently blinding targeted soldiers or civilians, the weapon was considered cruel, causing unnecessary suffering. However, that it could be prohibited so quickly and unanimously certainly had to do with the fact that it was regarded as having only marginal military potential. This point is underlined by a comparison with cluster bombs: the prohibition of this class of weapons is based on the observation that they entail a high risk of killing without distinction.

---

[27] See http://www.brad.ac.uk/acad/sbtwc/keytext/genprot.htm (accessed 5 May 2017).

[28] See https://www.opcw.org/chemical-weapons-convention/articles/ (accessed 5 May 2017).

But because they are regarded as having a considerable military potential, several powerful international actors, such as the United States, Russia and China, have not yet ratified the treaty.

As the conclusion of international treaties is a lengthy process, and its ultimate success is dependent on a subsequent, ideally universal, ratification process, international humanitarian law comprises a couple of generally valid principles, which are supposed to guarantee the law's ability to adapt to and keep up with technological progress. The principle of distinction, which forms the very core of this legal system, as well as the fundamental prohibition of causing unnecessary suffering, are applicable to any weapons system, regardless of its age. The International Court of Justice confirmed as much in its advisory opinion on nuclear weapons, a category of weapons that is not mentioned in the Geneva Conventions or their Additional Protocols.[29] These principles are sufficiently abstract and, therefore, timeless. They are applicable without regard to the kind of technology in question, which means that they can be used to govern even cutting edge technologies, such as cyberspace or autonomous weapons systems. At the same time, the high degree of abstractness implies that there is always room for discussion and re-interpretation, which is why it seems to be recommended to aim for a specific treaty which is better suited to tackle the weapon's specific details. Still, past experience shows that, more often than not, it was recourse to those general principles that would initiate negotiations between states in the first place.

Two other general principles found in international humanitarian law that aim at securing both the timelessness and the dynamic flexibility of the legal system are the so-called Martens Clause on the one hand and the obligation to test newly developed weapons as regards their compliance with the rules of international humanitarian law, as enshrined in Article 36 of the Additional Protocols to the Geneva Conventions on the other.

In its most recent phrasing, as enshrined in Article 1(2) of Additional Protocol I, the Martens Clause reads: 'In cases not covered by this Protocol or by other international agreements, civilians and combatants remain under the protection and authority of the principles of international law derived from established custom, from the principles of humanity and from the dictates of public conscience'. The ICRC holds

---

29 ICJ, Legality of the Threat or Use of Nuclear Weapons, Advisory Opinion, 8 July 1996, at 84 et seq., accessed 5 May 2017 at http://www.icj-cij.org/docket/files/95/7495.pdf.

the view that this rule implies that each party is under the obligation to test new weapons' compliance with said principles of humanity and dictates of public conscience prior to their use in an armed conflict.[30] Closely interlinked with this rule is Article 36 of Protocol I, which attaches conditions to the employment of newly developed weapons:

> In the study, development, acquisition or adoption of a new weapon, means or method of warfare, a High Contracting Party is under an obligation to determine whether its employment would, in some or all circumstances, be prohibited by this Protocol or by any other rule of international law applicable to the High Contracting Party.

In other words, the agreement stipulates the duty for all parties to test new weapons prior to their use, as regards their compliance with the rules of international humanitarian law or any other applicable rule of international law. Although Article 36's status as one of customary international law is in dispute, even some militarily significant states that are not party to Protocol I have recognized a general obligation to test weapons in this sense, and introduced a formal process of verification.[31] However, the consequences of a violation of Article 36 remain a contentious issue: does it entail a prohibition of the weapon in question by default, or is a further, specific treaty necessary to that end? The more convincing standpoint argues that a weapon cannot possibly be permitted if it contradicts the basic principles of international humanitarian law per se. However, as already shown above, those principles are themselves deliberately ambiguous and lack specification as to a tangible standard—and that standard must in any case be considered to be rather high. It follows that a weapon will rarely evidently be in violation of them.[32]

In regard to the development of autonomous weapons, experts have rightly pointed out that it is crucial to start with the verification concerning the listed principles early on and to closely monitor the entire process step by step. Otherwise—in case the weapons were built solely in line with technical possibilities—the considerable research and development costs put into their development would create strong incentives,

---

[30] ICRC, *A Guide to the Legal Review of New Weapons, Means and Methods of Warfare* (ICRC 2006) 17, accessed 5 May 2017 at https://www.icrc.org/eng/assets/files/other/icrc_002_0902.pdf.

[31] See e.g. for the United States, US Department of Defense Directive 5000 01: The Defense Acquisition System, Defense Acquisition Guidebook, E1.1.15 (Legal Compliance), 12 May 2003, accessed 5 May 2017 at https://acc.dau.mil/CommunityBrowser.aspx?id=314789.

[32] Geiß (n 25).

which would render the employment of the expensive systems all but factually inevitable. Given such a constellation, it is highly questionable if anyone could still conceivably come to the conclusion that the weapons are actually in violation of the principles stipulated by Article 36 of Protocol I.[33]

That aside, a more fundamental question arises as regards the verification process of new weapons. It is true of course that international humanitarian law, with its principles just mentioned, is sufficiently dynamic and capable of adapting to technological change.[34] However, it is hardly deniable that the rules are, at the same time, static, in the sense that they are still based on the same ethical principles, which were stipulated more than a century ago. Both the prohibition of unnecessary suffering and the principle of distinction (albeit in a very rudimentary version) already existed prior to World War I as rules of international law. Bearing that in mind, if one considers the development of autonomous weapons, not simply a mere evolution of weapons technology, but, rather, a true paradigm shift in the way wars are fought, then it is in any case up for discussion whether the traditional principles of international humanitarian law are indeed still sufficient to deal with this entirely new class of weapons, despite the otherwise undisputed continuing relevance of these fundamental principles.

4. AWS AS A PARADIGM SHIFT FOR INTERNATIONAL LAW

As already indicated, the potential introduction of autonomous weapons to future theaters of war or other situations of armed conflict raises numerous issues, both ethical and legal. To be sure, autonomy as such is not the problem: there is nothing inherently critical about the employment of autonomously operating minesweeping boats or robots that are used to dispose of bombs. It is rather the delegation of critical, potentially lethal functions to system components not controlled by humans that needs to be scrutinized.

---

33 Marco Sassòli, 'Autonomous Weapons and International Humanitarian Law: Advantages, Open Technical Questions and Legal Issues to be Clarified' (2014) 90 Intl L Stud 308, 322.

34 Dave Wallace and Shane R Reeves, 'Modern Weapons and the Law of Armed Conflict' in Geoffrey S Corn et al (eds), *U.S. Military Operations: Law, Policy, and Practice* (Oxford University Press 2015) 41, 66.

At the same time, even when focusing on lethal autonomous weapons, there are without doubt conceivable scenarios that appear to be less problematic: just imagine an armed conflict carried out solely by means of employing autonomous weapons by both or all conflicting parties. One might well argue that such a mode of conflict would have to be qualified as being inherently more ethical than traditional forms of armed conflict with human casualties.[35] However, such a scenario seems highly unlikely. Moreover, it has been convincingly pointed out that it would be impossible to predict how such a mode of conflict would play out in reality. Experts argue that the inherent incalculability of highly complex algorithms that confront each other might eventually lead to a military escalation that cannot be controlled.[36] Above all, however, it is not plausible that states that plan to employ AWS will focus their planning on such scenarios. On the contrary, it is to be expected that legally protected interests of humans will remain primarily affected in future wars. As political scientists have pointed out, it is precisely the asymmetrical constellation in most modern-day armed conflicts that is the foremost incentive to advance the development of autonomous weapons.[37] As long as armed conflicts are carried out in order to gain control over human populations or territories by means of armed force, purely virtual scenarios, in which harm to human beings is ruled out, will remain utopic.

Various contentious questions are currently being discussed as regards the development and deployment of AWS: whether armed conflicts between states or between states and non-state armed groups would become more likely; whether AWS are capable of complying with the rules of international humanitarian law; whether AWS inherently violate the dignity of their human targets in lethal operations; whether there is, therefore, a duty to program them to operate only in a non-lethal mode; and finally, how to solve problems of legal accountability for the actions of AWS. It is this latter issue upon which this chapter will lay emphasis in the following section. The other aspects will (briefly) be addressed thereafter.

---

[35] See e.g. Peter M Asaro, 'How Just Could a Robot War Be?' in Philip Brey et al (eds), *Current Issues in Computing and Philosophy* (IOS Press 2008) 50, 62.

[36] Schörnig (n 14).

[37] Münkler (n 26).

4.1 Problems of Accountability

Proponents of AWS claim that the machines will be able to obey the rules of warfare and other applicable norms such as those stemming from human rights law much better than human soldiers, mostly due to the absence of factors like fear, stress, anger, hate or exhaustion—autonomous systems are supposed to function in a consistent manner, as the obedience necessary to comply with the law is part of their very fabric, that is, the underlying code that controls their actions.

However, as already mentioned, complex scenarios of modern conflicts entail a moment of unpredictability that, at least potentially, undermines the notion of the failure-proof agency of autonomous weapons. As such, errors can never entirely be ruled out, even if the system otherwise works in a perfectly predictable manner. If errors happen which lead to a violation of an applicable rule of international law, the question is thus who is to be held accountable, and on the basis of which criteria? Legal responsibility is a fundamental precondition of the function of law, and, hence, of the protections guaranteed by the rules of international humanitarian law and human rights law.[38] Without identifiable and stable agency, the legal prescriptions lack an addressee.

Traditionally, models regarding accountability are usually premised on some manifestation of control or foreseeability of the results of an action. However, higher levels of autonomy in weapons systems imply lower levels of control and foreseeability. Therefore, it will be more difficult to establish accountability on the basis of such traditional models the more autonomous the respective weapons system is. This is not only a problem for AWS and international law. Indeed, the same challenge is faced by the makers and customers of other, civil, autonomous technology, such as self-driving cars.

Still, this observation does not mean that there is an inevitable or insurmountable 'accountability gap' when it comes to the employment of AWS. This chapter argues that, at least in the area of state responsibility, accountability challenges can be overcome by way of regulation and clarification of already existing rules. As such, there is no conceptual barrier for holding a state accountable for wrongful acts committed by a robot. Therefore, it is not necessary to devise a new legal category to appropriately capture AWS, such as 'e-persons', 'non-human actors', or

---

[38] Human Rights Watch 42.

'virtual legal entities'. The same goes, it is submitted, for the accountability of individual human beings, even if the challenges are greater as regards individual criminal responsibility.

Accountability for the actions of AWS will, in the following, be analyzed concerning state responsibility as well as criminal responsibility and liability under private law.

4.1.1 State responsibility

According to the general rules on state responsibility as found in customary international law, a state is responsible for all violations of the laws of armed conflict if they can be attributed to the state. The issue of attribution of the acts of autonomous systems seems thereby to be straightforward and does not pose any particular problems. At least as long as humans decide on the deployment of AWS—and for the time being, no other scenario is conceivable—accountability is to be determined on the basis of the well-established rules on attribution. The most fundamental of those rules of customary law is the one that holds states responsible for any conduct of their state organs. Thus, if a member of the armed forces—a state organ—of a certain state decides to deploy an autonomous weapon on a combat mission, all activities carried out by the system will be attributable to the state. This assessment is not altered or affected by the mere fact that the system has some autonomous capabilities.

The principal challenge in this context is not the question of attribution. In order to establish the responsibility of a state, it first needs to be determined whether an internationally wrongful act has been committed, that is, whether a (primary) norm of the laws of armed conflict has been violated. Of course, this assessment rests on the specific requirements and interpretation of the primary rule in question.[39] Here, an important distinction is to be made: some rules of international law are violated whenever and as soon as their objective requirements are met. Other rules, however, require an element of 'fault', that is, some form of intent, recklessness or negligence. Then again, it needs to be kept in mind that it is often not clear or in any case not entirely settled what exactly is required by a given primary rule of international humanitarian law.

Thus, accountability challenges arise whenever the primary norm under scrutiny requires such an element of 'fault'. Intent, recklessness, and negligence all denote mental states of human beings. They are by

---

[39] International Law Commission, Draft Articles on Responsibility of States for Internationally Wrongful Acts, with Commentaries, Article 2, at 3.

definition absent in autonomous (weapons) systems. Therefore, within this context, it can only matter whether the military commander deploying the system acted with intent, recklessness or negligence respectively. Of course, cases that involve a certain degree of premeditation or deliberation will not pose any particular problems. For instance, if civilians are killed because a state's military commander intentionally programmed an autonomous weapons system to attack civilians, there can be no doubt that the commander is individually (and criminally) responsible for the commission of a war crime and that the state is therefore responsible for a violation of the laws of armed conflict. There is no difference between deliberately shooting at a protected person and programming a robot to do so.

To be sure, while such situations are certainly conceivable, they are not, in themselves, critical for the legal debate concerning accountability for the conduct of AWS. Other and presumably more frequent scenarios involving autonomous systems will raise more difficult accountability challenges. The main concern here is based on the assumption that it will likely be unpredictable exactly how autonomous systems will operate under actual battlefield conditions. It remains a contentious issue among experts the level of predictability (or lack thereof) that will eventually remain. For the time being, this issue remains speculative. However, this also means that, for the sake of argumentation, legal scholars need to work under the assumption that unpredictable outcomes of an AWS mission cannot be ruled out.

Thus, a military commander may deploy a thoroughly tested and duly authorized autonomous weapons system that nevertheless ends up violating the laws of armed conflict while operating in a complex and dynamic battlefield environment. In such a scenario, there might indeed exist what may be called an 'accountability gap', because it will most likely be impossible to establish either intent, recklessness, or negligence on the part of the commander. Without the affirmation of these mental states, however, no one will be at 'fault', so the requirements of the primary norm of international humanitarian law will not be met if it indeed comprises such an element.

A further accountability challenge arises from the complexity of autonomous weapons systems. AWS could always have any number of hidden errors in their controlling code, which are both unknowable to the military commander who is in charge of the decision to deploy the system, and impossible to prove ex post facto in the case that something went wrong as a result of one of these software flaws.

In order to mitigate the mentioned risks concerning the operation of AWS, states could resort to a number of different solutions. Aside from indispensable tasks such as weapons reviews and thorough testing at the pre-deployment stage, or options for the deployment stage like human override or deactivation facilities, as a matter of policy states could also by default program robots to act in a (more) conservative manner, for example to always 'shoot second' or to 'double check' before using lethal force. Moreover, states could also agree to only deploy AWS in restricted environments, that is, under battlefield conditions that make encounters with protected persons impossible or at least highly unlikely.

In light of the foregoing, state responsibility for the conduct of lethal autonomous weapons should be established, following a quite simple rationale. It is to be based on the basic assumption that, while the deployment of AWS on the battlefield is not, per se, unlawful, in any case it is a high-risk activity—at least in scenarios that potentially involve the presence of persons protected under international humanitarian law, as argued above. As a novel, game-changing technology, its implications and the consequences of its deployment in the complex environment of modern-day battlefields are not yet fully understood. This means that there is an inherent element of predictable unpredictability. Therefore, a state that benefits from the various strategic and tactical gains associated with this new technology should be held responsible whenever the unpredictable risks inherent in this technology are realized.

Due diligence obligations As a general rule, prevention is surely better than cure. Therefore, identifying and specifying detailed due diligence obligations aiming at risk prevention and harm reduction are as important as and indeed complimentary to the question of liability. Such obligations are for instance already common in the area of international environmental law.[40] For example, Article 3 of the ILC Draft Articles on Prevention of Transboundary Harm from Hazardous Activities provides that '[t]he State of origin shall take all appropriate measures to prevent significant transboundary harm or at any event to minimize the risk thereof'. With respect to the laws of armed conflict, such obligations could for example be derived from common Article 1 of the four Geneva Conventions (and corresponding customary international law), which requires states to ensure respect for the laws of armed conflict in all circumstances. The problem hereby is not so much a lack of a legal basis, but, rather, the lack of clarity as to what exactly it is that the due

---

[40] See e.g., *Case Concerning Pulp Mills on the River Uruguay (Argentina v Uruguay)*, ICJ judgment, 20 April 2010, at 223; more generally see Robert P Barnidge Jr, 'The Due Diligence Principle Under International Law' (2006) 8 Intl Community L Rev 8.

diligence obligation to ensure respect requires with regard to autonomous weapons systems. In order to comply with a due diligence obligation, the usual standard is what a reasonable actor would do under the given circumstances. However, it is not easy to determine what should be considered reasonable in this sense, when dealing with an entirely new technology, for which clear standards, practical experiences, and benchmarks naturally cannot yet exist. It is obvious that, without such workable guidelines, due diligence obligations aimed at risk mitigation are bound to remain empty shells.

Therefore, the debate surrounding autonomous weapons systems should focus on the specification and clarification of these due diligence obligations. The inherent risks resulting from unpredictable behavior of robots could be mitigated in various ways. As a general rule, the higher the risk, the stricter the obligation to mitigate risks must be. There is thus a graduated set of risk mitigation obligations depending on the scenario of deployment, the range of tasks to be fulfilled, and the specific features of the weapons system at issue. In other words, risk mitigation obligations will be rather low or even negligible if an autonomous system is deployed in a predetermined area with no human beings present. Conversely, if a machine were to be deployed in a complex, highly dynamic, perhaps urban environment, the obligations would be very high.

Liability If the risk mitigation obligations could be established, as proposed above, then the responsibility of a state for an unlawful act committed by an autonomous weapons system could be linked to the failure to abide by these obligations. A state that fails to exercise the required due diligence in order to mitigate risks and prevent damage, could be held accountable for such failure to abide by relevant due diligence obligations. Conversely, in a case where a state exercised the required due diligence but where damage occurred nevertheless, it could not be held accountable for failure to abide by its due diligence obligations.

However, while this legal construction might be preferred by states—especially those promoting the development of autonomous weapons systems—in theory there are other ways conceivable as to how state responsibility for the actions of AWS could be construed de lege ferenda.

First, one might consider a system of strict liability.[41] Such a regime would not require any proof of fault on the part of the acting party. Strict

---

[41] See with regard to this argument also Rebecca Crootof, 'War Torts: Accountability for Autonomous Weapons' (2016) 164 U Penn L Rev 1347.

liability, which is sometimes also called 'absolute liability', 'presumed responsibility', 'risk liability', 'operator's liability', or '*responsabilité pour risqué crée*', means that the issue of 'fault' is removed from the legal assessment. Under a strict liability regime, responsibility is triggered by default as soon as the risks inherent in unpredictable robotic behavior are realized. Pursuant to such a regime of strict liability, it becomes irrelevant what the operator, the commander in charge of the mission, or the state behind the operation thought or expected in terms of the autonomous system's behavior. It is only the actual conduct of the machine that matters. Moreover, it likewise makes no difference whether the operational failure that led to the breach of the rule was due to technical failure of the system's sensors, unforeseen external interference, environmental conditions, programming errors, or other software defects, to name but a few possible causes.

Such strict liability regimes are not uncommon in regard to hazardous activity or processes of high complexity, albeit admittedly it seems rather unlikely that states would agree to adopt such a model in the domain of the laws of armed conflict. For instance, domestic product liability regimes frequently involve strict liability. It is therefore hardly surprising that such liability models are currently also being discussed on the domestic level with respect to civil uses of autonomous technology. The Swedish carmaker Volvo, for example, has pledged to be 'fully liable' for accidents caused by its self-driving technology.[42] In international law, three examples of strict liability are relevant for the consideration under scrutiny here, as they may serve as a model for the handling of AWS: Principle 4 of the ILC Draft Principles on the Allocation of Loss in the Case of Transboundary Harm Arising out of Hazardous Activities, Article VII of the so-called 1967 Outer Space Treaty, and Article II of the 1972 Space Liability Convention. According to Article II of the 1972 Space Liability Convention, for instance, '[a] launching State shall be absolutely liable to pay compensation for damage caused by its space object on the surface of the earth or to aircraft in flight'.

The similarities to the situation concerning AWS today are indeed striking: when the Outer Space Treaty and the Space Liability Convention were drafted and adopted during the 1960s and early 1970s, outer space technology was not yet fully understood. It was thus perceived as a high-risk technology with potentially unpredictable outcomes. This very same rationale applies to autonomous weapons systems today. One

---

[42] See http://europe.newsweek.com/volvo-will-accept-full-liability-self-driving-car-crashes-334333 (accessed 5 May 2017).

crucial distinction between the mentioned treaties and the regime envisaged for AWS needs to be addressed, however. The liability regimes laid out in both the Outer Space Treaty and the Space Liability Convention provide that a state shall be absolutely liable for *any damage* caused by its space objects on the surface of the earth. Such an arrangement, which links liability to damage alone, is not feasible in the context of autonomous weapons, as they are operating under the laws of war, which, by definition, means that there are various scenarios in which the causation of damage by those systems is explicitly permissible. Therefore, autonomous weapons systems require a more nuanced liability regime, one that is specifically tailored to the context of armed conflict. As soon as international humanitarian law applies due to the existence of an armed conflict, certain damages—for instance, the destruction of a military objective—are lawful and therefore do not lead to any state responsibility. Furthermore, the risks and uncertainties associated with autonomous weapons systems (and hence the need for a strict liability regime) might be considered more relevant in certain scenarios, such as if the system is deployed in an urban environment, than in others, like a naval context or in outer space.

These distinctions suggest the introduction of a graduated liability regime: strict liability could be imposed in certain scenarios or with regard to certain fundamental rules but not in all scenarios or in relation to the entire body of rules comprising the laws of armed conflict. Such a graduated liability regime, which combines strict liability with other forms of liability that require certain elements of fault, may be most appropriate in order to respond to the specific risks and uncertainties inherent in the deployment of autonomous weapons technologies.

As an alternative to the described strict liability regime, a future liability regime for autonomous weapons systems could also be designed as to (merely) reverse the burden of proof. Here, in the case of the breach of a rule of international humanitarian law or other applicable norm of international law by the autonomous system, the state would still have the legal possibility to exonerate itself by showing that it had indeed sufficiently complied with its obligations vis-à-vis the development, construction, deployment and monitoring of the machine, as well as the planning of the mission.

Finally, it bears mentioning that there exists a direct correlation between the issue of accountability, as elaborated on in the foregoing section, and the concept of 'meaningful human control'. This aspect will be dealt with in greater detail below.

4.1.2 Criminal responsibility

With regard to the question of who is to be held criminally responsible in a case where the actions of an autonomous weapons system have led to a breach of a rule of international law, two approaches seem possible. On the one hand, the manufacturer of the system and/or the coders of the controlling algorithms might be held criminally responsible. On the other hand, responsibility for the crime may lie with the commander in the field. Furthermore, depending on the facts of the case under scrutiny, superior generals or even political decision-makers that initiated the military campaign might be held criminally responsible.[43] To begin with, it is necessary to clarify that the state of the law is unambiguous as far as intentional breaches of the laws of armed conflict that amount to war crimes are concerned. If, for example, the programmer intentionally writes a code that makes the robot attack civilians on the battlefield then criminal responsibility is not in doubt. Similarly, if a commander is aware of certain software defects that provoke hazardous malfunctioning in an AWS and deploys the system in a densely populated area regardless and without caring about the likely loss of civilian lives, criminal responsibility could be established.[44] The application of the law, however, is particularly challenging if all involved human beings deployed the autonomous weapons system under the assumption that it works perfectly well and without acting negligently or with the intention of harming protected persons. Rather, the scenario at issue is one in which the unpredictability inherent in any autonomous weapons systems materializes in unlawful damage on the battlefield, that is, where an otherwise perfectly functioning robot unexpectedly kills a human being.

Autonomous weapons systems are extraordinarily complex. Even for the persons involved in the manufacturing process, it is not always easy to foresee all possible consequences of a deployment. This is inevitable, as the systems are built to be able to react autonomously to unforeseen situations. Indeed, taken the notion of 'autonomy' seriously, it is inherently impossible to test all possible actions and reactions in advance. On the contrary, it is to be expected that unexpected environmental conditions and external interferences will come up during an actual mission.[45] Such factors, however, have to be taken into consideration in a

---

43 Heyns (n 2) 77.

44 John Lewis, 'The Case for Regulating Fully Autonomous Weapons' (2015) 124 Yale L J 1309, 1324.

45 US Chief Air Force Scientist, Report on Technology Horizons. A Vision for Air Force Science and Technology During 2010–2030, 2010, 105 et seq.,

criminal trial, as they would most likely alleviate individual responsibility. Foreseeability is a precondition for culpability even with regard to offenses of negligence. Yet, constructed in this way, it becomes possible that any misconduct of autonomous systems is indeed equated with force majeure, amounting, in other words to an event beyond the influence of humans. This problem is further aggravated if we are dealing with a machine learning algorithms. Here, it will be even more difficult to anticipate the autonomous weapons system's future behavior.

The commander's responsibility is problematic as well. Article 28 of the Rome Statute of the International Criminal Court deals with the military commander's accountability under international criminal law, so an analogous application of the norm might be considered. However, the rule requires that the commander knew or ought to have known at a certain point in time that the subordinate was committing a crime or was about to do so. This precondition essentially entails two problems. For one, it is questionable whether the norm is indeed applicable analogously. It is based on the premise that the conduct involves acts by two entities with moral agency. This construction is, by definition, not transferable to the relationship between a human and a machine. If algorithms make the system's behavior unpredictable and, moreover, if they are learning, then it is difficult to ever usefully determine at what point a commander 'ought to have known' that the autonomous system was about to violate the rules of international humanitarian law.[46] On the other hand, if the commander observes the robot's actions and notices that the machine starts to commit war crimes due to insufficiencies in its software or other malfunctions, the legal assessment may differ. If the commander refuses to abort the mission immediately by deactivating the autonomous system, he could be held criminally liable for all transgressions from that point in time onwards—provided, of course, that he is actually 'on the loop', that is, if it is in his factual power to stop the machine.

The above theoretical scenario underlines the basic realization that responsibility decreases where the system's autonomy increases. This problem is structural: there is no responsibility without control. The more a system is capable of acting autonomously, the greater the (potential) accountability gap becomes. At a certain point on this scale, culpability

---

accessed 5 May 2017 at http://www.flightglobal.com/assets/getasset.aspx?ItemID=35525.

[46] Peter M Asaro, 'On Banning Autonomous Weapons Systems: Human Rights, Automation, and the Dehumanization of Lethal Decision-Making' (2012) 94 Intl Rev of the Red Cross 687, 693; Noel E Sharkey, 'The Evitability of Autonomous Robot Warfare' (2012) 94 Intl Rev of the Red Cross 787, 790.

can no longer be established. It is therefore not a valid argument to simply point to necessary human involvement at some point in time in order to circumvent the issue.[47] Furthermore, to simply state that individual criminal responsibility is overrated in any case, as it is merely one option among several to ensure compliance with the rules of international humanitarian law,[48] is not fully convincing either, as the contention disregards the fundamental importance of criminal responsibility in this regard. Still, chances to close resultant accountability gaps under criminal law in the near future appear slim. While, in theory, there are certain options for establishing criminal responsibility in relation to robotic activity, in practice, their implementation seems unrealistic. Thus, the best (theoretical) option to close accountability gaps would be to develop a regime of strict criminal liability akin to the strict liability regime in the area of state responsibility referred to above. In other words existing accountability gaps should be closed by creating statutory offenses that criminalize the dangers inherent in lethal autonomous weapons systems. Under such a model, a commander's criminal responsibility could be triggered by default whenever the dangers inherent in an AWS (for example, the danger of unpredictable behavior) materialize on the battlefield, provided that the commander had created the hazardous situation by deploying the system. But such a strict liability regime is unheard of in international criminal law (although there are examples of such a regime in certain domestic criminal law systems) and it would arguably be unjust to allocate criminal responsibility exclusively with the random military commander deciding to deploy an autonomous system.

4.2 Compliance with the Rules of Armed Conflict

Aside from questions of accountability, a large part of the debate surrounding autonomous weapons systems is concerned with the question of whether those machines are indeed capable of operating within the limits of the rules of international humanitarian law.

The relevant rules as established by treaty and custom have already been mentioned: first, the principle of distinction as enshrined in Article 51(2) of Additional Protocol I; second, the principle of proportionality, laid down mainly in Articles 51(5)(b) and 57(2)(a)(iii) of Additional

---

[47] That has been proposed by Schmitt & Thurnher (n 15) 277.

[48] See Kenneth Anderson and Matthew C Waxman, *Law and Ethics for Autonomous Weapons Systems: Why a Ban Won't Work and How the Laws of War Can* (The Hoover Institution 2013) 17.

Protocol I; and third, the principle of precautions in attack in accordance with Article 57(1) of Additional Protocol I.

Concerning all three rules, there is disagreement among experts as to whether robots will ever be able to take the necessary decisions in a complex battlefield environment that allow them to comply with those fundamental principles of international humanitarian law.[49] That civilians are never legitimate targets is beyond doubt. Whether autonomous weapons systems will reliably be able to distinguish a civilian from a legitimate military objective, however, remains questionable. First and foremost, to be sure, this is a technical question.[50] However, given the realities of modern conflicts, it is rather unlikely to expect clear-cut situations, which—for many observers makes it difficult to imagine algorithms ever to be up to the task. Particularly in urban warfare constellations, where fighters routinely blend in with the civilian population, ambiguous circumstances will be the rule rather than the exception. These scenarios will require more than just cutting-edge sensor technology, namely the ability to interpret human behavior in highly dynamic scenarios. In this context, it is of interest to note that the United States Army Soldier's Guide, when addressing the 'Ethical Reasoning Process' every soldier ought to undertake when deciding on a course of action in response to a given situation, explicitly includes a 'gut check' as a final criterion. By this the guidelines urge the soldier to question whether the course of action '"feel[s]" like it is the right thing to do'.[51] This final step of moral deliberation is by definition beyond the limits of machine 'intelligence', as even proponents of autonomous weapons systems acknowledge.[52] Similarly, experts doubt whether it could be possible to write the ability to hesitate or abort a lethal mission in the case of 'doubt' concerning a certain situation or a certain person's legal status under international humanitarian law.[53]

---

[49] See in this context also Michal Klincewicz, 'Autonomous Weapons Systems, the Frame Problem and Computer Security' (2015) 14 J of Mil Ethics 162, who argues that the software of truly autonomous weapons systems would need to be so complex that the system inevitably would become critically vulnerable to hacking, which undermines the advantages of AWS.

[50] See only Sharkey (n 46) 788.

[51] Quoted in Ronald C Arkin, 'Governing Lethal Behavior: Embedding Ethics in a Hybrid Deliberative/Reactive Robot Architecture' Technical Report GIT-GVU-07-11, 51, accessed 5 May 2017 at http://www.cc.gatech.edu/ai/robot-lab/online-publications/formalizationv35.pdf.

[52] See ibid.

[53] See Michael N Schmitt, 'Autonomous Weapon Systems and International Humanitarian Law: A Reply to the Critics' (2013) Harvard Natl Sec J Online 16,

On the other hand, proponents of the technology maintain that human soldiers do not necessarily have an advantage in making the distinction between combatants and protected persons as prescribed by Article 51(2) of Additional Protocol I. On the contrary, they are, unlike machines, confronted with emotional states such as stress, anger or fear. It is those indeed very human conditions that are responsible for triggering the majority of transgressions in combat situations. For experts in favor of AWS, this serves as the main argument: machines are even more capable of complying with the laws of armed conflict in the middle of the 'fog of war', emotionless as they are.[54] An autonomous system does not fear for its life, which, at least theoretically, means that it can afford to maintain the legally prescribed presumption that a person is a (protected) civilian right up until the moment when that person actually draws a weapon, turning him or her into a legitimate target. A human soldier, on the other hand, will, out of fear for his or her life, hardly ever wait that long, thereby vastly increasing the likelihood of incidentally harming a civilian. Moreover, the so-called 'scenario fulfillment' problem, which means the unconscious execution of a chain of previously rehearsed steps ending in the lethal use of force,[55] even though a proper assessment of the situation at hand would have indicated a different action, cannot happen to autonomous weapons systems.[56]

Aside from distinguishing between legitimate targets and protected persons, the principle of proportionality pursuant to Articles 51(5)(b) and 57(2)(a)(iii) of Additional Protocol I may also pose a serious obstacle for the deployment of autonomous weapons systems. The rule prohibits attacks that 'may be expected to cause incidental loss of civilian life, injury to civilians, damage to civilian objects, or a combination thereof, which would be excessive in relation to the concrete and direct military advantage anticipated'. Such a decision on the consequences of an attack comprising collateral damage requires a complex mental operation based on values and, in relation to the singular instance at hand, leading to an assessment of the overall circumstances. Again, the crucial question is whether algorithms could ever be expected to make such a complex and critically nuanced calculation. Could an autonomous system, for example, appropriately assess the expected military advantage of the

---

accessed 5 May 2017 at http://harvardnsj.org/2013/02/autonomous-weapon-systems-and-international-humanitarian-law-a-reply-to-the-critics/.

[54] Arkin (n 51) 6.

[55] See the shooting down of Iran Air 655 by the USS Vincennes in July of 1988 as one of the most striking examples of this problem.

[56] Arkin (n 51) 6.

isolated operation? In 2003, the International Criminal Tribunal for the Former Yugoslavia held that in order to determine whether an attack was proportional 'it is necessary to examine whether a reasonably well-informed person in the circumstances of the actual perpetrator, making reasonable use of the information available to him or her, could have expected excessive civilian casualties to result from the attack'.[57] Some experts raise doubts as to whether machines could be programmed so that they would be able to make such an evaluation, at least in the near future.[58]

However, others, again, counter this by suggesting that it would at least already be possible to write code that would enable the autonomous systems to make assessments of a given conflict situation that would be on par with the evaluation made by a human being.[59] Even more so, lacking the instinct of self-preservation, some experts argue that autonomous weapons systems would ultimately have less incentive to over-react and resort to an excessive use of force. This, in itself, would lead to a more reliable compliance with the principle of proportionality.[60] However, this assertion is of course, as such, incapable of countering the arguments of those experts who doubt that autonomous systems could ever be programmed in such a way in the first place.

The principle of precaution as enshrined in Article 57(1) of Additional Protocol I, which is closely linked to both the principle of distinction and the principle of proportionality, is the third issue to be examined in relation to the development and deployment of AWS. It prescribes that '[i]n the conduct of military operations, constant care shall be taken to spare the civilian population, civilians and civilian objects'. The duty to undertake precautionary measures in this sense concerns everyone involved and every phase from planning an attack until its execution, but equally already the construction and coding of the autonomous system.[61] Furthermore, the original planning must be still appropriate and relevant at the time the mission is actually being carried out. As many unforeseen events might happen during an ongoing deployment, some commentators argue that the precautionary principle implies the obligation to always keep a human soldier at least 'on the loop', so he or she can react to a

---

57 ICTY, *Prosecutor v Stanislav Galic*, Judgment (Trial Chamber) (Case No. IT-989-29-T), 5 December 2003, at 58.

58 William H Boothby, *Conflict Law: The Influence of New Weapons Technology* (Springer 2014) 110 et seq.

59 Schmitt (n 53) 19.

60 Arkin (n 51) 58.

61 Boothby (n 58) 115.

change of circumstances.[62] Others go even further, cautioning that, due to the fact that computer systems are capable of processing information much faster than humans, it is doubtful whether a human could actually effectively intervene if an AWS commences to transgress a rule of international humanitarian law on the battlefield.[63]

Considering the foregoing objections, it might be argued that proper 'precaution' means that a lethal autonomous system could only be lawfully deployed if the commander can guarantee that encounters with the civilian population are ruled out per se. In view of modern conflicts, such a scenario does not seem very realistic.

As regards all fundamental rules of international humanitarian law, this section has shown that most issues hinge on the question of whether or not it will be possible to develop sensor technologies and code algorithms, which, in combination, are indeed capable of enabling autonomous weapon systems to comply. This is mostly still a matter of speculation. That aside, it is clear that computers will, by definition, never achieve comprehensive, contextual intelligence: they are not able to 'think' outside of their algorithms. This, in itself, implies the inherent risk that the systems fail whenever something truly unforeseen occurs that would force the system to depart from the original mission plan.[64] In the words of Noel Sharkey, '[w]hen a machine goes wrong it can go really wrong in a way that no human ever would'.[65]

On the other hand, if it can be shown that AWS, once properly developed, will be, in fact, more capable of complying with the rules of armed conflict as compared to human soldiers, one might even argue that a commander would have the obligation to deploy machines instead of soldiers that are prone to transgressions, in order to adequately protect the civilian population. If the human is the weakest link in military command chains that attempt to abide by the law, due to emotions like fear or anger, then robots could indeed be considered superior. Moral reasoning, it may be maintained, is, in fact, overrated: the rules of international humanitarian law already incorporate ethical judgments made by the world society and which are thus universally applicable. Soldiers are not meant to come to their own moral conclusions on the

---

62 Geneva Academy (n 8) 16.

63 Philip Alston, 'Lethal Robotic Technologies: The Implications for Human Rights and International Humanitarian Law' (2011/2012) 21 Information & Science 36, 54.

64 Paul Scharre, 'Why Unmanned' (2011) 61 Joint Force Quarterly 89, 92.

65 Sharkey (n 46) 790.

battlefield. And, when it comes to the application of such predetermined rules, 'cold-blooded' algorithms are invariably more reliable.[66]

However, such a line of argumentation overlooks a much more fundamental consideration. Concerning international humanitarian law, it might well be claimed that the whole body of law is implicitly based on the premise that human beings, with their human and thus necessarily limited capacity of decision-making, are its sole addressees.[67] If human characteristics, such as stress, fatigue, a predisposition to err, or the instinct of self-preservation, are already factored into the ratio of the laws of armed conflict, then it is probably a priori misguided to ask whether autonomous weapons systems are capable of complying. The real question would rather be whether these rules are the 'right' rules when dealing with autonomous systems, or if new, stricter rules need to be found.

Indeed, under these premises, it may be concluded that, at the very least, autonomous weapons systems should abide by a much higher standard than that provided by the existing rules. This could mean, for example, that the principle of distinction as applied to machines prescribes the algorithms to be written in such a way that the AWS would only resort to lethal force in situations in which enemy combatants unambiguously act in an aggressive and offensive manner. As long as the situation is less clear-cut, the machine would not be allowed to harm its opponent.[68]

4.3 Political and Legal-philosophical Problems

As already indicated, the problems concerning accountability and responsibility are of course not the only issues in regard to the development and deployment of autonomous weapons systems.

4.3.1 An increase of armed conflicts?

On a political level, some experts warn that the proliferation of AWS could increase the likelihood of armed conflicts. If states were freed of the necessity to risk their own soldiers' lives in order to wage war against an opponent, so the argument goes, the politicians' and military commanders' inhibition threshold to resort to force would be significantly reduced.[69] The lack of risk of the loss of human lives may furthermore

---

66 Arkin (n 51) 55.
67 Asaro (n 46) 700.
68 Ibid 701.
69 Asaro (n 35) 62.

increase acceptance of the use of armed force among the state's population. Armed conflicts of the recent past, most prominently those in Afghanistan and Iraq, have once again shown that a population's attitude towards a military mission rapidly changes once the political and societal costs of the war start to rise. On the other hand, the continuing 'drone war' conducted mainly by the United States against Al Qaeda and other non-state armed groups connected to transnational terrorism seems to support the argument as protest from within the American population against the strategy remains negligible.[70] However, at the present moment, a reliable prediction in this regard remains impossible.

4.3.2 Autonomous weapons and human dignity

From a more ethical-legal perspective and on a very fundamental level, it has been argued that autonomous weapons systems ought to be banned outright, due to the assertion that the lethal use of force by machines that are controlled by algorithms inherently infringes human dignity. Philosophically, the concept of human dignity entails as a minimum that every human being is to be valued and treated as an individual that is both unique and as such irreplaceable. It is this aspect of the notion that needs to be considered when evaluating the deployment of autonomous weapons systems.

The crucial question in this context is hence whether it is invariably a violation of the uniqueness of every single human life if the decision to kill such a being is based solely on the absolute rationality of an algorithm. One may assert that this process is never entirely rational and should thus not be 'outsourced' to an entity, which has no moral agency. The inherent irrationality expressed in the act of killing a human being may be considered a basic precondition of a minimum of morality. Even if a soldier acts in accordance with the laws of armed conflict, and in spite of the commander's orders, the decision to kill an enemy combatant in a given situation remains a highly personal decision that requires the examination of conscience in a moral sense.[71] Machines are incapable of executing such an operation of human rationality, which comprises an element of power of judgment and compassion. Autonomous weapons systems take the 'decision' to kill with literally merciless precision, without any prior moral evaluation.[72] Therefore, the human being is

---

[70] Alston (n 63) 55.

[71] O'Connell (n 3) 231. See in this context also Dave Grossman, *On Killing: The Psychological Cost of Learning to Kill in War and Society* (Back Bay Books 1995).

[72] Asaro (n 46) 695.

indeed not considered as a unique individual but as a mere object of a mathematically calculated decision to use lethal force—in the words of UN Special Rapporteur, 'death by algorithm'.[73] Against this backdrop, it seems reasonable to maintain that it indeed constitutes a violation of the principle of human dignity if a computer system makes an autonomous decision on life and death—leaving aside the more technical question concerning the legal status of human dignity under international law.[74]

Furthermore, if lethal autonomous weapons systems are going to be deployed in areas with resident civilian populations, the latter may be impaired in their right to live in dignity. It has correctly been pointed out that the presence of unmanned systems among civilians is capable of inducing a general feeling of discomfort, anxiety, and even trauma.[75] This effect has been studied in detail within the context of the protracted deployment of drones by the United States in Pakistan, with disconcerting results.[76] Under such circumstances, a normal daily life becomes almost impossible. It is likely that the deployment of AWS could have a similar impact.

5. OUTLOOK: MAINTAINING MEANINGFUL HUMAN CONTROL

In order to tackle the issues analyzed in the foregoing sections, in particular the problems concerning accountability for unlawful acts and compliance with the rules of international humanitarian law, many experts as well as state representatives have come to the conclusion that autonomy in weapon systems should only be allowed to the degree that

---

[73] Comments by Christof Heyns, UN Special Rapporteur on extrajudicial, summary or arbitral executions, Informal Meeting of Experts on Lethal Autonomous Weapons: Conventional Weapons Convention, 16 April 2015, 5, accessed 5 May 2017 at http://www.unog.ch/80256EDD006B8954/(httpAssets)/1869331AFF45728BC1257E2D0050EFE0/$file/2015_LAWS_MX_Heyns_Transcript.pdf.

[74] See on this e.g., Niels Petersen, 'Human Dignity, International Protection' in Rüdiger Wolfrum (ed), *Max Planck Encyclopedia of Public International Law* (Oxford University Press 2012) 1.

[75] Heyns (n 2) 98.

[76] International Human Rights and Conflict Resolution Clinic at Stanford Law School and Global Justice Clinic at NYU School of Law, Living Under Drones: Death, Injury, and Trauma to Civilians From US Drone Practices in Pakistan, 2012, accessed 5 May 2017 at https://law.stanford.edu/publications/living-under-drones-death-injury-and-trauma-to-civilians-from-us-drone-practices-in-pakistan/.

'meaningful human control' over those systems is always retained, at least as far as decisions concerning the lethal use of force are concerned.[77]

Despite growing support, the definition of 'meaningful human control' remains a contentious issue.[78] However, considering the concept's wording, the lowest common denominator of all attempts must come down to a prohibition of full machine autonomy as regards certain critical functions or aspects.[79] If a human is in control, then there is per se no autonomy. In a certain sense, the very notion of 'meaningful human control' as a qualification of autonomous conduct by computer systems is thus an oxymoron. However, it may nevertheless serve as a useful starting point for the necessary debate on the two most critical questions: (1) which decisions should remain under human control and (2) what could such control look like. The required degree of control may concern different factors, such as the amount of time between the last decision made by a human and the execution of (lethal) force by the machine; the environment in which the machine is deployed, in particular as regards the question whether civilians are present; the purpose of the mission, that is, whether the machine is supposed to carry out defensive or offensive tasks; the question whether the machine is at all set up to use lethal force; the degree of training of those persons who are in charge of exerting control over the machine; the question to what extent the human soldiers are capable of intervening in the case of an emergency, and to abort the mission; how to guarantee the possibility of accountability, for example by arranging an uninterrupted recording of all of the robot's actions.

Along these lines, for the UK-based non-governmental organization Article 36, critical functions that ought to remain under meaningful

---

[77] See e.g. Human Rights Watch and International Human Rights Clinic, *Killer Robots and the Concept of Meaningful Human Control: Memorandum to Convention on Conventional Weapons (CCW) Delegates*, April 2016, accessed 5 May 2017 at https://www.hrw.org/sites/default/files/supporting_resources/robots_meaningful_human_control_final.pdf; International Committee of the Red Cross (ICRC), *Autonomous Weapons: Decisions to Kill and Destroy Are a Human Responsibility*, 11 April 2016, accessed 5 May 2017 at https://www.icrc.org/en/document/statement-icrc-lethal-autonomous-weapons-systems.

[78] See, for an overview on the different approaches, Rebecca Crootof, 'The Meaning of "Meaningful Human Control"' (2016) 30 Temple Intl & Comp L J.

[79] Statement of the International Committee of the Red Cross (ICRC), CCW Meeting of Experts on Lethal Autonomous Weapons Systems, Geneva, 13 April 2015, accessed 5 May 2017 at https://www.icrc.org/en/document/lethal-autonomous-weapons-systems-LAWS.

human control at all times concern '[t]he pre-programmed target parameters, the weapon's sensor-mechanism and the algorithms used to match sensor-input to target parameters; [and] [t]he geographic area within which and the time during which the weapon system operates independently of human control'.[80] Thus, control cannot be 'meaningful' if it is limited to some degree of mere supervision ('on the loop') or considered finished after a (however careful) planning or programming phase. Similarly, the ICRC, identifying human control over autonomous weapons as 'the overarching issue in this debate',[81] maintains that its need 'is consistent with legal obligations, military operational requirements and ethical considerations'.[82] Accordingly, the organization proposes that, as a way forward concerning autonomous weapons systems, states should 'develop the parameters of human control in light of the specific requirements under IHL and ethical considerations ... thereby establishing specific limits on autonomy in weapon systems'.[83]

However, despite a growing number of publications and opinions on the definition of 'meaningful human control' and despite growing consensus that is roughly in line with that of the ICRC, Crootof correctly observes that 'the grey area is wide, and full of complicated situations',[84] but suggests that the notion's lack of a clear definition might ultimately be a strength, as ambiguity makes it more likely that relevant stakeholders will be able to agree on backing the concept as such—subsequently, state practice could then substantiate the notion as a legally binding concept.[85] Such a formalistic approach notwithstanding, Scharre and Horowitz attempt to identify three key substantial components of 'meaningful human control' which in any case should inform the ongoing search for a workable definition:

> (1) Human operators are making informed, conscious decisions about the use of weapons; (2) Human operators have sufficient information to ensure the lawfulness of the action they are taking, given what they know about the target, the weapon, and the context for action; (3) The weapon is designed and

---

[80] Article 36, Killing By Machine: Key Issues for Understanding Meaningful Human Control, April 2015, 3, accessed 5 May 2017 at http://www.article36.org/wp-content/uploads/2013/06/KILLING_BY_MACHINE_6.4.15.pdf.

[81] ICRC, Decisions to Kill (n 77).

[82] Ibid.

[83] ICRC, Views of the International Committee of the Red Cross (ICRC) on Autonomous Weapon Systems, 11 April 2016, 6, accessed 5 May 2017 at https://www.icrc.org/en/document/views-icrc-autonomous-weapon-system.

[84] Crootof (n 78) 4 (draft pagination).

[85] Ibid 5 et seq.

> tested, and human operators are properly trained, to ensure effective control over the use of the weapon.[86]

If an agreement on the concept of 'meaningful human control' can be reached that comprises all of the three elements, then the risk of an allegedly insurmountable 'accountability gap' becomes a non-issue. This is the main reason why experts who are skeptical of the development and introduction of autonomous weapon systems promote the idea of comprehensive human control. Every time the system's deployment leads to the breach of a rule of international humanitarian law, a human could be held directly responsible for failing to exert sufficient control over the machine's actions. Moreover, if a commander decides to send the machine on the battlefield without ensuring meaningful human control at all times of the mission, he or she could potentially be held responsible for acting unlawfully, whether or not the autonomous weapon system itself actually committed an unlawful act while operating on the battlefield.[87]

---

[86] Michael C Horowitz and Paul Scharre, 'Meaningful Human Control in Weapon Systems: A Primer' CNAS Working Paper, March 2015, 14 et seq., accessed 5 May 2017 at https://www.cnas.org/publications/reports/meaningful-human-control-in-weapon-systems-a-primer.

[87] Human Rights Watch (n 77) 6.

13. Making autonomous weapons accountable: command responsibility for computer-guided lethal force in armed conflicts

*Peter Margulies**

In debates about autonomous weapons systems (AWS)[1] in armed conflict, narratives have their own momentum. One could frame AWS as a

* I thank my research assistant, Nicole LaCicero, law librarians Lucinda Harrison-Cox and Stephanie Edwards, and library access services coordinator Thelma Dzialo for their help, and Ken Anderson, Rebecca Crootof, Missy Cummings, and Gabor Rona for comments on a previous draft.

[1] An AWS (this chapter also uses the abbreviation, AWS, as a plural noun referring to more than one autonomous system) is a system in which a computer network makes and executes decisions about the conduct of an armed conflict without ex ante human authorization for each decision. Certain types of AWS have been used for years, for example, on naval vessels to detect and counter incoming enemy missiles. These applications have not inspired major controversy. See International Committee of the Red Cross, 'Autonomous Weapons Systems: Technical, Military, Legal and Humanitarian Aspects' (2014) Background Paper for Meeting of Experts 57 1, 65–66, accessed 5 May 2017 at https://www.icrc.org/en/document/report-icrc-meeting-autonomous-weapon-systems-26-28-march-2014; Chris Jenks, 'False Rubicons, Moral Panic & Conceptual Cul-De-Sacs: Critiquing and Reframing the Call to Ban Lethal Autonomous Weapons' (2016) XLIV(1) Pepperdine Law Review 21–22, accessed 5 May 2017 at http://ssrn.com/abstract=2736407. In contrast, the future development and deployment of AWS that would use lethal force to specifically target *human* adversaries has triggered fierce debate. See Hin-Yan Liu, 'Categorization and legality of autonomous and remote weapons systems' (2012) 94(886) International Review of the Red Cross 627, 632; Marco Sassòli, 'Autonomous Weapons and International Humanitarian Law: Advantages, Open Technical Questions and Legal Issues to be Clarified' (2014) 90 Naval War College International Law Studies 308; Michael N Schmitt and Jeffrey S Thurnher, '"Out of the Loop": Autonomous Weapons Systems and the Law of Armed Conflict' (2013) 4 Harvard National Security Journal 231. Critics of AWS express deep concern about their compliance with international humanitarian law (IHL). United Nations, *Report of the Special Rapporteur on Extrajudicial, Summary or Arbitrary Executions* (UN Doc A/HRC/23/47, 2013) para 55

variant of driverless cars: a means to reduce the havoc and mayhem caused by human error. Indeed, some commentators view AWS in armed conflict as a potential cure for defects in human perception and judgment.[2] In contrast, AWS opponents warn of killer robots going rogue, and urge a ban on development and deployment of AWS.[3] Proponents of a ban often also raise the specter of impunity, asserting that it will be impossible to hold a human accountable for the mistakes of a computer[4] (a 'machine' or 'agent', in data scientists' parlance).[5]

This chapter argues that a ban on AWS is unwise. Adaptations in current procedures for the deployment and use of weapons can ensure that any AWS used in the field complies with IHL. Solving the AWS accountability problem hinges on the doctrine of command responsibility, applied in a three-pronged approach that the chapter calls 'dynamic diligence'.

Dynamic diligence is a demanding standard. First, it requires continual adjustments in the machine-human interface, performed within a military command structure staffed by persons who possess specialized knowledge of AWS's risks and benefits. Second, dynamic diligence requires ongoing assessment of the AWS's compliance with IHL.[6] This assessment starts with validation at the weapons review stage, prior to an

(cautioning that AWS do not currently possess emotion of "compassion" in human sense of the term); Peter Asaro, 'On banning autonomous weapons systems: human rights, automation, and the dehumanization of lethal decision-making' (2012) 94(886) International Review of the Red Cross 687 (warning that AWS may decrease respect for human life).

2 Sassòli (n 1) 310.

3 Asaro (n 1); but see Michael A Newton, 'Back to the Future: Reflections on the Advent of Autonomous Weapons Systems' (2015) 47 Case Western Reserve Journal of International Law 5, 8–9 (arguing that weapons bans have often been counterproductive); see generally Eric Talbot Jensen, 'The Future of the Law of Armed Conflict: Ostriches, Butterflies, and Nanobots' (2014) 35 Michigan Journal of International Law 253, 294–5 (predicting that AWS will play increasing role and that they offer potential improvements in IHL compliance, while also presenting risks).

4 Asaro (n 1); cf Rebecca Crootof, 'War Torts: Accountability for Autonomous Weapons' (2016) University Pennsylvania Law Review (forthcoming), accessed 5 May 2017 at http://ssrn.com/abstract=2657680.

5 Peter Flach, *Machine Learning: The Art and Science of Algorithms that Make Sense of Data* (Cambridge University Press 2012).

6 The US Department of Defense has issued a directive that would require periodic review of AWS, although it is unclear from the language of the directive whether it would require review of AWS software, let alone the frequent, comprehensive review advocated in this chapter. See US Department of Defense,

AWS's deployment. It includes frequent, periodic assessments of an AWS's learning in the field, to ensure that field calculations enabled by the machine's software are IHL-compliant. Dynamic assessments also contemplate continual updates to the AWS's inputs, including databases such as terrorist watch lists. Third, dynamic diligence requires flexibility in the parameters governing the machine's operation, with a presumption favoring interpretability of the AWS's outputs. A state that practices dynamic diligence with respect to AWS can comply with IHL, ensure accountability for any IHL violations, and potentially conduct an armed conflict with greater safety for both civilians and its own combatants.

1. TARGETING AND IHL

Analysis of the status of AWS under international humanitarian law (IHL) should begin with the law itself. IHL emerges from the delicate balance of two values: military necessity, which recognizes that states have a legitimate interest in measures that give them a military advantage in an armed conflict, and humanity, which tempers necessity with concern for the safety of civilians, civilian objects, cultural treasures, and the environment.[7] Additional Protocol 1 of the Geneva Convention on the Protection of Victims of International Armed Conflicts (API)[8] sets out the

---

Autonomous Weapons System Directive 3000.09. (2012) Enc. 4: Para. 8(a)(6), accessed 5 May 2017 at www.dtic.mil/whs/directives/corres/pdf/300009p.pdf (requiring that military "establish and periodically review training, TTPs [tactics, techniques, and procedures], and doctrine" for AWS "to ensure operators and commanders understand the functioning, capabilities, and limitations of a system's autonomy in realistic operational conditions, including as a result of possible adversary actions").

[7] Michael N Schmitt, 'Military Necessity and Humanity in International Humanitarian Law: Preserving the Delicate Balance' (2010) 50(4) Virginia Journal of International Law 795, 796.

[8] International Committee of the Red Cross, Protocol Additional to the Geneva Conventions of 12 August 1949, and Relating to the Protection of Victims of International Armed Conflicts (Protocol I), *adopted* June 8, 1977, 1125 UNTS 3 (API).

three principles of IHL that balance these core values: distinction,[9] proportionality[10] and precautions in attack.[11]

The principle of distinction requires parties to an armed conflict to refrain from intentionally targeting civilians. Each party to an armed conflict can target the armed forces of the other side and other 'military objectives', including factories that produce armaments or other material necessary for war-fighting. An armed force cannot target civilians who have no direct role in combat. However, it can target civilians who directly participate in hostilities (DPH).[12] In a non-international armed conflict (NIAC), between a state and a non-state group, identification of non-state combatants and DPH civilians can be challenging, because non-state actors do not always heed API's mandate to 'distinguish

---

[9] API, Article 51(2); Yoram Dinstein, *The Conduct of Hostilities Under the Law of International Armed Conflict* (2nd edn, Cambridge University Press 2010) 89; cf Kenneth Anderson, Daniel Reisner and Matthew Waxman, 'Adapting the Law of Armed Conflict to Autonomous Weapons Systems' (2014) 90 International Law Studies 386, 401–5 (applying IHL principles to AWS).

[10] API, Article 51(5)(b); Schmitt and Thurnher (n 1) 253–4.

[11] API, Article 57(2)(a)(i); Schmitt and Thurnher (n 1) 259–60. While the United States has not ratified API, it accepts these principles as elements of customary international law (CIL) that bind all nations. Michael J Matheson, Remarks, 'Session One: The United States Position on the Relation of Customary International Law to the 1977 Protocols Additional to the 1949 Geneva Conventions' (1987) 2 American University of Journal of International Law and Policy 419, 420; cf Michael A Newton, 'Exceptional Engagement: Protocol I and a World United Against Terrorism' (2009) 45 Texas International Law Journal 323, 344–7 (discussing political agendas that drove enactment of Protocol I).

[12] For discussion of DPH, see Nils Melzer, International 'Interpretive Guidance on the Notion of Direct Participation in Hostilities Under International Law' (2009) International Committee of the Red Cross 33–6 [hereinafter ICRC Guidance], accessed 5 May 2017 at www.aco.nato.int/resources/20/Legal%20Conference/ICRC_002_0990.pdf (taking narrow view of DPH); but see Bill Boothby, 'And for such time as: The Time Dimension to Direct Participation in Hostilities' (2010) 42 New York University Journal of International Law and Policy 741, 753–5 (analyzing membership in an organized armed group as a basis for targeting); Michael N Schmitt, 'Deconstructing Direct Participation in Hostilities: The Constitutive Elements' (2010) 42 New York University Journal of International Law and Policy 697, 699 (critiquing the ICRC's guidance as failing "to fully appreciate the operational complexity of modern warfare"); Kenneth Watkin, 'Opportunity Lost: Organized Armed Groups and the ICRC "Direct Participation in Hostilities" Interpretive Guidance' (2010) 42 New York University Journal of International Law and Policy 641, 643–4 (arguing that the ICRC's guidance in effect favored non-state actors over states).

themselves from the civilian population'.[13] NIACs, including conflicts with terrorist groups such as ISIS or Al Qaeda, therefore pose particular complexities for compliance with both the principle of distinction and the principle of proportionality, which bars incidental harm (sometimes called 'collateral damage') to civilians that is 'excessive in relation to the concrete and direct military advantage anticipated'.

Finally, the requirement of precautions in attack obligates a state to 'take all feasible precautions in the choice of means and methods of attack' to minimize harm to civilians. The feasibility standard does not require that a state take any and all *possible* precautions. Rather, feasibility entails what is practicable under the circumstances. For example, the principle of precautions may require some kind of warning to civilians who live in an area that will soon be subject to attack.[14]

We can break down targeting into two different steps. One is target selection: determining which individuals are combatants or civilian DPH, and thus can be lawfully targeted. The second is implementation (sometimes called engagement): subsequent to target selection, using lethal force against a combatant or DPH in a manner that complies with the principles of distinction, proportionality and precautions in attack.

Target selection can turn on categorical, individual or situational factors. Categorical targeting involves identification of large groups of targetable personnel, who are identifiable because they wear insignia that distinguish them from civilians and carry arms openly. The classic example of categorical targeting involves the opposing uniformed armed forces of states in an international armed conflict (IAC) facing off in a traditional battle space such as a desert or field, located at some distance from civilian population centers. In such situations, targeting is consistent with IHL if an individual wearing such insignia is not *hors de combat*, that is, that individual has not surrendered or been grievously wounded.

Individual targeting, in contrast, focuses on specific, named persons. In an IAC, such specific targeting typically flows from that individual's status within a uniformed state force. For example, one state might specifically target an adversary's senior commander. In such a situation, targeting the individual commander poses no special problems under

---

13 API, Article 44(3).

14 Pnina Sharvit Baruch and Noam Neuman, 'Warning Civilians Prior to Attack under International Law: Theory and Practice' (2011) 87 International Law Studies 359; cf Geoffrey S Corn, 'War, Law, and the Oft Overlooked Value of Process as a Precautionary Measure' (2015) 42 Pepperdine Law Review 419, 423–4 (suggesting that principle of precautions in attack may have more concrete value than proportionality in ensuring IHL compliance).

IHL, since the individual is still being targeted because of her role in the uniformed force's chain of command. In a NIAC in which a state faces off against a non-state actor, state targeting of individuals requires an additional step. The absence of uniforms requires a state to gather evidence that a particular person's role has a sufficient nexus to hostilities to support targeting that person as a combatant or civilian DPH.

To that end, individual targeting decisions by the United States in NIACs against Al Qaeda or Al Qaeda in the Arabian Peninsula (AQAP) often entail a lengthy process of accumulating information and deliberating about whether the proposed target meets the legal standard. For example, in targeting Anwar al-Awlaki, a US citizen whom US officials believed to be head of AQAP, officials had to determine that al-Awlaki had participated not merely in propaganda directed at the United States, but in actual operational planning of attacks. Officials had to answer this latter question in the affirmative before they approved identification of al-Awlaki as a lawful target.[15]

The most difficult situation is what I call situational targeting. In this context, typically a NIAC featuring combatants for a terrorist group who do not wear uniforms or distinguishing insignia, an opposing force will target unknown individuals whose behavior indicates that those individuals are combatants or DPH. This approach typically involves inferences from behavior such as association with other, known individuals who are combatants or DPH, or travel in armed groups near the vicinity of known outposts of terrorist organizations. Sometimes called 'signature' strikes, this tactic has attracted criticism in some quarters.[16] In

---

[15] Daniel Klaidman, *Kill or Capture: The War on Terror and the Soul of the Obama Presidency* (Houghton Mifflin Harcourt 2012); Charlie Savage, *Power Wars: Inside Obama's Post-9/11 Presidency* (Little, Brown and Company 2015) 249–52; see also Robert Chesney, 'Who May Be Killed? Anwar Al-Awlaki as a Case Study in the International Legal Regulation of Lethal Force' (2010) 13 Yearbook of International Humanitarian Law 3–60 (discussing legal standard); Gregory S McNeal, 'Targeted Killing and Accountability' (2014) 102 Georgetown Law Journal 681, 685 (concluding based on document review and interviews that US process involved elaborate analysis engaged in by dozens or even 'hundreds' of officials); but see Mary Ellen O'Connell, 'Unlawful Killing with Combat Drones: A Case Study of Pakistan, 2004–2009' in Simon Bronitt, Miriam Gani and Saskia Hufnagel (eds), *Shooting to Kill: Socio-Legal Perspectives on the Use of Lethal Force in Context* (Hart Publishing 2012) 263 (criticizing use of remotely piloted drones to target individuals and groups outside of states in which clear armed conflict exists).

[16] Craig Martin, 'A Means-Methods Paradox and the Legality of Drone Strikes in Armed Conflict' (2015) 19 International Journal of Human Rights 142.

principle, targeting unknown individuals who do not wear uniforms but are in fact combatants or DPH is no different from categorical targeting of massed, uniform groups in an IAC; in the latter situation, as well, an opposing force typically does not know the names of particular members of the uniformed group. IHL does not require such knowledge. Indeed, such a requirement would make the conduct of military operations impossibly burdensome for both sides. That result would be inconsistent with IHL, which aims to balance military necessity and humanity, not outlaw armed conflict. However, situational targeting does require great care in the formulation and application of factors that constitute evidence of targetability.[17]

2. WEAPONS REVIEW AND AWS

Before a state can deploy a weapon for use in targeting adversaries, it must conduct a weapons review. In this process, the state makes a threshold determination about the weapon's consistency with IHL. That threshold test addresses the inherent nature of the weapon, not its effects in particular situations.[18] At some point in the future, an AWS may be able to meet this standard.

To be found lawful by a weapons review, a weapon cannot be indiscriminate by its very 'nature'.[19] In its normal, routine use, a weapon must be capable of discriminating between civilians and combatants, as the principle of distinction requires.[20] This is a low bar; few weapons fail this test.[21] Second, a weapon, consistent with IHL's balance between military necessity and humanity, cannot be of a 'nature' to engender 'unnecessary suffering or superfluous injury'.[22] This provision, which applies only to combatants (since other provisions like proportionality protect civilians), applies to weapons that produce needlessly cruel consequences. Third, a weapon can be illegal if its deleterious impacts

[17] Kevin Jon Heller, 'One Hell of a Killing Machine: Signature Strikes and International Law' (2013) 11(1) Journal of International Criminal Justice 89.

[18] Anderson, Reisner and Waxman (n 9) 399–400.

[19] William H Boothby, *Weapons and the Law of Armed Conflict* (Oxford University Press 2009) 78.

[20] Gary D Brown and Andrew O Metcalf, 'Easier Said Than Done: Legal Reviews of Cyber Weapons' (2014) 7 Journal of National Security Law and Policy 115.

[21] Anderson, Reisner and Waxman (n 9) 399.

[22] API, Article 35(2).

cannot be 'controlled'.[23] This principle would generally bar biological weapons, since the path of a virus cannot be guided once the virus is released[24]—the course of infection is determined by the virus's biochemistry and environmental factors such as wind, hygiene, and personal contact, not by tactical decisions made by the party using the weapon.

As noted above, a weapon does not fail any or all of these conditions merely because it might be *used* in an indiscriminate manner.[25] Any weapon, from a bayonet or machine-gun to a cruise missile, can be *used* to target civilians in a particular situation. However, that possibility does not render the weapon indiscriminate by its very 'nature'. If the mere possibility of misuse led to failing this condition, *no* weapon would be legal. That prospect might appeal to pacifists, but would not be consistent with the balance of military necessity and humanity at the heart of IHL.[26]

Since the weapons review standard is low, a properly validated AWS that operates in land warfare and targets humans may well be able to pass a weapons review at some future date. However, many intermediate steps will be necessary. In the short term, states will likely focus on using autonomous elements in systems that remain under direct and ongoing human control. An apt analogy here might be anti-collision, pedestrian detection, or self-parking programs in human-driven motor vehicles. Moreover, as a later section of this chapter notes, a passing grade at the weapons review stage does not mean that a state can deploy an AWS as a 'set and forget' weapon. A reasonable expectation that the AWS will not be indiscriminate will be premised on continuing, periodic human engagement.

---

[23] Schmitt and Thurnher (n 1) 250; see also API, Article 51(4)(c) (noting that 'method or means of combat' is illegal if its 'effects ... cannot be limited as required by this Protocol').

[24] Anderson, Reisner and Waxman (n 9) 400.

[25] Ibid 399–400.

[26] Cf Sean Watts, 'Regulation-Tolerant Weapons, Regulation-Resistant Weapons and the Law of War' (2015) 91 International Law Studies 540, 556–8 (arguing that Hague Convention's Martens Clause, which notes role in warfare of 'the laws of humanity and the requirements of public conscience', applies only to situations not otherwise covered by established norms of customary international law or treaties—norms in the latter category, including IHL principles relevant to an AWS, already build in the requisite balance of military necessity and humanity).

3. AWS ACCOUNTABILITY AND THE DOCTRINE OF COMMAND RESPONSIBILITY

While weapons review does not assess any and all possible uses of a weapon, certain uses may violate IHL. Persons responsible for those uses must be accountable. Autonomous weapons trigger questions in this regard, since they can make decisions without ex ante human authorization. That absence of human authorization need not create an 'accountability gap'. In dealing with AWS, the appropriate mechanism for accountability is the familiar doctrine of command responsibility.

Under command responsibility, a person in command is accountable for crimes committed by subordinates if the leader knew or should have known that subordinates were engaged in illegal activity and failed to take reasonable steps to prevent such acts.[27] The premise of command responsibility is two-fold. First, commanders, by virtue of their role, have supervisory responsibilities. Commanders accrue benefits from this role, as do states or other entities that rely on the chain of command. Commanders receive a pool of people and other instrumentalities that do the commander's bidding, backed up by a system of military discipline that attaches severe penalties to disobedience. States benefit from that efficiency in the projection of military force.

Since commanders enjoy these benefits, they should also shoulder the burdens of command—the responsibility for taking reasonable steps to ensure that subordinates comply with IHL.[28] On this view, command

[27] See *In re Yamashita* 327 U.S. 1 (1946); *Prosecutor v Aleksovski* Judgment Case No. IT-95-14/1-T, paras. 66–81 (International Criminal Tribunal for the Former Yugoslavia Trial Chamber I, 25 June 1999); Allison Marston Danner and Jenny S Martinez, 'Guilty Associations: Joint Criminal Enterprise, Command Responsibility, and the Development of International Criminal Law' (2005) 93 California Law Review 75, 120–30; Victor Hansen, 'What's Good for the Goose is Good for the Gander: Lessons from Abu Ghraib: Time for the United States to Adopt a Standard of Command Responsibility Toward its Own' (2006–07) 42 Gonzaga Law Review 33; cf Jens David Ohlin, 'The Combatant's Stance: Autonomous Weapons on the Battlefield' (2016) 92 International Law Studies 1, 23 (discussing command responsibility as recklessness, on theory that participation in armed conflict coupled with failure to exercise adequate control over instrumentalities using lethal weapons entails engaging in dangerous acts with knowledge of their potential consequences); Yuval Shany and Keren R Michaeli, 'The Case Against Ariel Sharon: Revisiting the Doctrine of Command Responsibility' (2002) 34 New York University Journal of International Law and Politics 797, 816–30 (discussing origins and limits of command responsibility doctrine).

[28] Shany and Michaeli (n 27) 833.

responsibility, while a relatively recent doctrine, is a logical outgrowth of age-old concerns about warfare. Just as the principles of distinction, proportionality and precautions in attack balance military necessity and humanity, command responsibility places this onus on the individual best equipped to bear the load: the commander, who has an opportunity to shape the strategy and tactics that subordinates execute.[29]

In addition, command responsibility is prophylactic in character. Without command responsibility, a commander could subtly or indirectly encourage subordinates to commit war crimes, but avoid responsibility because no standing orders required the commission of such illegal acts. Placing an affirmative legal duty on the commander structures incentives for the commander to train troops properly, monitor troops' conduct, and take prompt action to deter, punish and remedy IHL violations.[30] With

---

[29] For a discussion of the urgency of ensuring accountability for AWS violations of IHL, see Markus Wagner, 'The Dehumanization of International Humanitarian Law: Legal, Ethical, and Political Implications of Autonomous Weapons Systems' (2014) 47 Vanderbilt Journal of Transnational Law 1371, 1399–1401. A conceptual difficulty arises here, because command responsibility is a doctrine of vicarious responsibility that imposes liability based on the actions of human subordinates. Jack Beard, 'Autonomous Weapons and Human Responsibilities' (2014) 45 Georgetown International Law Journal 617, 655–60; Tim McFarland and Tim McCormack, 'Mind the Gap: Can Developers of Autonomous Weapons Systems be Liable for War Crimes?' (2014) 90 International Law Studies 361, 365–6. Those human subordinates would also be accountable directly. Shany and Michaeli (n 27) 831–2. The accountability gap for AWS arises precisely because an AWS would not be liable directly. This disparity between an AWS and a human subordinate demonstrates that the doctrine of command responsibility is not a perfect patch for the accountability issues that AWS critics pinpoint; Crootof (n 4). However, like other legal doctrines, command responsibility can evolve in the face of changed circumstances. That adaptation is particularly appropriate because, in a functional sense, an AWS performs like other combatants on the battlefield, and adversaries of the force deploying an AWS will seek to destroy or disable the AWS, as they would treat any other enemy instrumentality. Ohlin (n 27) 20; but see Sassòli (n 1) 324 (arguing that imposing direct, not vicarious, responsibility, on the commander is most appropriate approach, since AWS is closer to weapon whose use gives rise to direct responsibility for personnel employing weapon than it is to personnel who control the weapon).

[30] Shany and Michaeli (n 27) 834 (noting that commander, by virtue of her superior position and consequent access to information, is well situated to ensure that subordinates obey the law); but see Geoffrey S Corn, 'Autonomous Weapons Systems: Managing the Inevitability of "Taking the Man Out of the Loop"' (2014) 21, accessed 5 May 2017 at http://ssrn.com/abstract=2450640 (arguing that official in charge of initial development and procurement of AWS should

modest adaptations, the command responsibility doctrine is an appropriate vehicle for addressing the accountability problems posed by AWS.

4. AUTONOMOUS WEAPONS ANALYZED

The operation of an AWS typically entails decisions made by a computer network in the course of the identification and/or implementation phases of targeting. An armed force that deploys an AWS must have a reasonable belief that the AWS will comply with IHL in all of these decisions. An AWS is not a pre-programmed piece of software like Microsoft Office. Through what data scientists call 'machine learning', an AWS receives inputs and then develops approaches to analyzing and acting on fresh information. The following paragraphs describe AWS outputs based on 'machine learning'.

Using machine learning, a computer-guided agent makes decisions and takes actions without direct ex ante human intervention. Machine learning affects performance in four key domains for AWS: (1) probability assessment, which will aid in the selection of suspected targets; (2) pattern and visual recognition, which includes identifying faces and distinguishing combatants from civilians; (3) movement, which entails an awareness of the dynamic state caused by an AWS's movement and actions taken by the AWS in response to encounters in the field; and (4) interpretability, which would provide a commander supervising an AWS with a substantive, verbal explanation of factors identifying a target.

Before exploring these attributes, it is useful to briefly note the steps common to any form of machine learning: training, testing and validation.[31] These steps will also be essential in a weapons review of an AWS. In the first step, a programmer inputs data, such as data on the appearance of combatants from country X or non-state actor Y. Inputting data trains the machine on criteria that determine good performance on a

---

bear legal responsibility); McFarland and McCormack (n 29) 375–6 (exploring whether system developer, who performed initial programming of AWS should be responsible for subsequent violations of AWS); Tim McFarland, 'Factors Shaping the Legal Implications of Increasingly Autonomous Military Systems' (2015) 97(900) International Review of the Red Cross 1313, 1335 (suggesting that accountability would be appropriate for developers 'or those they answer to … [who] exercise control over the behaviour of the system').

[31] Stephanie Tuffery, *Data Mining and Statistics for Decision Making* (John Wiley and Sons 2011) 304.

task, such as ways to distinguish combatants from civilians. The training process enables the machine to not only recognize criteria that data scientists expect will be relevant, but to also discern other criteria that are analogous but may not have been spotted by the machine's human minders. To ascertain whether the machine's training is sound, the programmer will test the machine on a separate data set.[32] If the machine does well on the test, the programmer will then validate the results by varying the test set to ensure that no fortuitous overlap between the previous test set and the training set artificially inflated machine performance.[33]

Validation will hinge on an acceptable percentage of errors compared with accurate results.[34] Two types of errors are relevant. False positives occur when a machine mistakenly identifies a subject as possessing an attribute, when in fact the attribute is absent. A false positive in armed conflict would be a civilian mistakenly classified as a combatant. False negatives occur when the machine mistakenly judges the subject as lacking an attribute we are seeking, when in fact that attribute *is* present. In armed conflict, a false negative would be a combatant or civilian DPH mistakenly classified as a civilian. A weapons review of an AWS would have to weigh the incidence of false positives against the legal standard that a weapon cannot be indiscriminate by nature. If the machine did sufficiently well at measures employed in the validation stage, the machine could lawfully be deployed in the field.

A. Drawing Inferences Based on Probabilities

High on the list of machine learning attributes is probabilistic inference. In the fog of war and in most other areas of human endeavor, absolute certainty is often unattainable. Humans settle for a substitute: judgments about conditional probabilities, which weigh the relative probability of competing hypotheses based on evidence. Machines today are ubiquitous

---

[32] Lina Zhou and others, 'A Comparison of Classification Methods for Predicting Deception in Computer-Mediated Communication' (2004) 20(4) Journal of Management Information Systems 139, 152.

[33] Flach (n 5) 162 (discussing problem of 'overfitting' outputs, in which machine in essence memorizes training set, but develops no ability to generalize beyond training data); Stuart J Russell and Peter Norvig, *Artificial Intelligence: A Modern Approach* (3rd edn, Prentice Hall 2010).

[34] Flach (n 5) 56; Ian H Witten, Eibe Frank and Mark A Hall, *Data Mining: Practical Machine Learning Tools and Techniques* (3rd edn, Elsevier 2011) 174–7.

judges of conditional probabilities. Moreover, machines often outperform humans in such judgments.

Consider one common machine-learning tool for assessing conditional probability: Bayesian networks. Named after the great French mathematician, Bayesian networks take the mathematician's famous theorem as their premise. Under the theorem, new evidence may demonstrate that a particular hypothesis is more or less likely.[35] Some evidence is less probative because another hypothesis best explains that evidence. Bayesian networks allow us to assess how large amounts of data confirm or disprove a particular hypothesis.

As an example, consider the spam filter. Anyone who relies on email understands the importance of managing the deluge of spam messages that greet us each time we log on to our email accounts. Bayesian networks filter spam messages by making conditional probability assessments. Suppose that an email contains a term or phrase such as, 'Free!', 'only $9.95' (or any other monetary amount preceded by the word, 'only'), or 'be over 21'.[36] The first two terms make it more likely that the email's sender is trying to sell the recipient something, while the third term suggests that the merchandise in question is pornography, illegally distributed pharmaceuticals, or another commodity whose marketing to children would create additional legal risks that the spammer does not wish to bear. Suppose further that we add non-textual elements such as the domain type of the sender. Individuals or entities with an '.edu' domain are selflessly devoted to scholarship and teaching (or so we inform our deans). Academics rarely send spam, so a spam filter can typically permit these messages to pass through to the intended recipient. While it is possible that a message from a '.com' domain containing a term such as 'Free!' or 'only $9.95' is from a friend, relative, colleague or other legitimate source, the conditional probability of this hypothesis being true is far lower than the conditional probability that the message is spam.[37]

---

[35] Pedro Domingos, *The Master Algorithm* (Basic Books 2015) 151–2.

[36] Meheran Sahami, Susan Dumais, David Heckerman, and Eric Horvitz, 'A Bayesian Approach to Filtering Junk E-Mail' 3, accessed 5 May 2017 at ftp://ftp.research.microsoft.com/pub/ejh/junkfilter.pdf; cf Domingos (n 35) 151–2 (explaining conditional probabilities in spam filtering); Russell and Norvig (n 33) 865–6 (same).

[37] In another familiar machine tool based on conditional probabilities, a voice recognition system such as SIRI uses a variant of a Markov chain, named after the Russian mathematician who judged the probability that one letter would follow the one preceding it in Alexander Pushkin's epic poem, *Eugene Onegin*. A

Conditional probabilities can also aid in the identification of terrorists. Although critics of the use of machine learning in counterterrorism sometimes argue that there is far too little data on terrorism to allow machine learning to work, the use of conditional probabilities can ferret out useful patterns.[38] Consider the following data points:[39] suppose US officials know that a French national who until two weeks ago was physically located in Paris has participated in jihadist chat-rooms, expressing straightforward or overt support for violence against states that are fighting ISIS. That individual also recently reported a lost passport (in ISIS trade-craft, this would be consistent with an attempt to conceal a recent visit to Syria or Iraq). In addition, the same individual has attempted to use an encryption technique frequently recommended by ISIS's on-line 'help desk'. Moreover, a facial recognition program that is 95 per cent accurate used public video surveillance to ascertain that this individual had recently patronized a store specializing in pre-paid cell phones. Yesterday, a raid on an ISIS safe house in Syria disclosed that an individual with the same name as our 'person of interest' had stayed in the guest house within the last week.

Viewed in isolation, none of these facts would provide proof that an individual was a member of ISIS. Large numbers of people unaffiliated with a terrorist group surely engage in a number of these activities, including participation in chat-rooms, use of encryption, and purchase of a pre-paid cell phone. However, the convergence of *all* of the disparate

---

speech recognition system like Siri uses hidden states that stand for written words, linked together with conditional probabilities that certain sounds stand for combinations of letters and that subsequent sounds observed will form phrases and sentences used by an aggregate of Siri's many users or by the individual user. For example, if Siri hears sounds that conform to the combination of letters comprising the word, 'nice', it will then use a Markov chain to infer that the next word is likely to be a term such as 'work', 'job', or 'suit'. Domingos (n 35) 161–2; Russell and Norvig (n 33) 912–13.

[38] Domingos (n 35) 232–3; cf Nat'l Research Council of the Nat'l Academies, *Protecting Individual Privacy in the Struggle Against Terrorists: A Framework for Program Assessment* (National Academies Press 2008) 4, 77 (discussing risks and possible benefits of counterterrorism data mining); but see Bruce Schneier, *Data and Goliath: The Hidden Battles to Collect Your Data and Control Your World* (WW Norton and Company 2015) 136–9 (arguing that low incidence of terrorist attacks makes it impossible to gather sufficient data to enable machine predictions).

[39] This example is from Peter Margulies, 'Surveillance by Algorithm: The NSA, Computerized Intelligence Collection, and Human Rights' (2016) 68 Florida Law Review 1045, accessed 5 May 2017 at http://ssrn.com/abstract=2657619.

factors listed above markedly lowers the probability that the activities are innocent. The DC Circuit has employed probabilistic reasoning to determine the legality of detentions of suspected terrorists;[40] targeting decisions could rely on similar logic.

Machine learning can address the objection that even the confluence of probabilities above leaves too much room for false positives. Bayesian networks, like any form of machine learning, can be weighted to accommodate the disparate costs of certain types of errors. Which kind of error is more serious depends on context.[41] Spam filters strive to avoid false positives: mislabeling an email from a friend, relative or colleague might block an important message that the recipient would want to see immediately.[42] Calculating the costs of errors in armed conflicts is more complex. Both false negatives and false positives have dire consequences. A false negative such as the failure to identify a terrorist plotting a catastrophic attack can result in massive harm. Yet, false positives are also very serious. The negligent or reckless classification of an innocent civilian as a combatant or DPH could violate the IHL principles of distinction, proportionality and precautions in attack.

Fortunately, the relative seriousness of the error at issue can play a role in machine learning. For example, the model can be trained with a data set that multiplies false positives by a factor of 10.[43] Since models are rated on the accuracy of their outputs, weighting the training set toward false positives will act as a penalty on committing such errors. Because of this penalty, the model will tend to make decisions that avoid false positives. This approach could be used in training AWS to minimize the mistaken identification of civilians as combatants or DPH.[44]

---

[40] *Hussain v Obama* 718 F3d 964, 68 (DC Cir 2013); cf Matthew C Waxman, 'Detention as Targeting: Standards of Certainty and Detention of Suspected Terrorists' (2008) 108 Columbia Law Review 1365 (drawing analogy between detention and targeting of suspected terrorists).

[41] Witten, Frank and Hall (n 34) 163.

[42] Russell and Norvig (n 33) 710.

[43] Witten, Frank and Hall (n 34) 167.

[44] Similarly, the US military now uses rules of engagement (ROE), advice from military lawyers, and other measures to structure the parameters of an attack to minimize civilian casualties. McNeal (n 15) 685; see also Michael N Schmitt and John J Merriam, 'The Tyranny of Context: Israeli Targeting Practices in Legal Perspective' (2015) 37 University of Pennsylvania Journal of International Law 53, 73–5 (discussing role of Israeli military lawyers). Data scientists could include such constraints in the software for AWS; cf Russell and Norvig (n 33) 202 (discussing constraints); Domingos (n 35) 193.

Having either found facts based on probabilistic assessments or received facts at the input stage, machines can make normative decisions in a manner that complies with legal, ethical and moral rules.[45] In dilemmas at the intersection of law and medicine, data scientists have proved that machines can learn to treat each individual patient as having distinct rights, as a theory of deontological ethics would hold.[46] In such a theory, rights trump crude utilitarianism. For example, a machine can reject sacrificing one hospital patient to provide organs to several others.[47] Similarly, a machine can decide to administer a needed pain-killer to a patient even though the hospital is short of resources. To engineer such ethical, rule-based logics, data scientists quantify the alternatives, attaching a higher value for alternatives that are favored under a deontological approach, and a lower value for options favored under a strictly utilitarian calculus. If we can formulate and implement such logical rules for the bioethics context, we should in theory be able to do the same for IHL. The IHL context may be more fluid and dynamic, but that is a problem of application, not of theory.

Moreover, although the above approach relied on specific, 'hand-coded' human instructions, researchers have also been successful in using inductive methods that allow machines to learn results for themselves.[48] In one study, researchers used inductive logic programming (ILP), which relies on numerical values to help machines learn about competing ethical principles.[49] The machine then learns through analogy about resolutions for other dilemmas.

For example, consider the problem that arises in medical ethics when an adult patient with a life-threatening condition refuses a clinically sound treatment. The patient then opts for a course of treatment that the vast majority of medical professionals would agree is unavailing. Data scientists asked a machine whether a physician should seek to change the patient's mind. An answer must include an inquiry into the cost to patient autonomy imposed by the physician's persistence versus the cost to the

---

[45] Ronald C Arkin, 'Governing Lethal Behavior: Embedding Ethics in a Hybrid Deliberative/Reactive Robot Architecture' (2011) 43–6, accessed 5 May 2017 at www.cc.gatech.edu/ai/robot-lab/online-publications/formalizationv35.pdf.

[46] Selmer Bringsjord, Konstantine Arkoudas and Paul Bello, 'Toward a General Logicist Methodology for Engineering Ethically Correct Robots' (July–August 2006) 21(4) IEEE Intelligent Systems 38.

[47] Ibid.

[48] Michael Anderson and Susan Leigh Anderson, 'Machine Ethics: Creating an Ethical Intelligent Agent' (2007) 28(4) AI Magazine 15, 23.

[49] Ibid 21–2.

patient's health imposed by declining to engage the patient in further discussion. Moreover, an ethical physician would ask whether the patient exhibited any indicia of questionable decisional capacity, or whether other individuals such as family members were prodding the patient to take a stance that conflicted with the best available clinical recommendation. The researcher's machine learning model was able to sort out competing values and formulate ethically appropriate answers to the question presented.[50] This suggests that machines could actually contribute to more precise resolution of notoriously diffuse questions such as the content of the proportionality principle, if translating such principles into guidance for machines yielded a more objective, specific comparison of the respective value of civilian lives and military advantage.[51]

A machine's judgments can also outstrip human efforts. Humans suffer from a dizzying array of cognitive flaws that affect the layperson and expert alike. Probability assessments are a particular problem.[52] For example, in choosing which of two scenarios is accurate, humans rely on their own or another's experience and disregard base rates that establish each scenario's overall frequency. As a result, humans are likely to exaggerate the frequency of scenarios that are actually quite rare. Suppose that a human subject was told by psychological researchers that an expert had an 80 per cent accuracy rate in finding terrorists. That expert, the researchers informed the subject, had identified one individual in a randomized pool of 100 individuals as an ISIS fighter. Suppose further that researchers told their subject that approximately one person in 50,000 is a terrorist. Most subjects would find a significant probability that the expert was correct. However, because the overall percentage of terrorists is tiny, a random pool of 100 individuals has a vanishingly

---

[50] Ibid 23. Other data scientists have suggested that the ILP model, with its reliance on single-integer values, is too limiting. These experts have proposed richer, case-based models that incorporate more detail into their inputs and hypotheticals. Bruce M McLaren, 'Computational Models of Ethical Reasoning: Challenges, Initial Steps, and Future Directions' in Michael Anderson and Susan Leigh Anderson (eds), *Machine Ethics* (Cambridge University Press 2011). In a cautionary note on the possibility of machines making autonomous normative choices, data scientists who have developed case-based models view their work as *assisting* humans, who will continue to make the ultimate decisions. Ibid 298.

[51] Sassòli (n 1) 334.

[52] Dale Griffin and Amos Tversky, 'The Weighing of Evidence and the Determinants of Confidence' (2002) in Thomas Gilovich, Dale Griffin and Daniel Kahneman (eds), *Heuristics and Biases: The Psychology of Intuitive Judgment* (Cambridge University Press 2002) 230, 231 (noting that many human predictions clash with rules of statistics and probability).

small chance of containing even a single ISIS fighter, whatever the expert's acumen.[53] Humans should factor such information into their probability estimates, but they often ignore such information altogether, or fail to give it sufficient weight.

As the above example demonstrates, humans' neglect of base rates skews probabilities. Cognitive flaws are not confined to laypersons; experts are subject to the same infirmities of judgment.[54] A machine, in contrast, could be far more accurate.[55] In an armed conflict, that accuracy would protect civilians and promote compliance with IHL.

B. Pattern Recognition

Machines are very good at recognizing patterns, links, and associations. Even critics of using machine learning techniques in counterterrorism acknowledge that such techniques can be useful in mapping affinities that can ripen into terrorist affiliations.[56] Those technologies, like the probabilistic methods discussed above, can be immensely helpful in identifying

---

[53] For an illustration of the human base-rate fallacy, see Ernest Fehr and Jean-Robert Tyran, 'Individual Irrationality and Aggregate Outcomes' (2005) 19 Journal of Economic Perspectives 43, 57–8.

[54] Mark R Waser, 'Implementation Fundamentals for Ethical Medical Agents', in Simon P van Rysewyk and Matthijs Pontier (eds), *Machine Medical Ethics* (Springer 2015) 55–6. Unstructured expert predictions of dangerousness often have higher error rates than more systematic methods. Cf Christopher Slobogin, 'Risk Assessment and Risk Management in Juvenile Justice' (2013) 27 Criminal Justice 10, 12–13 (suggesting that actuarial techniques for predicting recidivism among offenders that rely on a common list of factors are typically more accurate than unstructured clinical assessments).

[55] Machines will still encounter formidable problems in the fluid context of armed conflict. For example, a machine will have to integrate many sources of data, some of which may contain errors. State officials and data scientists will have to work together to ensure that databases are accurate and include machine-readable markings of their currency (a bit like the 'use-by' dates on perishable foods or the 'last-visited' term in scholarly citations to sources on the Internet). See nn 91–93 and accompanying text; see also M L Cummings, 'The Human Role in Autonomous Weapons Design and Deployment' (2014), accessed 5 May 2017 at https://www.law.upenn.edu/live/files/3884-cummings-the-human-role-in-autonomous-weapons (cautioning that machine analysis based on other databases will compound the errors in that data).

[56] Schneier (n 38) 138; Rachel Levinson-Waldman, 'IBM's Terrorist-Hunting Software Raises Troubling Questions' (2016) Just Security, accessed 5 May 2017 at www.justsecurity.org/29131/ibms-terrorist-hunting-software-raises-troubling-questions/.

previously unknown followers of ISIS or other groups and implementing a targeting plan.

For example, artificial neural networks are useful for discerning patterns between individuals, groups, and objects or behaviors.[57] Neural networks mimic the functioning of the human brain.[58] The human brain works through neurons that are connected through fibers. As in the human brain, neurons in an artificial network are interconnected but distinct, enabling the neurons to break down complex data into more manageable parts.[59]

As with any form of machine learning, neural networks are first trained with examples that the network learns with the aid of a learning algorithm. Neural networks then produce outputs that find patterns between a new stimulus and other data inputs. The principal architectural innovation of neural nets is the presence of one or more hidden layers between inputs and outputs.[60] Multiple hidden layers divide the sorting of data into steps, facilitating greater precision. Each layer sorts data from the previous step.[61]

---

[57] This discussion is based on material in Margulies, 'Surveillance By Algorithm' (n 39).

[58] Russell and Norvig (n 33) 728; Michael Aikenhead, 'The Uses and Abuses of Neural Networks in Law' (1996) 12 Santa Clara Computer and High Tech Law Journal 31.

[59] Witten, Frank and Hall (n 34) 232.

[60] Zhou (n 32) 149–50.

[61] Support vector machines (SVMs) are another method for finding intersections between variables. See Domingos (n 35) 190–96; Witten, Frank and Hall (n 34) 191–2. SVMs efficiently analyze data with many variables. An SVM expresses intersecting variables as 'dimensions' that an SVM can plot in space, using hyperplanes that cleanly separate disparate groups. Linear modes of data analysis such as graphs can only plot two variables at a time, such as the intersection of mass and acceleration to compute force or age and education to help predict an individual's likelihood of voting. In contrast, the hyperplanes derived by SVMs can separate groups along 20 or more variables. Using hidden layers, SVMs discern relationships between variables that humans would miss.

To illustrate how an SVM could aid in counterterrorism, consider the identification of ISIS recruits through the relationship of a large number of variables, each of which alone might be useless. For example, ISIS recruits might cite particular commentaries or interpretations of sacred texts as authorizing violence. In addition, ISIS recruits might refer to such religious commentaries or to operational and logistical details using code. A directed search would find only codes already known to government officials. In contrast, an SVM could also find uses of language that were analogous to known codes in syntax, frequency of word choice, and similar factors. The SVM might detect new codes

Consider an artificial neural network approach to facial recognition that would assist in the selection of terrorist targets and the implementation of that selection decision. After being trained on a large number of images, a neural network would analyze an even larger set of inputted photographs.[62] A neural network subdivides that task, first searching for facial features, such as eyes and noses. To detect eyes, the network will use at least two hidden layers. The first layer will search for short edges. Eyes have short edges, compared with larger inanimate or inorganic objects, such as buildings or trucks. Another layer searches through the images of short edges and spots the curved short edges that characterize an eye. A third layer could identify facial types, and a fourth could identify specific faces.

To aid in the target selection phase, a neural net tasked with facial recognition could inspect public video surveillance feeds of individuals associating with known combatants or DPH civilians. It could then look for the same face in images of state identification documents, such as passports. The neural net could also look for the face in video from ISIS or Al Qaeda training camps. In the implementation phase, a neural net receiving a video feed from a drone sent to the proposed target's assumed location would confirm that the face of the individual at the location matched the face of the nominated target. A neural net could perform these tasks with accuracy and efficiency that was equal to or greater than a human being.[63]

---

by grouping the incidence of certain word choices with other individual behaviors, such as frequent visiting of chat rooms advocating violent extremism, use of specific kinds of encryption, or patronage of stores selling burner phones. In this fashion, ISIS recruits that would escape detection by humans, directed searches, or other autonomous searches would 'pop' in SVM outputs.

[62] See 'Human Face Recognition Found in Neural Networks Based on Monkey Brains: A neural network that simulates the way monkeys recognize faces produces many of the idiosyncratic behaviors found humans, say computer scientists' (2015) MIT Technology Review, accessed 5 May 2017 at www.technologyreview.com/view/535176/human-face-recognition-found-in-neural-network-based-on-monkey-brains/; see also Laura K Donohue, 'Technological Leap, Statutory Gap, and Constitutional Abyss: Remote Biometric Identification Comes of Age' (2012) 97 Minnesota Law Review 407, 543–8 (discussing facial recognition and expressing concern about abuse of this technology).

[63] The neural net would have to be cross-validated with an array of modified test sets before it was tasked with this mission, and would also be subject to regular, frequent human monitoring and assessment.

Machines have also improved greatly in related areas such as perceiving images in a landscape. In the implementation phase of targeting, distinguishing moving human figures from objects and animals is crucial. Assessments of the size of those figures, their distance from the AWS, and the speed of their movements are of equal importance. Image perception and reconstruction techniques can assist in these tasks, which are vital for both perceiving combatants and DPH civilians and distinguishing those lawful targets from protected persons.

Consider a task such as distinguishing people in the immediate vicinity of a surface AWS. An AWS can use machine learning to distinguish people near a moving AWS, whom the human driver of an ordinary civilian vehicle would classify as pedestrians. Using video cameras to provide inputs, an autonomous vehicle would create a graphic[64] that depicts edges within a scene. Using machine learning, the graphic highlights edges that correspond to outlines of a person, and mutes all others.[65] The AWS can then avoid collision with innocent pedestrians.

Deep neural nets, which sort out a large number of variables through many hidden layers, have dramatically enhanced machine capabilities for accurately mapping the exact locations and dimensions of structures and roads in landscapes.[66] Ordinary maps, including detailed renderings produced for expert use, often omit objects or mistake their location by a small unit of distance. Using aerial photographs, a machine can model a landscape with pinpoint accuracy. Pinpointing landmarks like structures and roads equips an AWS with an updated map to locate moving figures in a landscape. Using an updated map overlaid with GPS capability, an aerial or surface AWS can pinpoint its own location within fractions of an inch.[67] An AWS can then discern the distance and size of figures in a landscape using its updated map and an image attribute called optical flow. Optical flow cues visual perspective: from the standpoint of a person or AWS moving in space, objects that are closer to the horizon and appear to move more slowly than other objects are more distant.

---

[64] The graphic is called a histogram of oriented gradients (HOG). Russell and Norvig (n 33) 946.

[65] Machine learning experts will use a support vector machine for optimal separation of edges corresponding to pedestrians from other edges. Russell and Norvig (n 33) 946.

[66] Volodymyr Mnih and Geoffrey E Hinton, 'Learning to Label Aerial Images from Noisy Data' (2012) Proceedings of the 29th International Conference on Machine Learning (ICML-12), accessed 5 May 2017 at www.cs.toronto.edu/~vmnih/docs/noisy_maps.pdf.

[67] Russell and Norvig (n 33) 965.

This combination of capabilities aids the AWS's IHL compliance. If distant objects are near landmarks such as buildings or roads depicted in detail on the AWS's updated map, an AWS with GPS capability can also calculate those distant objects' size. The size of figures will provide important clues about the figures' status: figures that are small in size are more likely to be children who cannot be targeted, while the observation of a group of larger figures is one building block in identification of lawful targets.[68]

C. Movement

Machine learning has also been instrumental in enhancing the movement abilities of autonomous agents that perform tasks such as driving. Many agents engaged in complex tasks, such as operating land or aerial vehicles, use reinforcement learning. Reinforcement learning, which uses the probability-based calculations pioneered by Bayes and Markov, hinges on integrating a robot's knowledge about its current physical context[69] with its assessment of the likely effects of subsequent actions.[70] To complete these calculations efficiently, machine learning will use layers of the kind we saw in our earlier discussion of artificial neural networks.

As an example, consider a driverless car learning to obey traffic laws. Machine learning will provide the agent with awareness that its state is dynamic, changing as the vehicle moves. The agent's sensors will be color-sensitive, so the agent will recognize when a green light turns red. Ensuring that the machine learns what to do entails a bifurcated approach to probabilities.

The first track allows the agent to learn from its own experience in a real or simulated environment. The agent can use trial and error, much as humans do. In other words, it can first stop at a red light, and observe the results. Next, the agent can ignore the red light, move through the intersection, and observe these results, as well. Of course, these scenarios reveal the risk of exclusive reliance on learning from experience. The agent may disregard several red lights without adverse effects, depending

---

[68] Estimates of figure size are most useful in identifying true negatives—persons who are hors de combat. Figure size should never be the sole basis for labeling a figure as positive for targeting purposes, since many males are also civilians whom IHL protects.

[69] Russell and Norvig (n 33) 658.

[70] Domingos (n 35) 217–22.

on traffic at each intersection.[71] If the agent's experience were its only input, it would assess the probability of safely disregarding future red lights as being higher than it is in practice (to say nothing of the legal ramifications if a police car—driverless or not—is in the vicinity).

The second probability track comes to the rescue, anchoring the agent in an objective Bayesian estimate of the *overall* risk of disregarding a red light. As we discussed in addressing utility functions, we can also weight these risks according to their gravity. Humans would typically (we hope) attach high importance to the risk of a fatal car accident or even getting a traffic summons, and dismiss the trivial gain in time realized by running a red light. Programmers could group the probability estimate with a utility function that would express this disparity in numerical terms, or could weight the agent's training set heavily toward the worst case scenarios of a traffic accident or summons.

Because it is complex to keep track of the agent's own state and its environment while assessing the probable effects of various acts, the agent will need several layers to do all the calculations required. Some agents will use three layers: a reactive layer to understand the environment, an executive layer to take actions in response to that environment, and a deliberative layer to analyze those actions and adjust probabilities for the future.[72] Moving through the three layers sequentially can be time-consuming. For this reason, data scientists have developed a so-called 'pipeline' approach that puts all of the layers on parallel tracks, working simultaneously.[73] This approach preserves the accuracy of the three-layer method, while substantially enhancing its efficiency.

Using such approaches, data scientists have developed agents that learn many complex tasks far more quickly than humans. For example, autonomous agents can absorb inputs from humans performing sophisticated helicopter maneuvers,[74] including flips, and maneuvers performed by fixed-wing aircraft, such as nose-in circles.[75] Autonomous agents can learn in less than an hour maneuvers that humans take months to master. If machines can learn tasks this complex, it seems foolhardy to insist that they are inherently incapable of learning to comply with IHL.

---

71 Russell and Norvig (n 33) 835.
72 Ibid 1004.
73 Ibid 1006.
74 Domingos (n 35) 222.
75 Russell and Norvig (n 33) 1004–6.

D. Interpretability

One problem with methods such as artificial neural networks is that they generate outputs that are not readily interpretable—that is, the outputs indicate robust links among many disparate variables, but do not provide a substantive, verbal explanation of those links. This lack of interpretability complicates the use of neural nets and similar methods in the target selection phase. A commander charged with an IHL violation because of a machine's mistake will not satisfy the tribunal by discussing a neural net's hidden layers. Fortunately, there are methods that increase the interpretability of machine learning.

Consider decision trees, a graphical form of machine learning that breaks down factors that contribute to a decision or drive the classification of an individual or activity.[76] A leaf in a decision tree represents a causal factor, depicting how that factor contributed to a particular outcome.[77] In a classic example, a decision on waiting for a table at a crowded restaurant might depend on factors such as the day of the week (on Friday and Saturday, alternative establishments might also be crowded), the type of cuisine served, the aspiring diners' hunger, the weather (pouring rain would raise the appeal of staying put), and whether the restaurant had a bar (which could lessen the sting of waiting). In order to manage the large amounts of data that disparate variables can produce, a decision tree will generate an explanation that is as simple as possible, given the data, with leaves pruned away if they are unnecessary for prediction. For example, if our prospective diners cared less about the type of cuisine at the restaurant than the presence of a bar (on the theory that imbibing sufficiently would make any cuisine taste sublime), a machine would prune the cuisine leaf. Moreover, unlike some other forms of machine learning, decision trees are interpretable and transparent—a human can trace the leaves of a decision tree, and readily gain a substantive, verbal understanding of a particular tree's components.

Decision trees can help analyze past events and model attributes possessed by terrorists who might play a role in future catastrophes. A decision tree analyzing the Titanic shipwreck of 1912 would find that the confluence of gender and ticket class was the best predictor of survival.[78]

---

[76] This discussion also relies on Margulies, 'Surveillance by Algorithm' (n 39).

[77] Russell and Norvig (n 33) 757.

[78] Tuffery (n 31) 314–15. Gender mattered because the Titanic's captain ordered that women and children move first to the vessel's lifeboats. Ticket class

In modeling attributes of ISIS recruits, a useful decision tree would move beyond the simplistic claim that recruits were more likely to be young, male and Islamic. These factors would include an unacceptable level of both false negatives and false positives. Women have also become ISIS recruits[79] and the vast majority of young Islamic males have resisted ISIS's lure. To more effectively find actual recruits, a decision tree might include leaves for other factors, such as employment history (to detect the disaffected or alienated), information from facial recognition software regarding attendance at events sponsored by violent groups, search history data showing visits to extremist sites, and social media posts, texts, emails or phone calls with known ISIS personnel. A leaf representing travel to a state such as Syria or Iraq, in which ISIS is currently a party to an armed conflict, would add to the tree's predictive value for targeting under IHL.

While neural nets can be more adept than decision trees at finding connections, data scientists have also developed ways to extract intelligible rules from neural networks.[80] Suppose that a neural net studies possible links between individuals in a given city. In seeking out overlapping data points based on inputted common variables, the neural net discovers that several individuals have received parking tickets in an area around a large athletic stadium while no events were scheduled at the venue.[81] The individuals do not live, work or attend school in the area. While the neural net has merely outputted the individuals' names,

---

mattered because first class passengers had berths closest to the ship's main deck, where lifeboats were located.

[79] See Katrin Bennhold, 'Religion Meets Rebellion: How ISIS Lured 3 Friends' *New York Times* (18 August 2015) A1 (describing ISIS's recruitment of three teenage girls in Britain).

[80] Rudy Setiono, Bart Baesens and Christophe Mues, 'Rule Extraction from Minimal Neural Networks for Credit Card Screening' (2011) 21 International Journal of Neural Systems 265; Rudy Setiono and others, 'Automatic Knowledge Extraction from Survey Data: Learning M-of-N Constructs using a Hybrid Approach' (2005) 56 Journal of the Operational Research Society 3; Phanida Phukoetphim, Asaad Y Shamseldin and Bruce W Melville, 'Knowledge Extraction from Artificial Neural Networks for Rainfall-Runoff Model Combination Systems' (2014) 19 Journal of Hydrologic Engineering 1422.

[81] This example was part of a demonstration of a new IBM program that seeks to detect terrorists before they engage in attacks. Patricia Tucker, 'Refugee or Terrorist? IBM Thinks Its Software Has the Answer' (2016) Defense One, accessed 5 May 2017 at www.defenseone.com/technology/2016/01/refugee-or-terrorist-ibm-thinks-its-software-has-answer/125484/. Of course, the claims of any software vendor merit close scrutiny. Levinson-Waldman (n 56).

rule extraction may be able to prod data scientists to study other areas for anomalous parking tickets or other unexplained variables. Data scientists can then incorporate this feature into training examples for other machines. In a target selection process based on conditional probabilities of combatant or civilian DPH status, such examples of shared associations and activities would be relevant, as they would be in determining after the fact that a subsequent drone strike on any or all of the parking ticket recipients was consistent with IHL.[82]

E. Summary

The account of AWS attributes above should not be mistaken for unbridled enthusiasm. While AWS have extraordinary capabilities, experts have not yet knit those strengths together into a system that will pass a weapons review and be suitable for deployment against humans in an armed conflict. Moreover, deployment will surely come first in a traditional IAC arena, in which target selection is typically categorical in nature, based on an enemy force's uniforms or emblem-laden armored

---

[82] To coordinate inputs from a wide range of sources, including documents, online communications, and video feeds, a data scientist would use a suite of machine 'learners'. Driverless cars use coordination mechanisms of this kind. Russell and Norvig (n 33) 1005. To ensure that a single learner's output is not distorted by skewed inputs, a data scientist uses ensemble learning—sometimes called 'metalearning'—which employs multiple machine learners, each with a revised data set, and then combines the outcomes. This process is also part of the validation of a machine-generated hypothesis. Ibid 748–9; Witten, Frank and Hall (n 34) 354–8; see also Domingos (n 35) 238 (discussing metalearning, including techniques used by Netflix that combine hundreds of models for movie recommendations). Combined outputs derived from metalearning can be difficult to interpret, since they are often designed to generate a numerical average or other aggregate calculation of individual learners without an overarching substantive verbal rationale for their findings. Domingos (n 35) 239. Fortunately, data scientists are developing more interpretable ensemble learners such as MetaCost, which applies combined probability estimates to training data. Using those probability estimates, MetaCost generates new cost calculations, and weights training examples accordingly. Witten, Frank and Hall (n 33), 356. Imagine a combined 'vote' of learners that try to predict likely candidates for ISIS recruitment. Suppose that the ensemble vote indicates that a particular factor in a predictive hypothesis, such as a possible candidate's use of a specific encryption technique, has only a modest effect on the conditional probability that a given subject is an actual ISIS recruit. Metacost would relabel this example in training data to downgrade its significance.

divisions.[83] Even if results of that deployment indicate that an AWS can be both accurate and effective, decision-makers should be conservative in deploying an AWS for more complex individual and situational target selection tasks, and for implementation in untraditional battle spaces, such as urban areas.

Furthermore, even if an AWS's actions in such more challenging deployments would comply with IHL, a state may question whether deployment is wise. In target selection, for example, the image of a responsible official such as Harold Koh deliberating over a suspected terrorist's dossier may be reassuring to both domestic and global audiences. Although evidence might demonstrate that machines are more accurate in this task, avoiding false negatives and positives alike, a state might rightfully be reticent. However, a state that took a more aggressive stance would still have substantial duties under IHL. It is to these duties that the next subsection turns, with special attention to the problem of accountability.

5. ENSURING THAT AWS COMPLY WITH IHL: EXERCISING DYNAMIC DILIGENCE

To adequately deal with the doctrinal and practical problems of accountability for AWS, IHL should require an approach of dynamic diligence. This approach casts commanders as being active, informed and engaged regarding an AWS's past, present and future performance. Approval of an AWS in the weapons review phase should be contingent on substantial ongoing human engagement with the weapons system. That human engagement has limits: A dynamic diligence approach will not require human ex ante authorization of AWS targeting decisions. However, this regime will require frequent, periodic assessment and, where necessary, adjustment of AWS inputs, outputs and interface with human service members.[84]

---

[83] Michael N Schmitt, 'Autonomous Weapons Systems and International Humanitarian Law: A Reply to the Critics' (2013) 4 Harvard National Security Journal 1, 17.

[84] Some have suggested that such review will be so burdensome that no state will undertake it. Wendell Wallach and Colin Allen, 'Framing robot arms control' (2013) 15 Ethics and Information Technology 125, 133. That may be so, although technological progress will ease these concerns over time. Ongoing remote assessment of an AWS's learning will also be complicated by concerns about hacking by hostile forces or private groups. These concerns, however, are

This significant human participation is consistent with a broad understanding of 'meaningful human control'.[85] That broad understanding would preclude the deployment of 'set and forget' AWS designed for the use of lethal force. The 'set and forget' approach fails to pay sufficient heed to the primary virtue of AWS: their ability to learn independently. It also fails to recognize that this virtue, left unattended, can morph into a curse that ushers in violations of IHL.

While the dynamic diligence approach is consistent with this broader understanding of meaningful human control, the approach taken in this chapter rejects a narrow definition of meaningful human control. That narrow approach would preclude autonomous targeting and requires ex ante human authorization of targeting decisions. The narrower conception fails to acknowledge that humans, including supposed experts, are prone to a legion of errors. A vision of IHL that locks in such errors intensifies the fog of war. Dynamic diligence retains what is best about human involvement, while transcending the errors that human involvement often yields.

A. The Human-machine Interface

The human-machine interface is crucial to AWS' compliance with IHL. Adjustments in the interface between an AWS and human service members will help ensure accountability. Those adjustments will require attention to both the structure of command and the tactical environment for use of such weapons.

The structure of command will have to accommodate the unique capabilities of AWS. An AWS cannot merely be treated like another weapon at the disposal of an ordinary infantry unit, used with the routine ease of a bazooka or machine-gun. The demands of IHL will not allow a military unit to use an 'off the shelf' AWS. Because of the extraordinary technical demands posed by AWS, deploying these weapons will entail a dedicated command structure.

---

not confined to AWS. In an age in which weapons are influenced by the 'Internet of Things', foiling hacking is a pervasive concern in weapons development. This chapter acknowledges such difficulties, which states will have to address. The chapter argues only that these difficulties do not justify the ban on AWS proposed by some commentators.

[85] Michael C Horowitz and Paul Scharre, 'Meaningful Human Control in Weapons Systems: A Primer' (2015) Center for New American Security, accessed 5 May 2017 at https://www.cnas.org/publications/reports/meaningful-human-control-in-weapon-systems-a-primer.

A dedicated command structure for particularly complex weapons is nothing new in modern militaries. Many of today's weapons or instrumentalities of warfare have special needs for operation and maintenance that require a degree of specialization in command that was unknown centuries ago. While medieval armies may not have had a 'Crossbow Command', that situation has changed. Air power, armored forces and submarines all have a dedicated command structure. In the United States, cyber has its own command. A separate AWS command would be consistent with this trend.

AWS commanders will have significant responsibilities. They should be conversant with machine learning, and work with fellow officers and civilian support staff who share this technical expertise. A commander who is familiar with AWS and has adequate support staff to cope with the challenges expected should be 'on the loop' for any use of AWS in the field.[86] When a case arises in which an AWS has engaged in action that appears to violate IHL, that dedicated commander should be accountable. If a state deploys AWS without such a dedicated command structure and colorable IHL violations occur because of AWS action, senior commanders in that state's armed forces, as well as civilian leadership in the chain of command, should be accountable for that omission.

Moreover, AWS should have a dynamic tactical interface with humans, to address changes in the targeting environment. For example, while an AWS may be able to operate fully autonomously in urban areas, real-time human monitoring or ex ante human review of targeting decisions may be appropriate in that setting, given the myriad variables in an urban setting and the greater likelihood of encountering substantial numbers of civilians.[87] An AWS should have the capability of requesting such review, based on its own assessment of the situation. Similarly, human service members should be able to override an AWS's machine learning protocol. In the event of a mistake that constituted a colorable IHL violation, the

---

[86] Cf Duncan B Hollis, 'Setting the Stage: Autonomous Legal Reasoning in International Humanitarian Law' (2016) Temple International and Comparative Law Journal 1, 4 (forthcoming), accessed 5 May 2017 at http://ssrn.com/abstract=2711304 (discussing human in the loop systems, which would require human ex ante authorization of targeting, human on the loop systems, which require periodic human engagement, and human out of the loop systems, which have a far more limited human role).

[87] Cf Benjamin Wittes and Gabriela Blum, *The Future of Violence: Robots and Germs, Hackers and Drones* (Basic Books 2015) 32 (noting that degree and scope of autonomy is flexible and hinging on context, not "binary" choice).

lack of such an adaptable interface would constitute prima facie evidence that a violation had occurred.

However, while an AWS must have the *capability* for greater human interface, IHL would not *require* human intervention, if an AWS could do the job as well or better. For example, given this assumption, the principle of precautions in attack would generally not require human ex ante approval for each individual AWS targeting selection. First, particularly when time is of the essence, a human authorization might not be 'feasible'. While states like the United States have up to now required human authorization for the selection of targets,[88] one could argue that this practice emerged because no realistic alternative existed. A state might demonstrate in the future that an autonomous system that relied on post hoc human review of target selection would be more efficient. In certain cases, the time expended in layers of human review, from military lawyers to a senior official such as Harold Koh, might permit prospective targets to 'go to ground' and hide their whereabouts. If such cases predominated, officials could argue that human review was not 'feasible' because it interfered with achievement of an expected military objective.

Moreover, whether human involvement is a precaution against civilian casualties or an added risk factor is an empirical question. The distortions in judgment caused by human anger, fear and cognitive flaws may exacerbate errors in the targeting process. While it may be reassuring to view human involvement as necessarily benign, that rosy sentiment may be merely another symptom of humans' cognitive faults: that is, optimism about humans' targeting abilities may be part of the *problem*, not the solution.[89]

B. Dynamic Assessment

Assessment should be dynamic regarding the effect of the AWS's experience in the field. We expect commanders to engage with humans serving under their command and discern changes caused by contact with an enemy force. While duties for AWS commanders may be more technical, they stem from the same principle of command responsibility.

---

88 McNeal (n 15).

89 Cf David Dunning, Judith A Meyerowitz and Amy D Holzberg, 'Ambiguity in Self-Evaluation: The Role of Idiosyncratic Trait Definitions in Self-Serving Assessments of Ability' (1989) 57(6) Journal of Personality and Social Psychology 1082 (discussing self-serving bias in human cognition and motivation).

To that end, human analysts should at periodic and frequent intervals review the AWS's outputs.[90] The ability to elicit video, audio, text and coded feedback from the AWS is vital to this task. Dynamic assessment should entail running a new test set to ascertain that the AWS is still observing parameters, such as compliance with the principle of distinction and a low level of permissible collateral damage, that the analysts have imposed. A dynamic assessment should also entail the ability to promptly revise AWS software if problems of compliance appear.

Dynamic assessment will involve another task: confirmation that a competent state official has updated databases used by an AWS to make target selection and implementation decisions. For example, an AWS may view a possible target's association with another individual listed on the terrorist watch list as raising the conditional probability that the possible target is also part of ISIS. However, the reliability of that conditional probability assessment depends on the accuracy of the database. US case law suggests that such lists contain numerous errors, including false positives.[91] Moreover, although procedures for giving individuals

---

[90] Cf US Department of Defense (n 6) (discussing importance of reviewing TTPs for use of AWS); US Department of Defense, Defense Science Board, *Summer Study on Autonomy* (2016) 15 (noting that autonomous systems will learn once they are deployed in the field and will 'outgrow their initial verification and validation'; consequently they will 'require more dynamic methods to perform effectively'); Lt Col Christopher M Ford, Stockton Center for the Study of International Law, *Remarks at the 2016 Informal Meeting of Experts*, UN Office in Geneva (2016) 4, accessed 5 May 2017 at http://www.unog.ch/80256EDD006B8954/(httpAssets)/D4FCD1D20DB21431C1257F9B0050B318/$file/2016_LAWS+MX_presentations_challengestoIHL_fordnotes.pdf (noting that an AWS 'should be re-reviewed periodically based upon feedback on how the weapon is functioning'); Sassòli (n 1) 332 (discussing updating of AWS inputs to allow system to weigh military advantage against harm to civilians, as the principle of proportionality requires).

[91] *Latif v Holder* 28 F Supp 3d 1134 (D OR 2014); *Tanvir v Lynch* 2015 US Dist Lexis 117661 (SD NY 2015); *Ibrahim v Dep't of Homeland Security* 62 F Supp 3d 909 (ND CA 2014); see also Irina D Manta and Cassandra Burke Robertson, 'Secret Jurisdiction' (2016) 65 Emory Law Journal 1313, accessed 5 May 2017 at http://ssrn.com/abstract=2647779 (discussing litigation over no-fly lists); cf *Abdelfattah v Dep't of Homeland Security* 787 F3d 524, 529–31, 536–39 (DC Cir 2015) (describing US lawful resident plaintiff's subjection to repeated security checks that may have been triggered by erroneous information in government databases and holding that plaintiff could seek relief under US law, but denying relief because plaintiff had not demonstrated basis, including tangible harm, for remedy sought).

recourse to correct such errors are improving, they are still less robust than they should be.[92] To support autonomous targeting, a database must be frequently and periodically updated.

A commander would not be obliged to personally update a database, but would merely have a ministerial duty to confirm the currency of a state certification. The duty to update is arguably a state obligation under human rights law, which bars the arbitrary taking of human life.[93] If a state expects that a commander will input databases into an AWS to facilitate targeting, the state should regularly certify the database as being up to date. A commander need only confirm that the state's certification is current. If the AWS commander cannot confirm that the certification is current, the commander should order a pause in AWS targeting. Under dynamic diligence, failure to order a pause would constitute a failure of command responsibility, if that omission led to an AWS targeting error.

C. Dynamic Parameters

In addition to a dynamic human-machine interface and periodic assessments, operation of the AWS will require dynamic parameters governing

---

[92] Commercial firms should also regularly update their own data, although sometimes commercial updating hinges on business factors, rather than on fairness to customers. Danielle Keats Citron and Frank Pasquale, 'The Scored Society: Due Process for Automated Predictions' (2014) 89 Washington Law Review 1.

[93] *Hassan v United Kingdom* App No 29750/09, para 104 (ECtHR, 16 September 2004) (observing that in an international armed conflict, human rights norms 'continue to apply, albeit interpreted against the background' of IHL); cf Legality of the Threat or Use of Nuclear Weapons, Advisory Opinion, 1996 ICJ 226, 262 (8 July 1996) (noting that human rights law applies in armed conflicts); Gerald L Neuman, 'Understanding Global Due Process' (2009) 23 Georgetown Immigration Law Journal 365, 387 (endorsing 'modifying the content of … treaty norms … by importing relevant rules (if any exist) from the law of armed conflict'). The precise interaction of IHL and human rights law is a complex and evolving area beyond the scope of this chapter. Unfortunately, the US Supreme Court took a more deferential stance when it held in 2009 that the good-faith exception to the exclusionary rule governing searches under the US Constitution's Fourth Amendment covered a search conducted in reliance on a flawed database. *Herring v United States* 555 US 135, 145–7 (2009); but see ibid at 153–6 (Ginsburg, J dissenting) (arguing that exclusionary rule should provide greater incentive to law enforcement to update databases); cf Jennifer E Laurin, 'Trawling for Herring: Lessons in Doctrinal Borrowing and Convergence' (2011) 111 Columbia Law Review 670 (discussing ways in which *Herring* dovetailed with other lines of precedent that reduced checks on law enforcement).

its use. These parameters will include limits on time, distance and maximum expected collateral damage. Parameters should also address the interpretability of the machine learning techniques used and whether a particular machine-learning model will be used in isolation or as part of an ensemble with other models.

To lower the probability of an AWS going rogue, each AWS should include time and distance defaults. The AWS should be programmed to move into a default hibernation mode after a relatively short, discrete period, which might be 24–96 hours. During this period, a human with remote access to the AWS's software could override the default and authorize continued operation for an additional increment of time. Absent a human override of the default, the AWS would shut down until a human intervened. Similarly, the AWS should be programmed to default to hibernation if it traveled more than a fixed, relatively short distance from the site at which the implementation phase began. For example, if an AWS had selected a target who was a senior ISIS leader in Syria, this parameter would force the AWS into hibernation mode if it traveled more than a particular distance—say 10–15 miles—from the location that the AWS had pinpointed for its target. As with time of operation, a human could override this default.

AWS should also be programmed with different maximum collateral damage parameters depending on the context of their deployment. If a state deploys an AWS to target a particular senior figure in ISIS or Al Qaeda, greater collateral damage as a consequence of that targeting would still comply with the proportionality principle. In contrast, a mission targeting lower-level ISIS figures in an urban setting would call for a lower ceiling on collateral damage.[94] US forces operate under such constraints today, complying with standing and mission-specific rules of engagement.[95] AWS should be governed by similar constraints. Such limits may mean that the AWS errs on the side of more false negatives, and therefore fails to target some combatants or DPH civilians. Nevertheless, imposing this limit, which might take the form of a penalty engineered into the AWS's calculations for even clearly proportional

---

[94] Data scientists can include comparable parameters in the AWS to compute military advantage. Schmitt, 'Autonomous Weapons Systems' (n 83) 20–21.

[95] Gary P Corn, 'Should the Best Offense Ever Be a Good Defense? The Public Authority to Use Force in Military Operations: Recalibrating the Use of Force Rules in the Standing Rules of Engagement' (2016) 49 Vanderbilt Journal of Transnational Law (forthcoming), accessed 5 May 2017 at http://ssrn.com/abstract=2709803.

harm to civilians, is appropriate in light of IHL's balance of military necessity and humanity.

In addition, target selection decisions by AWS should be interpretable and transparent.[96] If a target selection decision is mistaken and raises questions about compliance with IHL principles, a state should be able to present a clear account of the AWS's calculation to the tribunal adjudicating war crimes charges. In a war crimes tribunal, finders of fact need to review the accuracy of the targeting process used in the case at hand. Only a substantive, verbal explanation will suffice—an explanation of the method's scientific validity is both too susceptible to manipulation and too general to answer the specific questions of the tribunal.

The interpretability criterion for target selection decisions would require a model that is transparent, such as a decision tree or rule extraction that can lend greater precision to outputs from a neural network or another model opaque mode of machine learning. Each of these options increases the costs of target selection by an AWS. A single decision tree is not necessarily as accurate as other models. Using a rule-extraction approach that combines the outputs of more than one model is more complicated, time-consuming and expensive. However, the need for interpretability justifies these added costs.

Consider how interpretability interacts with time and with dynamic assessment in a target selection that is based on situational factors. Recall that in target selection based on situational factors, neither the machine nor a human has discovered the name or other personal identifiable information of a possible target.[97] Suppose that the visual feed of a drone connected to an autonomous targeting model indicates that the possible targets are currently meeting with two known low-level ISIS fighters in a village in Iraq. A decision tree or Bayesian network would represent this fact as raising the conditional probability that the possible target was also an ISIS fighter or DPH civilian. However, physical co-location with individuals identified as ISIS fighters might not in itself provide a reasonable basis to believe that the possible target was also part of ISIS. The co-location might have resulted from another factor, for example, attendance at a non-conflict event such as a wedding or a meeting with tribal elders.

Moreover, even the initial conditional probability estimate leans hard on the positive identification of the possible target's acquaintances as

---

[96] Dustin A Lewis, Gabriella Blum and Naz K Modirzadeh, 'War-Algorithm Accountability' (2016) 98, accessed 5 May 2017 at http://ssrn.com/abstract=2832734 (arguing for audit logs to promote interpretability of AWS decisions).

[97] For a critique of situational or 'signature' strikes, see Martin (n 16).

ISIS figures. That initial estimate is reliable only if dynamic assessment has assured continual updating of relevant databases. An interpretable model would also list the latest update of the relevant database. An update that was less recent would further decrease the conditional likelihood of IHL-compliant targetability.

However, even a recent update would still leave questions about the explanation for the co-location of the ISIS figures and the possible target. To resolve these questions and raise conditional probability to levels that complied with IHL, the AWS would need a dynamic temporal parameter: the machine, like a drone remotely piloted by a human, would have to maintain visual contact for an additional period that would either increase or diminish the conditional probability of IHL-compliant targetability. If the meeting between the possible target and known ISIS figures was short in duration, that would diminish the conditional probability of targetability, since it would raise the likelihood that the meeting was innocent and casual, or for a non-conflict purpose. However, if the possible target got into a vehicle with known ISIS figures, traveled more than a mile in the vehicle, and spent the next five to seven hours in the company of the ISIS fighters, that would raise the conditional likelihood of targetability, since it would make an innocent explanation less likely.

Because of a drone's capabilities, such extended surveillance of a possible target is feasible. The additional time might therefore be required by the principle of precautions in attack. It could also be required to comply with the principle of proportionality, since killing two or more presumed civilians just to kill two low-level fighters might well be excessive in light of the military advantage expected.

Without interpretability and the other dynamic attributes listed above, a tribunal examining this incident could find that targeting violated IHL. Suppose the prosecution in a war crimes tribunal presented evidence from residents of the village that all of the men observed were guests at a wedding. According to the witnesses, none were ISIS fighters. Instead, they were local farmers and tradesmen. Suppose as well that the tribunal lacked an interpretable record of the AWS's calculations, proof that the ISIS database had been recently updated, or evidence that the AWS had taken additional time to observe the targets and their acquaintances. Lacking this evidence, a tribunal might find that the targeting decision had not complied with IHL.[98]

---

[98] This chapter leaves for another day the question of who should bear the burden of proof on whether a commander has fulfilled her responsibility for the actions of an AWS. Lex lata indicates that the prosecution bears the burden of

Moreover, suppose that, due to human error, the AWS had not been operating under a constraint that required additional time if a preliminary targeting assessment was based solely on co-location. If the AWS's outputs were interpretable, a tribunal would be able to discover this fact. It would then be able to impose liability on the AWS commander under a command responsibility theory. If the outputs were not interpretable, the tribunal might be left in the fog of war. Interpretability penetrates that fog, enabling accountability.

6. CONCLUSION

Of all the challenges to IHL posed by AWS, accountability is perhaps the most serious. Putting a machine in the dock at a war crimes trial would create a sorry spectacle. Humans who shrugged off their own role in alleged IHL violations by an AWS would inspire even greater revulsion. Fortunately, with modest revisions the familiar doctrine of command responsibility provides a vessel for the accountability that AWS critics rightly demand.

Application of the doctrine of command responsibility would encourage states to harness an AWS's potential for IHL compliance. An AWS has abilities that humans can only envy to assess probabilities and find patterns based on new evidence. These strengths, coupled with a growing capacity for movement, promise heightened compliance with IHL principles such as distinction, proportionality and precautions in attack. This potential may not be enough to ensure today that an AWS can pass a weapons review. However, that day will come, probably within a shorter period than the time required to move the Internet from a scientist's brainstorm to a fixture of everyday life. Nevertheless, deployment of AWS will require safeguards. Command responsibility ensures the integrity of those protections.

Command responsibility is admittedly not a perfect fit for filling the AWS accountability gap. As a doctrine of vicarious liability, command

---

proof. In that event, it might still be difficult to find a commander liable on these facts. However, one could interpret command responsibility as making the interpretability of an AWS's outputs or the presence of a dynamic temporal parameter substantive components of a commander's duties. On this view, a commander's failure to ensure that AWS outputs were interpretable and that the AWS was subject to a dynamic temporal parameter would be dispositive evidence of a violation, given an apparent AWS mistake regarding the principles of distinction, proportionality or precautions in attack.

responsibility generally is predicated on a negligent or reckless failure to curb willful acts by human subordinates. Under traditional notions of command responsibility, those human subordinates would also be accountable. Application of command responsibility to the actions of AWS requires a modest revision of the doctrine, which would extend its application to the acts of machines. That extension is a logical outgrowth of an evolution in the conduct of hostilities, in which more warfighting is done autonomously. As circumstances on the ground change, IHL should evolve, if it is to continue to preserve the balance of military necessity and humanity. Moreover, a modest revision to command responsibility is far superior to either devising an entirely new doctrine or foregoing the use of weapons that may actually comply with IHL more faithfully and efficiently than humans can, given humans' propensity for anger, fear and cognitive flaws.

Ensuring AWS compliance with IHL requires dynamic diligence that is the antithesis of a 'set it and forget it' approach. Dynamic diligence mandates attention to the machine/human interface, frequent and periodic assessments, and flexible operating parameters. The human interface with AWS should entail a dedicated AWS command, staffed by officers familiar with the capabilities and weaknesses of autonomous weapons. The interface should enable humans to step in immediately and override AWS's decisions, particularly on difficult issues involving the selection of NIAC targets or implementation of targeting in an urban area. Dynamic assessment should include both regular reviews of the AWS's learning process and confirmation that evidence relied on by the AWS, including terrorist watch lists, is up to date. Dynamic parameters include a presumption favoring interpretability of AWS findings in target selection.

In a sense, dynamic diligence is a variation on the theme of 'human on the loop' suggested by some commentators. Humans cannot go on their merry way as AWS conduct the grim business of war. At the same time, a rigid requirement that a human always be 'in the loop' would stifle innovation and curtail the promise of AWS for IHL compliance. Put another way, dynamic diligence is a practical version of what 'meaningful human control' would look like, if that phrase were deployed to permit autonomy while preserving checks on autonomy's excesses.

Dynamic diligence will not be easy. Keeping up with developments in this fluid field is a daunting challenge. However, the approach urged here also provides space for commanders to 'own' IHL compliance in a fashion that is both new and linked to long-standing military practice. A ban on AWS would seek to blink away the future of war. In contrast, dynamic diligence bets both on the future and on continued fidelity to core IHL principles.

14. The strategic implications of lethal autonomous weapons

*Michael W Meier**

1. INTRODUCTION

Over the past 15 years the public has become more aware of the issues surrounding autonomy in weapons systems. Unfortunately, much of the information comes from sensationalized portrayals in the media. Long before the current debate, an original *Star Trek* episode from the 1960s, 'Taste of Armageddon',[1] looked at automation and warfare. In that episode, two worlds, Eminiar and Vendikar, had been engaged in an armed conflict for over 500 years using a sophisticated computer program, which simulated attacks and designated casualties on both sides. Once designated by the computer as a casualty, the person was required to report to a disintegration chamber. When Captain Kirk and his party arrive on Eminiar to meet with their leaders, he is told that the computer has determined that the Enterprise was destroyed in an attack and the entire crew must report to the disintegration chambers. Anan, their leader, informs Captain Kirk that if one side fails to have their casualties report, then actual weapons will be used. Captain Kirk refuses to send his crew to their deaths and admonishes Anan that they have made war so neat and painless that neither side has any need to stop it. He tells them it is the horror of war that makes it something to be avoided and by eliminating this aspect, the parties have simply allowed the war to go on for over 500 years. After Captain Kirk blows up the computer, he tells Anan that he has given them back a war with actual destruction and they now have a reason to finally stop it.[2]

This episode represents an extreme example of remote warfare as there are no armies, weapons or actual fighting. Instead, the computer makes

* The views expressed in this chapter are those of the author in his personal capacity and should not be understood as representing those of the Department of State, Department of Defense, or any other United States government entity.

[1] *Star Trek:* 'A Taste of Armageddon' (NBC television broadcast 23 February 1967).

[2] Ibid.

life and death decisions without human involvement. We have not reached that stage yet for armed conflict, and hopefully never will, but the potential development of Lethal Autonomous Weapons Systems (LAWS), which is sometimes defined as a weapons system that can select and engage a target without human intervention,[3] raise important questions about how far does society want to go with autonomy in weapons systems, in particular with life and death decisions. Are LAWS simply the next evolution in remote warfare, or are they, as some claim, the next revolution of military affairs after gunpowder and nuclear weapons? Would the development of LAWS mean we are 'Crossing the Rubicon' where machines and not humans determine when to take a life in armed conflict? Science fiction aside, LAWS are receiving serious attention in the international community through the framework of the Convention on Certain Conventional Weapons (CCW).[4]

There has been much written on the legal and ethical issues with the potential development of LAWS.[5] This chapter instead will focus on

---

[3] See US Dept of Defense, Directive 3000.09, Autonomy in Weapons Systems (21 November 2012) [hereinafter DoD Directive 3000.09]. The Directive defines an 'autonomous weapon system' as 'a weapons system that, once activated, can select and engage targets without further intervention by a human operator …'.

[4] UN Convention on Prohibitions or Restrictions on the Use of Certain Conventional Weapons which may be deemed to be Excessively Injurious or to have Indiscriminate Effects, 10 October 1980, 1342 UNTS 137, 163. It was adopted in 1980 and entered into force in 1983. Ibid. The Convention on Certain Conventional Weapons (CCW) was negotiated under the auspices of the United Nations in 1979 and 1980 and builds upon long-standing rules related to armed conflict, including the principle of distinction and the prohibition of weapons that are deemed to be excessively injurious or have indiscriminate effects. Ibid. There are five protocols to CCW, including non-detectable fragments (Protocol I); Mines, Booby-traps and Other Devices as amended on 3 May 1996 (Amended Protocol II); Incendiary Weapons (Protocol III); Blinding Lasers (Protocol IV); and Explosive Remnants of War (Protocol V). Ibid 168–72; Additional Protocol to the Convention on Prohibitions or Restrictions on the Use of Certain Conventional Weapons Which May be Deemed to be Excessively Injurious or to Have Indiscriminate Effects (Protocol IV, entitled Protocol on Blinding Laser Weapons), 30 July 1998, 2024 UNTS 163, 167; Protocol on Explosive Remnants of War to the Convention on Prohibitions or Restrictions on the Use of Certain Conventional Weapons Which May be Deemed to be Excessively Injurious or to Have Indiscriminate Effects (Protocol V), 28 Nov 2003, 2399 UNTS 100, 126.

[5] *Background – Lethal Autonomous Weapons Systems*, UNOG, accessed 5 May 2017 at http://www.unog.ch/80256EE600585943/(httpPages)/8FA3C2562A60FF81C1257CE600393DF6?OpenDocument.

the strategic and national security implications that have been raised with respect to LAWS by looking at four commonly cited concerns: (1) LAWS, if developed and used, will lower the threshold for armed conflict; (2) LAWS will have a negative impact on global and regional security through unintended engagements; (3) the use of LAWS will lead to more asymmetric warfare, including resorting to possible terrorist attacks; and (4) the development of LAWS will proliferate and could cause an arms race as states will feel compelled to develop or acquire them in order to maintain regional stability and balance of power. Finally, this chapter will propose actions the United States should take internationally within the CCW framework, as well as on a national basis to mitigate these strategic stability concerns.

2. OVERVIEW OF LAWS DISCUSSION WITHIN THE CONVENTION ON CERTAIN CONVENTIONAL WEAPONS

LAWS have been the subject of three informal meetings of experts at the CCW in Geneva, Switzerland.[6] Even though there remains no agreement or definition of what would constitute LAWS, the Campaign to Ban Killer Robots, a coalition of civil society organizations, and certain states have called for a pre-emptive ban or moratorium on the development and fielding of LAWS.[7] Although there remain many divergent views with respect to lethal autonomous weapons systems, the one thing that is clear is that this debate will continue for the foreseeable future—and it should

---

6 *2016 Meeting of Experts on LAWS*, UNOG, accessed 5 May 2016 at http://www.unog.ch/80256EE600585943/(httpPages)/37D51189AC4FB6E1C1257F4D004CAFB2?OpenDocument.

7 Human Rights Watch and the Campaign To Stop Killer Robots, for example, called for pre-emptive bans on the development and use of autonomous weapons. Statement, Stephen Goose, Director, Arms Division, Human Rights Watch, Statement by Human Rights Watch to the Convention on Certain Conventional Weapons Informal Meeting of Experts on Lethal Autonomous Weapons Systems (13 May 2014), accessed 5 May 2017 at http://www.unog.ch/80256EDD006B8954/(httpAssets)/6CF465B62841F177C1257CE8004F9E6B/$file/NGOHRW_LAWS_GenStatement_2014.pdf; Statement, Mary Wareham, Human Rights Watch, Campaign to Stop Killer Robots Statement to the Convention on Certain Conventional Weapons Informal Meeting of Experts on Lethal Autonomous Weapons Systems (13 May 2014), accessed 5 May 2017 at http://www.unog.ch/80256EDD006B8954/(httpAssets)/33AFAF2B1AFFFB3CC1257CD7006AAB67/$file/NGO+Campaign+Killer+Robots+MX+LAWS.pdf.

continue—as states, international organizations, civil society and others have begun to recognize and understand the complex issues that need to be addressed.

The first meeting of informal experts, chaired by Ambassador Jean-Huges Simon-Michel of France, took place in May 2014 with a mandate to discuss 'the questions related to emerging technologies in the area of lethal autonomous weapons systems'.[8] The four-day substantive sessions addressed the legal, technical, ethical and operational and military aspects of LAWS.[9] As part of the Chair's report to the Meeting of High Contracting Parties, Ambassador Simon-Michel noted the 'impact of LAWS on international peace and security was discussed. The consequences on arms control were also raised'.[10] In 2015, Ambassador Michael Biontino of Germany chaired the second informal meeting of experts under the same mandate.[11] The meeting looked to build upon the work of the 2014 informal session by delving deeper into the issues surrounding the legal, technical, ethical and operational and military aspects of LAWS.[12] It also included a session on general security issues matters, including strategic implications for global and regional security, potential for an arms race, asymmetric warfare and lowering the threshold for applying force.[13]

---

[8] These 'informal experts' included several individuals specializing in the technical, legal, sociological/ethical, and military/operational issues surrounding LAWS. United Nations Office at Geneva, Convention on Certain Conventional Weapons, Report of the 2014 Informal Meeting of Experts on Lethal Autonomous Weapons Systems (LAWS) ¶ 1, CCW/MSP/2014/3 (11 June 2014), accessed 5 May 2017 at http://www.unog.ch/80256EDD006B8954/(%20httpAssets)/350D9ABED1AFA515C1257CF30047A8C7/$file/Report_AdvancedVersion_10June.pdf [hereinafter 2014 Chairs Report]; see 2014 Meeting of Experts on LAWS, Presentations and Statements from the Meeting of Experts, UNOG, accessed 5 May 2017 at http://www.unog.ch/__80256ee600585943.nsf/(httpPages)/a038dea1da906f9dc1257dd90042e261?OpenDocument&ExpandSection=1#_Section1 (listing various participants and providing their prepared remarks).

[9] 2014 Chairs Report (n 8) ¶ 1.

[10] Ibid ¶ 38.

[11] See United Nations Office at Geneva, Convention on Certain Conventional Weapons, Report of the 2015 Informal Meeting of Experts on Lethal Autonomous Weapons Systems (LAWS) ¶ 1, CCW/MSP/2015/3 (2 June 2015), accessed 5 May 2017 at http://www.unog.ch/80256EDD006B8954/(httpAssets)/587A415BEF5CA08BC1257EE0005808FE/$file/CCW+MSP+2015-03+E.pdf [hereinafter 2015 Chairs Report].

[12] 2015 Chairs Report (n 11) ¶ 9.

[13] Ibid ¶ 67.

In April 2016, Ambassador Biontino chaired the third informal meeting of experts, which saw an increased focus on the strategic implications of LAWS as two sessions were devoted to this topic.[14] Many delegations noted the challenges arising from the development and use of LAWS, to include risk of proliferation both to states and non-state actors, the prospect of an arms race, the risk of lowering the threshold for the use of force and the impact LAWS might have on global and regional stability.[15] Moving forward, one of the main issues that will need to be considered is whether the benefit that LAWS may provide to the commander in the way of greater capability and flexibility on the battlefield, is outweighed by the risk of proliferation and global and regional destabilization.[16]

3. THE IMPACT ON STRATEGIC STABILITY BY LAWS

(a) Lowering the Threshold for Armed Conflict

We have seen throughout history how the development and use of new weapons systems can transform the way states fight wars, such as through the crossbow or submarines.[17] One important factor that should be considered with the development of new weapons systems is the impact it will have on strategic stability.[18] Strategic stability is generally 'the condition that exists when two potential adversaries recognize that

---

[14] Statement, H E Ambassador Michael Biontino, Introductory Statement to Informal Meeting of Experts on Lethal Autonomous Weapons Systems, The Convention on Certain Conventional Weapons (CCW) Informal Meeting of Experts on Lethal Autonomous Weapons Systems (11 April 2016), accessed 5 May 2017 at http://www.unog.ch/80256EDD006B8954/(httpAssets)/6796F4DBA5B2F0D6C1257F9A00441922/$file/2016_LAWS+MX_GeneralExchange_Statements_Germany.pdf.

[15] Report of the 2016 Informal Meeting of Experts on Lethal Autonomous Weapons Systems (LAWS) (Advanced Version) [hereinafter 2016 Chairs Report] (10 June 2016) (on file with author).

[16] 2016 Chairs Report (n 15) ¶ 71.

[17] Peter B Postma, 'Regulating Lethal Autonomous Robots in Unconventional Warfare' (2014) 11 U St Thomas L J 300, 316–18.

[18] Jean-Marc Rickli, Some Considerations of the Impact of LAWS on International Security: Strategic Stability, Non-State Actors and Future Prospects, Presentation to CCW Informal Meeting of Experts (16 April 2015), accessed 5 May 2017 at http://www.unog.ch/80256EDD006B8954/(httpAssets)/B6E6B974512402BEC1257E2E0036AAF1/$file/2015_LAWS_MX_Rickli_Corr.pdf.

neither would gain an advantage if it were to begin a conflict with the other'.[19] The rapid advances that are being made with technology and autonomy in weapons systems have been described as the next revolution in military affairs on par with gunpowder and nuclear weapons.[20] One of the primary concerns expressed, especially by those opposed to the development of LAWS, is that these weapons systems may lower the threshold for states to engage in armed conflict. For example, the 2009 mission statement of the International Committee for Robot Arms Control (ICRAC) recommended the international discussion of autonomous weapons should consider '[t]heir potential to lower the threshold of armed conflict'.[21] Human Rights Watch in its 2012 report, *Losing Humanity*, stated:

> [T]he gradual replacement of humans with fully autonomous weapons could make decisions to go to war easier and shift the burden of armed conflict from soldiers to civilians in battle zones.
>
> While technological advances promising to reduce military casualties are laudable, removing humans from combat entirely could be a step too far. Warfare will inevitably result in human casualties, whether combatant or civilian. Evaluating the human cost of warfare should therefore be a calculation political leaders always make before resorting to the use of military force. Leaders might be less reluctant to go to war, however, if the threat to their own troops were decreased or eliminated.[22]

There are two aspects to consider with respect to the question of whether LAWS, if such systems are developed, fielded and used, will lower the threshold for engaging in armed conflict. The first is whether LAWS will negatively affect the jus ad bellum standard required for a state to use force, while the second is the political or policy calculation for engaging in the use of force.

---

19 Rickli (n 18) 1, quoting Steven Hildreth and Amy Woolf, *Ballistic Missile Defence and Offensive Arms Reductions: A Review of the Historical Record* (Congressional Research Service 2010) 4.

20 Peter Singer, *Wired for War: The Robotics Revolution and Conflict in the 21st Century* (Penguin 2009) 203.

21 Noel Sharkey, 'ICRAC Celebrates Successful Fulfillment of its 2009 Mission', International Committee for Robot Arms Control (18 May 2014), accessed 5 May 2017 at http://icrac.net/2014/05/icrac-celebrates-successful-fulfillment-of-its-2009-mission/.

22 Human Rights Watch, *Losing Humanity: The Case Against Killer Robots* (2012), accessed 5 May 2017 at https://www.hrw.org/sites/default/files/reports/arms1112_ForUpload.pdf (discussing the use of fully autonomous weapons and implications in lessening the threshold for going to war).

(1) Legal perspective

The starting point for determining whether a particular action constitutes a use of force can be found in Article 2(4) of the Charter of the United Nations, which provides that 'All Members shall refrain in their international relations from the threat or use of force against the territorial integrity or political independence of any state, or in any other manner inconsistent with the Purposes of the United Nations'.[23] Accordingly, any resort to the use of force by a state must have a legal basis that is assessed in light of the facts and circumstances to overcome this prohibition.[24] This prohibition is not absolute; there are several recognized exceptions to this prohibition on the use of force.

The first exception to the prohibition on the use of force would be an action under Chapter VII of the Charter of the United Nations.[25] Article 42 provides that the UN Security Council may take such action by air, sea or land forces as may be necessary to maintain or restore international peace and security, including demonstrations, blockades or other military operations.[26] The Security Council has used Chapter VII numerous times, but the resolutions generally do not specify the type of weapons that its member states would need to use to carry out the action.[27] Accordingly, the Security Council could authorize a Chapter VII action where a member state would deploy and use LAWS.

The second exception to the prohibition on the use of force is when it is undertaken with the consent of the territorial state.[28] In his April 2016 speech at the American Society of International Law (ASIL), Brian Egan, the Legal Adviser of the Department of State stated:

> As a matter of international law, the United States has relied on … consent … in its use of force against ISIL. Let's start with ISIL's ground offensive and capture of Iraqi territory in June 2014 and the resulting decision by the United States and other States to assist with the military response. Beginning in the

---

[23] UN Charter Article 2(4).

[24] Office of General Counsel, US Dept of Defense, Dept of Defense Law of War Manual (2015) § 1.11.3, accessed 5 May 2017 at http://archive.defense.gov/pubs/Law-of-War-Manual-June-2015.pdf [hereinafter DoD Law of War Manual].

[25] UN Charter ch 7.

[26] UN Charter Article 42; see DoD Law of War Manual (n 24) § 1.11.4.2.

[27] See Actions with Respect to Threats to the Peace, Breaches of the Peace, and Acts of Aggression (Chapter VII), Repertoire of the Practice of the Security Council (2012–2013) Article 42, accessed 5 May 2017 at http://www.un.org/en/sc/repertoire/actions.shtml; see also Geoffrey Corn et al, *The Law of Armed Conflict: An Operational Approach* (Wolters Kluwer 2012) 18–19.

[28] DoD Law of War Manual (n 24) § 1.11.4.3.

> summer of 2014, the United States' actions in Iraq against ISIL have been premised on Iraq's request for, and consent to, U.S. and coalition military action against ISIL on Iraq's territory in order to help Iraq prosecute the armed conflict against the terrorist group.[29]

Accordingly, the use of force in the territory of another state does not violate Article 2(4)'s prohibition against the use of force when the territorial state consents to it, even if the state used LAWS.[30]

Finally, a state is authorized to use force in self-defense pursuant to Article 51 of the Charter of the United Nations, which provides that 'nothing in the present Charter shall impair the inherent right of individual or collective self-defense if an armed attack occurs against a Member of the United Nations, until the Security Council has taken measures necessary to maintain international peace and security'.[31] The right of self-defense under the UN Charter does not supersede a state's inherent right of individual or collective self-defense under customary international law.[32] The state's inherent right to self-defense under customary international law is not unlimited, but rather a state's actions must be necessary and proportionate to the threat being addressed and all reasonable peaceful alternatives must be exhausted.[33]

In applying these jus ad bellum rules of international humanitarian law (IHL), consider the following scenario: Country A has developed and fielded an aerial lethal autonomous weapons system that is programmed to search for armored personnel carriers and, when found, it can then select and engage that target under specific programming parameters. Tensions between Country A and Country B have been rising over the past several months so Country A decides to deploy the LAWS along its border with Country B. During its operation, the LAWS locates an armored personnel carrier located in Country B and it engages and destroys it resulting in the deaths of the soldiers inside the vehicle. In reviewing this incident, there are two questions that need to be addressed:

---

[29] Brian Egan, 'International Law, Legal Diplomacy, and the Counter-ISIL Campaign: Some Observations' (2016) 92 Intl L Stud 235, 238.

[30] DoD Law of War Manual (n 24) § 1.11.4.3.

[31] UN Charter Article 51; see DoD Law of War Manual (n 24) § 1.11.5.

[32] DoD Law of War Manual (n 24) § 1.11.5.

[33] Ibid.

(1) Would the use of LAWS by Country A constitute a 'use of force';
(2) Would such a use of force trigger the right of self-defense by Country B?[34]

The first question would seem easy enough to answer. The strike by the LAWS would amount to a use of force under Article 2(4), which does not require a large-scale military operation, but covers all uses of 'force'.[35] The second question is more complicated because while every threat or use of 'force' would be prohibited under Article 2(4) of the UN Charter, not every use of 'force' will rise to the level of an 'armed attack' that triggers the right of self-defense.[36]

Some states, in accordance with the plain reading of Article 51, assert that the right of self-defense is triggered only when a state has suffered an 'armed attack'.[37] Other states, including the United States, take the view that self-defense is available against any illegal use of force.[38] There is no agreed upon definition of what constitutes an 'armed attack', but rather it is a consideration of various factors.[39] In the *Nicaragua* case in the International Court of Justice, that court found that only the 'most grave forms of the use of force' would constitute an armed attack and there must be a significant scale of violence above 'mere frontier incidents'.[40] This does not mean that an 'armed attack' requires a large scale attack, as even a single attack can rise to the level of an armed attack.[41] The International Court of Justice also noted it would be dangerous to unnecessarily restrict a state's right to self-defense as it

---

[34] Steven J Barela, *Legitimacy and Drones: Investigating the Legality, Morality and Efficacy of UCAVs* (Routledge 2015).

[35] Ibid.

[36] Ibid 40.

[37] Molly McNab and Megan Matthews, 'Clarifying the Law Relating to Unmanned Drones and the Use of Force: The Relationship between Human Rights, Self-Defense, Armed Conflict, and International Humanitarian Law' (2011) 39 Denv J Intl L & Poly 661, 675.

[38] DoD Law of War Manual (n 24) § 1.11.5.2; Harold Hongju Koh, 'The Obama Administration and International Law', Speech to the Annual Meeting of the American Society International Law, 25 March 2010, accessed 5 May 2017 at http://www.cfr.org/international-law/legal-adviser-kohs-speech-obama-administration-international-law-march-2010/p22300.

[39] McNab and Matthews (n 37) 675.

[40] Ibid 676.

[41] Ibid.

could limit that state's ability to legally respond to threats to its sovereignty.[42]

A state's right of self-defense is not unlimited as it requires that the actions taken be necessary to address the threat that authorized its use.[43] Further, self-defense does not automatically justify 'all-out' armed conflict to destroy the enemy, but permits those actions necessary to defend the state from the continuation of attack or imminent attacks.[44] Finally, the action taken in self-defense must be proportionate to the use of force that preceded it.[45] Applying these principles to the above scenario, states may reach different conclusions on whether the use of force by Country A constitutes an 'armed attack'.

The important consideration, whether or not the use of force by Country A would authorize Country B to respond in self-defense, is that irrespective of whether the use of force in this case was undertaken by a manned system, unmanned system or LAWS, the legal threshold will remain the same. Any unjustified threat of force or use of force will remain prohibited no matter what type of weapons system is used.[46] Just because a state uses an autonomous weapons system, the threshold for the use of force required for self-defense or when an armed conflict is taking place that would trigger applicable IHL principles is not different or lower. That does not mean, however, that the use of LAWS will not raise other questions. For example, the use of LAWS may make it harder to determine a state's intent, as questions will arise about whether the use of force was an intentional act or whether it was a malfunction on the part of the system. This may lead to escalations in tensions and possible confrontations, as states may take more aggressive or riskier actions if there is no threat to their own personnel, and make diplomatic relations between states more problematic.

(2) Policy perspective

In May 2016, the United States conducted a UAV strike in Pakistan that killed Taliban leader Mullah Akhtar Mohammad Monsour. President Obama remarked that the death of Mansour marked an 'important

---

[42] Ibid.

[43] DoD Law of War Manual (n 24) § 1.11.5.

[44] Committee Report, 'Use of Force: Report on Aggression and the Use of Force' (2014) 76 Intl L Assn Conf 648, 657.

[45] Ibid; cf *Oil Platforms (Iran v US)*, 2003 ICJ ¶ 72.

[46] DoD Law of War Manual (n 24) § 1.11.3.

milestone in [United States] longstanding effort to bring peace and prosperity to Afghanistan'.[47] Secretary of State John Kerry noted:

> We have had longstanding conversations with Pakistan and Afghanistan about this objective with respect to Mullah Mansour, and both countries leaders were notified of the airstrike. And it is important for people to understand that Mullah Mansour, as I said a moment ago, has been actively involved in planning attacks in Kabul, across Afghanistan, presenting a threat to Afghan civilians and to the coalition forces that are there.[48]

Pakistan officials, however claimed, 'the drone attack was a violation of its sovereignty, an issue which has been raised with the United States in the past as well'.[49] Although UAV strikes are considered a violation of Pakistani sovereignty, Pakistan reiterated both countries have 'good cooperation' on military and intelligence.[50] It is unclear whether this airstrike would have taken place without the ability to use an unmanned system. As this UAV incident illustrates, one of the principal arguments against the development and use of LAWS is that it will be the next evolution in remote warfare and these types of incidents will rise. Decision-makers may use LAWS in ways that they would not with manned systems, or even UAVs, thereby resulting in escalating international tensions and confrontations. It is hard to know exactly how LAWS may be used, since they do not yet exist; but there are lessons that can be learned from how UAVs are being used today.

In 2015, the Center for New American Studies (CNAS) launched its World of Proliferated Drones project.[51] As part of the project, CNAS commissioned essays from authors from 10 countries seeking their views on their country's use of UAVs.[52] Apart from use in a counterterrorism context, the authors viewed UAVs as providing greater strategic and

---

[47] Nic Roberston and Jamie Crawford, 'Obama: Taliban Leader's death marks "milestone"', CNN.com, accessed 5 May 2017 at http://www.cnn.com/2016/05/21/politics/u-s-conducted-airstrike-against-taliban-leader-mullah-mansour/index.html.

[48] Ibid.

[49] Ibid.

[50] Ibid.

[51] See Center for a New American Security, *A World of Proliferated Drones* (2015), accessed 5 May 2017 at https://www.cnas.org/research/future-of-warfare-initiative/proliferated-drones.

[52] Kelly Sayler, Ben FitzGerald, Dr Michael C Horowitz, and Paul Scharre, *Global Perspectives: A Drone Saturated Future* (2016), accessed 5 May 2017 at http://drones.cnas.org/reports/global-perspectives/#1460563103267-1900a533-b5dd.

operational flexibility for states and some expressed the view that their particular state would have a lower threshold for deploying these types of systems.[53] In Germany, a country that is traditionally risk-averse, UAVs allow them to contribute more frequently to international missions.[54] Israel also has a low tolerance for casualties and the use of unmanned systems could limit or eliminate casualties amongst its soldiers.[55]

In addition to providing greater operational flexibility, the various authors expressed the view that in a contested or sensitive environment, it was likely that unmanned systems would be used in place of manned systems. For example, the Republic of Korea would view manned flights as 'too risky' and that 'UAV missions should become the rule rather than the exception'.[56] It is believed that Germany would likely be willing to accept more risk in having a drone shot down than a manned system.[57] India has used unmanned systems in contested areas of Jammu and Kashmir, one of which was allegedly shot down by Pakistan.[58] Other states, however, such as France, would not be willing to use an unmanned system, instead of a manned system, in a contested area.[59]

In addition to providing greater flexibility for use in contested areas, states differed on how they would respond to an unmanned system crossing their borders.[60] Some states would not view a border incursion in the same way they would view a manned aircraft.[61] One author indicated that Vietnam would be unlikely to shoot down a drone that crossed its airspace due to concerns about political stability.[62] Another indicated that Singapore would likely take action if the intrusion was part of a broader effort to 'infringe upon Singapore's national airspace'.[63] However, the authors indicated that both Russia and France would take responsive action to include shooting down the unmanned system, in particular if it was thought to be an armed system.[64] This demonstrates that there are many differing views on how unmanned systems would be used and how other countries would respond.

---

53 Ibid 7.
54 Ibid.
55 Ibid.
56 Ibid 9.
57 Ibid.
58 Ibid.
59 Ibid.
60 Ibid 10.
61 Ibid.
62 Ibid.
63 Ibid.
64 Ibid.

Unlike UAVs, which have a human operator at all times, the development of LAWS would further remove direct human involvement from these engagements. It would seem likely that LAWS would continue the pattern of UAVs and be used in ways that differ from manned systems.[65] A concern is that, once LAWS are developed, their use will make it even less costly politically to use these systems in high risk operations as a state's own military forces are not at risk.[66] Decision-makers, no longer burdened about their forces becoming casualties, may be more inclined to use force, continue or even escalate an armed conflict without worry about domestic political reaction since their own population is not directly at risk.[67] The reduced political price political decision-makers face for using force may also lead politicians to fail to consider other nonviolent alternatives short of armed conflict.[68]

One counter argument to the suggestion that LAWS will make it politically less costly to engage in armed conflict is that no state is going to rely on LAWS alone.[69] LAWS are not going to replace humans on the battlefield so there will still be a risk of casualties.[70] A state's national defense will never rest solely with one weapons system, but rather a variety of weapons systems that complement various capabilities.[71] A single weapons system, even LAWS, may help secure a successful military campaign, but relying on it is more likely to guarantee failure.[72]

A second counter argument is that deliberately failing to develop new means and methods of warfare that offer greater protections to your civilian population and military forces would be equating those individuals to hostages as a way to pressure politicians to avoid engaging in

---

65 Ibid 12.

66 Markus Wagner, 'The Dehumanization of International Humanitarian Law: Legal, Ethical, and Political Implications of Autonomous Weapons Systems' (2014) 47 Vand J Transnatl L 1371, 1418.

67 Ibid 1419.

68 Ibid 1420.

69 See Eneken Tikks-Ringas, Notes to Presentation to Informal Meeting of Experts to CCW on LAWS (April 2016), accessed 5 May 2017 at http://www.unog.ch/80256EDD006B8954/(httpAssets)/DD2D25775C75786BC1257F9A00476DAF/$file/2016_LAWS+MX+Presentations_SecurityIssues_Eneken+Tikk-Ringas+note.pdf.

70 Ibid.

71 Ibid.

72 Ibid.

armed conflict.[73] As has been the case with UAVs, states have developed these new systems using new technology as a way to protect one's forces and civilians. There does not appear to be a logical explanation for suddenly abandoning this principle with the development of LAWS.[74]

It is clear that as UAVs continue to proliferate and the types of missions become more diverse, there remains the potential for misperceptions and miscalculations in their use.[75] It is not an unreasonable argument that the development and use of LAWS would continue this pattern of using unmanned systems for high-risk operations. Once developed, they could also take on an increased role, which would also increase the risk of unintended engagements.

(b) Unintended Engagements

Turning to the second argument, another concern with the development of LAWS is that since the system does not operate under direct human control, it can malfunction due to software errors, be hacked by the enemy or simply take unexpected actions as it senses and interprets inputs from its operating environment. If a malfunction or error occurs and results in an unintended engagement, the outcome has the potential to be disastrous. To date, the United States remains the only state that has a published directive dealing with autonomous weapons, which, in part, addresses unintended engagements.

One of the stated purposes of Department of Defense Directive (DoD Directive) 3000.09 is to 'minimize the probability and consequences of failures in autonomous and semi-autonomous weapons systems that could lead to unintended engagements'.[76] The Directive defines an unintended engagement to be:

> The use of force resulting in damage to persons or objects that human operators did not intend to be targets of U.S. military operations, including unacceptable levels of collateral damage beyond those consistent with the laws of war, [Rules of Engagement], and commander's intent.[77]

---

[73] Kenneth Anderson and Matthew Waxman, *Law and Ethics for Autonomous Weapon Systems: Why a Ban Won't Work and How the Law of War Can* (Hoover 2013) 18.

[74] Ibid.

[75] Sayler et al (n 52) 11–12.

[76] DoD Directive 3000.09 (n 3) § 1b.

[77] DoD Directive 3000.09 (n 3) glossary, part II, definitions; see Paul Scharre, *Autonomous Weapons and Operational Risk* (CNAS Ethical Autonomy Project 2016) 18.

In February 2016, Paul Scharre in his CNAS publication, *Autonomous Weapons and Operational Risk*, sets forth four key variables that can help determine the risk of an unintended engagement.[78]

The first variable is determining the inherent hazard with respect to the system.[79] This would require looking at the types of targets the autonomous weapons system is designed to engage and the munitions the system would employ. As Scharre notes, an autonomous weapons system that is designed to target combatants poses a greater risk to civilians than one that targets vehicles or is a defensive weapons system.[80] Further, if an autonomous system is armed with a large bomb it carries a higher risk than one that is unarmed or has non-lethal weapons.[81]

The second variable is how much of a time delay is there between a failure of the weapons system and the ability to take corrective action by the human operators.[82] One of the primary differences between an unmanned system and LAWS is the ability of the human operator to take action in real-time. With an unmanned system, the operator is the one who will select and engage a target. A fully autonomous weapons system does not have that human operator to take that corrective action while it is in operation, which likely will result in a time delay before corrective action can be taken.[83]

The third variable looks to the damage potential of the autonomous system. For example, this variable gauges how much damage the LAWS can cause before corrective action can be taken. Scharre suggests that the damage potential depends on a variety of factors, such as its inherent hazard, the time between the malfunction and corrective action, how quickly it can move between engagements, if at all, and the time and distance over which the system can operate and the number of munitions available to it.[84] A system that can operate over several days and several hundred miles and carries multiple large bombs carries more risk than a small autonomous system that is immobile and designed to protect against border incursions.

The final variable is determining the aggregate damage potential of all autonomous systems that may be in operation at one time.[85] Scharre

---

78 Scharre (n 77) 18.
79 Ibid.
80 Ibid.
81 Ibid.
82 Ibid.
83 Ibid.
84 Ibid 19.
85 Ibid.

postulates that if there is a flaw in one LAWS, such as a software error, then it is likely that the flaw will be present in the other weapons systems. The potential for damage can be significant if all such LAWS fail simultaneously.[86]

In reviewing all of these variables together, Scharre reaches a couple of conclusions. The first is that the intrinsic nature of LAWS creates a higher potential for damage and risk than an equivalent semi-autonomous system. Second, a key difference between LAWS and semi-autonomous systems comes down to the difference between machines and humans.[87] Although it is indisputable that humans are not infallible and certainly make mistakes that lead to civilian casualties, Scharre points out that they are 'idiosyncratic'.[88] Humans are all different and faced with the same circumstances will often reach different results or conclusions, but it is unlikely that multiple humans will make the same mistake in the same way. Autonomous weapons will continue to follow their programming, which could lead to mass failures.[89]

It is precisely because of this increased potential for greater damage that states will need to carefully consider how to manage and minimize the risk with LAWS, especially those systems that may have a long loitering time and continue to operate until they run out of ammunition, or a LAWS that may take an unexpected action across the border of a another country during a period of heightened tension that a human operator would not take. DoD Directive 3000.09, *Autonomy in Weapons Systems*, considers these risks and has sets out certain requirements on how the United States will ensure that the appropriate level of human judgment will be exercised over any LAWS that may be developed.[90]

DoD Directive 3000.09 requires that '[a]utonomous and semi-autonomous weapon systems shall be designed to allow commanders and operators to exercise appropriate levels of human judgment over the use of force'.[91] The Directive then sets out specific requirements that supplement this general requirement. There are three aspects that collectively help to ensure that if LAWS are developed, they can be used with the appropriate level of human judgment over the use of force. First, there must be reliable testing that ensures the LAWS are engineered to

86 Ibid.
87 Ibid 23.
88 Ibid.
89 Ibid.
90 DoD Directive 3000.09 (n 3).
91 DoD Directive 3000.09 (n 3) § 4a.

perform as expected.[92] Second, established training, doctrine and procedures must be in place for the operators.[93] Finally, there must be clear and readily understandable interfaces between the LAWS and the operators.[94]

One important aspect of ensuring that appropriate levels of human judgment over the use of force may be exercised is having the confidence that LAWS will operate in a predictable manner for its users and in accordance with legal and policy requirements. This predictability would be accomplished through rigorous hardware and software verification and validation and, prior to fielding, subjecting the weapons system to realistic system development and operational testing and evaluation.[95] Such measures are integral to ensuring that those weapon systems: (1) '[f]unction as anticipated in realistic operational environments against adaptive adversaries'; (2) '[c]omplete engagements in a timeframe consistent with commander and operator intentions and, if unable to do so, terminate engagements or seek additional human operator input before continuing the engagement'; and (3) '[a]re sufficiently robust to minimize failures that could lead to unintended engagements or to loss of control of the system to unauthorized parties'.[96] In order to allow for an appropriate level of human judgment to be exercised over the use of LAWS, they must operate in a predictable manner and in accordance with legal, policy and mission requirements.

A second important aspect of the DoD Directive to ensure appropriate levels of human judgment in the use of force is to require that the LAWS operators are properly trained.[97] This is a lesson the United States learned from its study of incidents in which weapons with autonomous capabilities have resulted in unintended engagements, such as when PATRIOT air defense batteries shot down a US jet and an allied jet in 2003.[98] As those cases indicated, the risk of an unintended engagement results not only from errors in programming, but also from operators who do not

---

[92] Ibid § 4a(1)(a).

[93] Ibid § 4a(1).

[94] Ibid § 4a(2)(b).

[95] Ibid § 4a(1).

[96] Ibid § 4a(1)(a)–(c).

[97] Ibid enclosure 3, § 1b(1).

[98] Michael W Meier, US Delegation Statement on 'Appropriate Levels of Human Judgment', Convention on Certain Conventional Weapons (CCW) Informal Meeting of Experts on Lethal Autonomous Weapons Systems (12 April 2016), [hereinafter Meier, Statement], accessed 5 May 2017 at https://geneva.usmission.gov/2016/04/12/u-s-delegation-statement-on-appropriate-levels-of-human-judgment/.

understand how to use the weapons system properly.[99] Accordingly, the Directive requires that appropriate training, doctrine, and tactics, techniques and procedures for the use of such weapons be established.[100]

It is important that the human operator fully understand the issues involving autonomy in weapon systems.[101] It can be argued that the interface between humans and machines is as important as determining the types of weapons systems to be developed.[102] Thus, the DoD Directive requires that '[i]n order for operators to make informed and appropriate decisions in engaging targets, the interface between people and machines for autonomous and semi-autonomous weapon systems shall: [b]e readily understandable to trained operators'; '[p]rovide traceable feedback on systems status'; and '[p]rovide clear procedures for trained operators to activate and deactivate system functions'.[103]

This is not to suggest that the DoD Directive is a panacea for eliminating unintended engagements by LAWS. No weapons system, even the simplest ones, will function correctly every time. A state must ensure that a weapons system can be used in accordance with applicable legal, policy and mission requirements, but a state must recognize the important differences between human operated systems and LAWS and take all appropriate action to mitigate the risks of unintended engagements.

(c) Increase in Asymmetric Warfare

The third argument often raised regarding the potential impact on strategic stability with respect to LAWS is that their development and use will cause those states without LAWS to engage in asymmetric warfare. During the general exchange of views at the CCW April 2016 informal meeting of experts, Ambassador Tehmina Janjua of Pakistan said:

> [t]he use of LAWS in the battlefield would amount to a situation of one-sided killing. Besides depriving the combatants of the targeted state the protection offered to them by the international law of armed conflict, LAWS would also risk the lives of civilians and non-combatants. The unavailability of a

---

99 Ibid.

100 DoD Directive 3000.09 (n 3) § 4a(1).

101 Meier (n 98).

102 Ibid.

103 Ibid § 4a(3)(a)–(c).

> legitimate human target of the LAWS user State on the ground could lead to reprisals on its civilians including through terrorist acts.[104]

Her statement raises two of the major concerns expressed by opponents of LAWS. The first is that states that do not possess LAWS will be unable to respond on the battlefield since there simply will not be an army there to fight, which will result in 'one-sided killing'.[105] The second is that the risk to the civilian population will increase as the boundaries of the battlefield will grow, as those states or armed groups that do not have LAWS will be forced to engage in attacks away from the battlefield or through the use of terrorist attacks.[106]

The potential development of LAWS is the continuation of the desire to add physical distance between combatants and their adversaries on the battlefield.[107] Similar to other weapons like the crossbow or submarines, states have sought to develop new weapons that not only effectively engage the enemy, but also protect its own forces from harm by continuing to move farther from direct combat.[108] However, the argument against the development and use of LAWS is that these systems are something much different than past technological advances.[109] The development of a LAWS and its use in armed conflict will create such a wide chasm between those states that have them and those states that do not that it will 'destroy … the idea of war as a contest that, while potentially uneven, is essentially still a contest between (human) equals'.[110] LAWS threaten to breakdown the established norm of what it means to be in an armed conflict as it will devolve into one state being able to attack combatants of another state at will without risk.[111] It will essentially result in one side paying with their lives while the other

---

104 Ambassador Tehmina Janjua, Opening Statement to Informal Meeting of Experts on Lethal Autonomous Weapons Systems, Convention on Certain Conventional Weapons (CCW) Informal Meeting of Experts on Lethal Autonomous Weapons Systems (11 April 2016).

105 Ibid.

106 Ibid.

107 William Boothby, 'Some Legal Challenges Posed by Remote Attack' (2012) Intl Rev of the Red Cross 886, 593.

108 Ibid.

109 Postma (n 17) 321.

110 See Frédéric Mégret, 'The Humanitarian Problem with Drones' (2013) 2013 Utah L. Rev. 1283, 1310.

111 Ibid.

merely pays with economic costs of developing and using the weapons system.[112]

There is, however, no prohibition against one state trying to obtain technological superiority over another state or trying to protect its forces from harm.[113] Armed conflict is not a sporting event. There is no principle of international law that requires armed conflict, like sports, to be fought with each opposing side using an equal number of combatants and being limited to the same weapons.[114] IHL does not require a fair fight, but only that states 'fight fairly'.[115] It would not only be unreasonable, but impossible, to force the technologically advanced state to change the way it fights or to try and prohibit it from continuing to develop the weapons it deems necessary to address its particular national security concerns.[116] The broader concern is that LAWS and the technology gap it may create may negatively impact the ability and willingness of the state or armed group that do not have LAWS to continue to comply with its IHL obligations.[117] Frederic Mégret notes that in an armed conflict where one group is being killed by its opponent without any risk of death for its own forces, there may be less incentive for the group getting killed to abide by the rules.[118] Essentially, there is really no IHL obligation to comply with since they never see the opposing forces.[119] Even if that group does manage to destroy the LAWS, its destruction likely does not inflict any real damage on the other side.[120] To compensate for the disproportionate technological advantage, the state or armed group without LAWS may look to bend or break its IHL obligations and resort to asymmetric warfare or engaging in terrorist acts.[121] It is also likely that the development of LAWS will have an effect on what constitutes the 'battlefield',[122] as the state or armed group

---

112 Ibid 1311.
113 Ibid 1313.
114 Ibid.
115 Ibid.
116 Ibid 1312.
117 Ibid.
118 Ibid.
119 Ibid.
120 Ibid.
121 Ibid.
122 See Michael W Lewis, 'Drones and the Boundaries of the Battlefield' (2012) 47 Tex Intl L J 293, 301 (There are some who argue that armed conflict today is subject to an increasingly strict geographical limitation. Christopher Greenwood argues 'it cannot be assumed—as in the past—that a state engaged in armed conflict is free to attack its adversary anywhere in the area of war'. Mary

will look to engage the state with LAWS in ways that they did not anticipate.[123]

Currently, there are members of the armed forces operating UAVs from locations within the United States versus being physically present on the 'battlefield'.[124] This raises the possibility that in an armed conflict, in particular with a state that may possess similar capabilities as the United States, an opposing force may engage these targets within the borders of the United States to disrupt or eliminate this capability.[125] One of the most effective ways to impact the United States would be for an adversary to engage in attacks inside the United States.[126] While some adversaries may look to engage in terrorist type attacks, others may engage what they consider to be legitimate military targets in locations such as the military bases where such operations are being conducted.[127] Further, they may seek to target service members even when they are not on duty, but engaging in day-to-day activities, such as pumping gas or shopping for groceries, merely because of their status as a member of the armed forces.[128]

Although there are legitimate concerns that the development may increase the risk of asymmetric warfare, there are also advantages to the development of these systems that cannot be overlooked. States develop advanced capabilities as an important part of military deterrence. Such systems can result in the use of less lethal force and destruction on the battlefield.[129] LAWS may continue the trend that we have seen with autonomy where these systems will likely lower casualty rates because it

---

Ellen O'Connell contends that the shooting down of Admiral Yamamoto's plane over Bouganville by US fighter jets in WWII would be considered illegal today because it occurred far from the battlefield. However, the argument that there is a legal restriction on the use of force in armed conflict based on the distance from the front lines is contrary to practice. For example, General Douglas McArthur made the Inchon landing over 150 miles from the fighting. During the First Gulf War, the coalition engaged in attacks over 500 miles away from Kuwait. The use of force away from the front lines would not only be legal, but is a routine part of armed conflict. Under the above analysis, a UAV operator or user of LAWS would be a legitimate military target even in the United States).

[123] Daniel Sukman, 'Lethal Autonomous Systems and the Future of Warfare' (2015) 16 Can Mil J 44, 48.

[124] Ibid 49.

[125] Ibid 48.

[126] Ibid 49.

[127] Ibid.

[128] Ibid.

[129] Postma (n 17) 314.

is expected they will be more accurate.[130] This increased accuracy, however, means they will likely be used more frequently and it will also likely reduce the margin for human error.[131] Those states will need to ensure that any LAWS that may be developed are used in accordance with the laws of armed conflict.

The argument that the development and use of LAWS means a state will not be risking its own forces is overstated.[132] No state will try to guarantee its security by relying on one weapon system.[133] It is more likely that LAWS will be another weapons system available to a state during armed conflict. LAWS will not replace humans, but rather complement them on the battlefield.[134] Importantly, states that develop LAWS will remain accountable under existing international law and their use will not happen in a vacuum.

Although a state may rely on asymmetric warfare tactics, once this technology becomes available and states successfully use it, it is more likely that states and armed groups that do not possess LAWS will seek to obtain them or develop their own LAWS in order to maintain global and regional stability, which will lead to the proliferation of these weapons, such as is now being seen in unmanned systems.

(d) Proliferation

The fourth argument related to strategic stability is that once LAWS are developed and fielded the technology will rapidly proliferate not just to states that are friends and allies to the United States, but to adversarial states, in particular rogue states and non-state armed groups. As former Deputy Secretary of Defense William J Lynn III said, 'few weapons in the history of warfare, once created, have gone unused'.[135] States will likely continue to take advantage of the most advanced technologies in military operations.[136] Although it is impossible to accurately predict whether or how LAWS and its related technology will proliferate since

---

[130] Ibid.

[131] Ibid.

[132] Anderson and Waxman (n 73) 7.

[133] See Tikks-Ringas (n 69).

[134] Nickolaus Hines, 'Autonomous Robots and Military A.I. Won't Fight Wars Alone, Pentagon Says' Inverse (27 April 2016), accessed 5 May 2017 at https://www.inverse.com/article/14916-autonomous-robots-and-military-a-i-won-t-fight-wars-alone-pentagon-says.

[135] Eric Talbot Jensen, 'Future War, Future Law' (2013) 22 Minn J Intl L 282, 315.

[136] See Tikks-Ringas (n 69).

none have yet been developed or fielded, the prospect of proliferation cannot be ignored. It is instructional to look at how UAVs have proliferated to better understand how LAWS might do the same.

The use of UAVs is not new and predates the events of 11 September that led to the armed conflict in Afghanistan.[137] It was not until 2001, however, that UAVs were armed and used in armed conflict by the United States in Afghanistan. Since that time, the use of armed UAVs has grown extensively, in particular with counter-terrorism operations increasing from about 50 strikes between 2001 and 2008 to approximately 450 between 2009 and 2014.[138] Since 2001, the number of states seeking to obtain this technology has increased; there are approximately 90 states, almost half in the world, which now have UAVs.[139] Although 90 countries may have UAVs, the number that have advanced UAVs, which are able to operate for at least 20 hours at an altitude of at least 16,000 feet, and have a maximum takeoff weight of at least 1,320 pounds, is less than a third of those countries, with only 27 states possessing this capability. Further, there are currently only 10 countries that possess armed drones, with another 12 seeking to acquire lethal systems.[140]

The argument against the proliferation of UAVs, similar to the argument being made by those who are seeking to pre-emptively ban LAWS, is that this is a transformative technology that will revolutionize armed conflict.[141] States will be able to conduct strikes without risking the lives of their own armed forces, which will minimize their casualties. The political threshold for using force will lower and states will be more willing to deploy military assets. The cost of crossing the border of another state will be less and since the cost, both in lives and money, is less there is less concern if they are shot down. This may have the effect of escalating an already tense situation between states.[142] Further, there is a concern about how these systems could be used once they are acquired by non-state actors. In January of 2016, the Oxford Research Group's Remote Control Project reported that Islamic State (ISIS) is 'obsessed with launching a synchronized multi-drone attack on large numbers of

137 Michael C Horowitz, Sarah E Kreps, and Matthew Fuhrmann, *The Consequences of Drone Proliferation: Separating Fact from Fiction*, 5, accessed 5 May 2017 at http://ssrn.com/abstract=2722311 (25 January 2016).

138 Ibid 2.

139 Ibid 5.

140 Ibid 6.

141 Ibid 8.

142 Ibid 9.

people in order to recreate the horrors of 9/11' and that 'drones are a game-changer in the wrong hands'.[143]

The counter to this argument is that UAVs are simply another weapons system that can provide technological advantage in armed conflict because these systems are able to perform tasks that manned aircraft cannot. UAVs are not invincible, as they have vulnerabilities that can be countered by other states. Andrea Gilli and Mauro Gilli assert that the proliferation of UAVs is unlikely to have the catastrophic consequences that many scholars and analysts predict because 'drone warfare is more than simply possessing drones ... [it] is about the employment in military operations of remotely piloted or autonomous aircraft'.[144] The use of armed UAVs on a strategic level requires advanced and reliable platforms that involve extensive logistical and infrastructure support that very few states will ever be able to possess.[145] This is not to say that UAVs cannot or will not ever be used by less technically advanced states or that such UAVs will not be able to conduct an armed attack, but that those particular states' or non-state actors' use of such systems will not lead to regional instability or armed conflict or affect the balance of military power on an international level.[146]

With respect to LAWS, it would be naïve to believe that once a state develops and successfully uses an autonomous weapons system in armed conflict there will not be a push by other states to develop them.[147] It will likely result in not only the development of more systems, but in an increase of the capabilities of such systems. The risk of proliferation will increase as states face pressure to protect their own armed forces and deploy these systems in their place.[148] Accordingly, states will need to ensure there are appropriate processes in place to prevent the diversion of such technology to rogue states and non-state armed groups.

---

143 James Bamford, 'Terrorists have Drones Now. Thanks, Obama' Foreign Policy, 28 April 2016, accessed 5 May 2017 at http://foreignpolicy.com/2016/04/28/terrorists-have-drones-now-thanks-obama-warfare-isis-syria-terrorism/

144 Andrea Gilli and Mauro Gilli, 'Why Concerns Over Drone Proliferation are Overblown' *The Diplomat*, 19 May 2016, accessed 5 May 2017 at http://thediplomat.com/2016/05/why-concerns-over-drones-are-overblown/.

145 Ibid.

146 Ibid.

147 Postma (n 17) 331.

148 Ibid.

4. IMPLICATIONS FOR UNITED STATES POLICY

The debate regarding LAWS will continue to generate questions regarding the legal, technical, ethical, and military and operational issues raised by these systems, as well as the issues surrounding their impact on strategic stability. The Campaign to Ban Killer Robots, a coalition of civil society groups, seeks a pre-emptive ban on the development of LAWS, through a new CCW protocol or other binding instrument.[149] However, states, civil society and other experts hold divergent views with respect to LAWS[150] and it is clear that this debate will continue for the foreseeable future. The United States will need to show leadership in this area. Although it is not inevitable LAWS will be developed, if such systems are it will take a concerted effort to ensure that these systems would be used in a manner in accordance with international humanitarian law and that such weapons systems will not negatively impact strategic stability. Therefore, it will be important for the United States to take a leading role with respect to LAWS, both internationally through continued engagement in CCW and domestically through establishment of a LAWS policy similar to the 2015 UAV export policy.

(a) International Discussions within the Convention on Certain Conventional Weapons

The United States has been a strong supporter of the LAWS discussions in CCW. There have been three informal meetings of experts within CCW since 2014 and the future of LAWS discussions was a primary focus at the Review Conference in December 2016. The High Contracting Parties agreed to establish a Group of Governmental Experts to meet

---

[149] See Mary Wareham, Coordinator, Human Rights Watch, *Campaign to Stop Killer Robots Statement to the UN General Assembly First Committee on Disarmament and International Security* (26 October 2015) 1–2, accessed 5 May 2017 at http://www.stopkillerrobots.org/wp-content/uploads/2015/10/KRC_StatementUNGA1_16Oct2015.pdf (promoting a new protocol including a pre-emptive ban on autonomous weapons in the CCW).

[150] See Michael W Meier, Statement, Head of Delegation Geneva, US Mission to the United Nations and Other Intl Orgs in Geneva, *US Delegation Closing Statement and the Way Ahead, Closing Statement Before the CCW Informal Meeting of Experts on LAWS* (17 April 2015), accessed 5 May 2017 at https://geneva.usmission.gov/2015/05/08/ccw-laws-meeting-u-s-closing-statement-and-the-way-ahead/. ('[I]t is clear that there remain many unanswered questions and divergent views on a wide variety of issues'.)

over 10 days in 2017.[151] As the discussions move forward in 2017, the United States should focus its efforts on the following three areas: (1) work with Switzerland and other like-minded delegations on a 'compliance based approach' to LAWS; (2) continue efforts to work on a non-binding document related to the weapons review process; and (3) encourage other High Contracting Parties to establish national policies on LAWS while discussions continue within CCW.

(1) Adopting a 'compliance-based' approach to LAWS

On 30 March 2016, Switzerland submitted a working paper, 'Towards a "compliance-based" approach to LAWS' for use during the April 2016 informal meeting of experts.[152] The working paper, in part, calls for considering the issues surrounding LAWS under a 'compliance-based' approach as a way to consider concerns related to LAWS through three criteria.[153] The United States should work with Switzerland and other like-minded delegations on this proposal as it provides a logical and comprehensive way to further the LAWS discussion in 2017 and is similar to proposals the United States has made. The Swiss proposal calls for assessing existing systems with limited autonomy in the targeting cycle; reaffirming and spelling out applicable international law, in particular IHL; and identifying best practices, technical measures and policy measures that would secure and facilitate compliance.[154]

The first criteria seeks to review certain existing weapons systems that use autonomy in the targeting process to determine how autonomy is incorporated and how human-machine interface ensures it can be used in

---

151 Campaign to Ban Killer Robots, *Moving Forward in 2016*, accessed 5 June 2017 at https://www.stopkillerrobots.org/2016/12/moving-forward-in-2016/ (The most significant development for the Campaign to Stop Killer Robots in 2016 came at the very end of the year when countries agreed to formalize and dedicate more time to their deliberations on lethal autonomous weapons systems. The move came after states held informal discussions on the matter since 2014. The final document of the five-year Review Conference of the Convention on Conventional Weapons (CCW) contains the decision to establish a Group of Governmental Experts to meet over 10 days in 2017 and then report-back to the CCW's annual meeting on 22–24 November).

152 Informal Working Paper by Switzerland, *Towards a Compliance-Based Approach to LAWS* (30 March 2016), accessed 5 May 2017 at http://www.unog.ch/80256EDD006B8954/(httpAssets)/D2D66A9C427958D6C1257F8700415473/$file/2016_LAWS+MX_CountryPaper+Switzerland.pdf.

153 Ibid ¶ 30.

154 Ibid ¶ 31.

compliance with IHL.[155] By determining how autonomy has been incorporated in existing systems, states would better understand how weapons systems that will incorporate with higher levels of autonomy would be able to be used in compliance with IHL.[156] The second criteria would reaffirm applicable international law, in particular IHL as they apply to LAWS.[157] By understanding the applicable IHL rules, especially when combined with a comprehensive weapons review, such as discussed below, would help guarantee there is the appropriate combination of human-machine interface with the LAWS that would ensure it could be used in compliance with IHL.[158] The third criteria would identify best practices, technical standards and policy measures that would complement, promote and reinforce the applicable legal obligations.[159]

Each of these criteria support positions the United States delegation has taken within CCW over the various sessions.[160] IHL does not specifically prohibit or restrict the use of autonomy to aid in the operation of weapons.[161] The use of autonomy can enhance the way IHL principles are implemented in military operations.[162] For example, some munitions have homing functions that enable the user to strike military objectives with greater discrimination and less risk of incidental harm.[163] Improving the performance of a weapon system not only improves military effectiveness, but provides a humanitarian benefit.[164] The United

---

[155] Ibid.

[156] Ibid ¶ 32.

[157] Ibid.

[158] Ibid ¶ 33.

[159] Ibid.

[160] See Statement, *US Delegation Statement on Overarching Issues* (16 April 2015), accessed 5 May 2017 at https://geneva.usmission.gov/2015/04/16/ccw-informal-meeting-of-experts-on-lethal-autonomous-weapons-systems-laws/ ('[The United States'] policy for many years has required legal review of the intended acquisition of a weapons system to ensure its development and use is consistent with applicable law, including IHL').

[161] DoD Law of War Manual (n 24) § 6.5.9.2; Statement, Michael W Meier, *US Delegation Statement on 'Weapons Reviews'*, The Convention on Certain Conventional Weapons (CCW) Informal Meeting of Experts on Lethal Autonomous Weapons Systems (13 April 2016), [hereinafter Meier, Statement, Weapons Reviews], accessed 5 May 2017 at https://geneva.usmission.gov/2016/04/13/u-s-statement-at-the-ccw-informal-meeting-of-experts-on-lethal-autonomous-weapons-systems/. ('IHL does not specifically prohibit or restrict the use of autonomy to aid in the operation of weapons.')

[162] Ibid.

[163] DoD Law of War Manual (n 24) § 6.5.9.2.

[164] DoD Law of War Manual (n 24) § 6.5.9.1.

States uses weapon systems with autonomous capabilities designed to defend against time-critical or saturation attacks.[165] These weapon systems have included the Aegis ship defense system and the Counter-Rocket, Artillery, and Mortar (C-RAM) system.[166] Demonstrating how autonomy makes these weapons more effective would be beneficial in allaying concerns that new systems that incorporate enhanced autonomy are something to be feared.

(2) Comprehensive Weapons Review

At the April 2016 CCW expert meeting, the United States Delegation again proposed that CCW should, as an interim step to the LAWS discussion, begin to document 'best practices' with respect to a comprehensive weapons review process that would apply if a state were considering the development or acquisition of LAWS.[167] Such a best practices document would assist in establishing a common understanding and approach and would help identify any specific issues related to evaluating LAWS.[168]

The comprehensive weapons review process would include not only the required legal review, but the policy, ethical, and technical and operational issues related to LAWS and its potential use.[169] Additional limiting factors, beyond IHL, related to the use of these weapons could be added to address concerns about how autonomy functions in the

---

[165] Ibid.

[166] Ibid.

[167] Michael W Meier, Head of Delegation, *US Delegation Opening Statement to the Convention on Certain Conventional Weapons (CCW)*, Informal Meeting of Experts on Lethal Autonomous Weapons (LAWS) (11 April 2016), accessed 5 May 2017 at https://geneva.usmission.gov/2016/04/11/laws/.

[168] Ibid. The statement, in part, provided as follows:

> Finally, we have consistently heard in the CCW interest expressed on the weapons review process and about the requirement to conduct a legal review of all new weapon systems, including LAWS. We believe this is an area on which we should focus as an interim step as we continue our consideration of LAWS in CCW. The United States would like to see the Fifth Review Conference agree to begin work, as part of the overall mandate on LAWS, on a non-legally binding outcome document that describes a comprehensive weapons review process, including the policy, technical, legal, and operational best practices that States could consider using if they decide to develop ALWS or any other weapon system that used advanced technology.

[169] Ibid.

specific system and its intended operational uses.[170] The comprehensive weapons review will permit an examination of how the autonomous functions of the system fit into the overall decision-making process and how it would carry out a commander's or operator's intent in compliance with IHL and applicable rules of engagement.[171]

Although some states have expressed skepticism that weapons reviews,[172] which are conducted on a national level, would provide sufficient guarantees with respect to LAWS, adopting a document that sets forth what constitutes a comprehensive weapons review process would be a tangible step towards ensuring some consistency and quality in the review of new weapons by states.[173] With respect to LAWS, conducting such a review to ensure it is sufficiently predictable and reliable is especially important. Appropriate system design and safeties, conducting rigorous hardware and software verification and validation and realistic operational testing could enhance IHL compliance.[174] It would minimize the probability of an unintentional engagement or loss of control over the system and help maintain strategic stability.[175]

(3) Encourage other states to implement national policies on LAWS

As the United States delegation said in its opening statement in CCW in April 2016, 'we would welcome more States coming forward with their national views and policies about the appropriateness of using technology to improve the performance of weapons systems'.[176] However, to date there are only two states that have developed a national policy with respect to LAWS—the United States and the United Kingdom.[177] As the

---

[170] Michael W Meier, *US Delegation Statement on Weapons Review to the Convention on Certain Conventional Weapons (CCW)*, Informal Meeting of Experts on Lethal Autonomous Weapons Systems (LAWS) (13 April 2016), accessed 5 May 2017 at https://geneva.usmission.gov/2016/04/13/u-s-statement-at-the-ccw-informal-meeting-of-experts-on-lethal-autonomous-weapons-systems/.

[171] Ibid.

[172] 2016 Chairs Report (n 15) ¶ 50.

[173] Michael W Meier, *US Delegation Closing Statement and the Way Ahead,* Informal Meeting of Experts on Lethal Autonomous Weapons Systems (LAWS) (17 April 2015), accessed 5 May 2017 at https://geneva.usmission.gov/2015/05/08/ccw-laws-meeting-u-s-closing-statement-and-the-way-ahead/.

[174] Meier, Statement on Weapons Review (n 170).

[175] Ibid.

[176] Meier, Opening Statement (n 167).

[177] Captain Cindy Koa, 'Autonomous Weapons Systems, International Law and Meaningful Human Control' XII Australian Army J 21, 30.

LAWS discussion within CCW continues, the United States should encourage other states to consider developing national policy on LAWS.

As discussed previously in this chapter, the United States has put in place DoD Directive 3000.09 that provides guidelines on reviews that must be accomplished for semi-autonomous and autonomous systems before formal development and again before fielding.[178] The Directive does not establish a US position on the potential future development of lethal autonomous weapon systems—it neither encourages nor prohibits the development of such future systems.[179] It requires for certain types of autonomous weapon systems an additional review by senior DoD officials before formal development and fielding.[180] It also provides additional requirements that are designed to minimize the probability and consequences of failure in autonomous and semi-autonomous weapon systems that could lead to unintended engagements and ensure appropriate levels of human judgment over the use of force.[181]

The United Kingdom is the only other country that has publicly stated its national policy with respect to LAWS, which differs from the US policy.[182] The United Kingdom, while considering that existing IHL is sufficient to regulate the use of LAWS, states that 'the autonomous release of weapons' will not be permitted and that 'operation of weapons systems will always be under human control'.[183] At the most recent CCW session, the United Kingdom reiterated its policy on LAWS, stating the 'UK believes that LAWS do not, and may never, exist. Furthermore, we have no intention of ever developing systems that could operate without *any* human control. The UK is committed to ensuring its weapons remain under human control. We encourage other States to share their national policy and approach on LAWS'.[184] They have suggested that as autonomy increases, it will be supported by human operators who will

---

178 Meier, Opening Statement (n 167).

179 Ibid.

180 DoD Directive 3000.09 (n 3) § 4d; see Meier, Statement on Weapons Review (n 170).

181 DoD Directive 3000.09 (n 3) enclosure 2; see Meier, Statement on Weapons Review (n 170).

182 Koa (n 177) 31.

183 Ibid.

184 United Kingdom of Great Britain and Northern Ireland, Statement to the Informal Meeting of Experts on Lethal Autonomous Weapons Systems' (11 April 2016), accessed 5 May 2017 at http://www.unog.ch/80256EDD006B8954/(httpAssets)/49456EB7B5AC3769C1257F920057D1FE/$file/2016_LAWS+MX_GeneralExchange_Statements_United+Kingdom.pdf.

set the parameters.[185] They view this as an 'intelligent partnership' between the weapons system and the operator, which will need to continue to evolve as autonomy progresses.[186]

A third state, Australia, may be considering whether to develop a policy with respect to LAWS.[187] In June 2015, there was a report from a Senate inquiry into the 'Use of unmanned air, maritime and land platforms by the Australian Defence Force', which included autonomous weapons.[188] One of the conclusions of this study was that '[having noted the U.S. Department of Defense policy directive on AWS] the committee considers the [Australian Defense Force] should review its own policy directives to assess whether a similar policy directive on AWS, or amendments to existing policies, are required'.[189] It recommended that 'the Australian Defense force review the adequacy of its existing policies in relation to autonomous weapon systems'.[190] However, to date Australia has yet to publish such a policy, but did state at the April 2016 CCW session that 'Australia is carefully considering the many aspects of the question of weaponisation of increasingly autonomous systems and the potential application of artificial intelligence to existing systems, and where this might lead'.[191]

As this discussion moves forward, additional states will need to develop their own national policies with respect to LAWS. The United States must continue its efforts to encourage other states to develop a national policy on the developing LAWS through diplomatic means, such as a demarche, as those states begin to prepare for the 2016 Review Conference in December and likely further work on LAWS in 2017.

---

[185] United Kingdom of Great Britain and Northern Ireland, 'Working Towards a Definition of LAWS', Statement to the Informal Meeting of Experts on Lethal Autonomous Weapons Systems (12 April 2016), accessed 5 May 2017 at http://www.unog.ch/80256EDD006B8954/(httpAssets)/44E4700A0A8CED0EC1257F940053FE3B/$file/2016_LAWS+MX_Towardaworkingdefinition_Statements_United+Kindgom.pdf.

[186] Ibid.

[187] Koa (n 177) 31.

[188] Koh (n 177) 22.

[189] Ibid 29–30.

[190] Ibid 30.

[191] Statement, Australia, *General Debate Exchange of Views*, Statement to the Informal Meeting of Experts on Lethal Autonomous Weapons Systems (11 April 2016), accessed 5 May 2017 at http://www.unog.ch/80256EDD006B8954/(httpAssets)/008A00242684E78FC1257F920057BD3C/$file/2016_LAWS+MX_GeneralExchange_Statements_Australia.pdf.

(b) Establish an Overarching United States Policy on LAWS

It is likely that the work in CCW will continue for the foreseeable future, but it is also important for the United States to demonstrate its leadership on this issue and take action on a national level to help address the strategic stability concerns with respect to LAWS. Accordingly, the United States should adopt an overarching policy that addresses the development and fielding, sale and transfer, and use of LAWS and related technology, similar to its UAV policy that was announced in 2015.

Turning first to the development and fielding of LAWs, the Department of Defense, through DoD Directive 3000.09, has put in place a process to consider proposals to develop and field autonomous weapon systems. Under the Directive, any proposal to develop and field certain autonomous weapon systems, for example, a system that would select humans as targets, will require an additional review by senior DoD officials before formal development and fielding.[192] Although the Directive provides important guidance as part of the acquisition process, it has its limitations. Most importantly, the Directive only addresses the DoD acquisition process and sets out additional measures prior to the development of LAWS.[193] There are other aspects related to LAWS that are not addressed in the Directive, such as whether LAWS should even be developed. It is also important to address the potential for proliferation to, and use by, other states of this technology, which directly impacts on the issue of strategic stability.

The first question that needs to be answered is the question of whether the United States should develop LAWS. This should be done through the interagency planning process with all the relevant stakeholders present, such as the Department of Defense, Department of State, the intelligence community, and other agencies that might have an equity in this issue.[194] As we have seen, many notable scientists and other groups

---

192 DoD Directive 3000.09 (n 3) § 4d.

193 Ibid.

194 See e.g., Dr John Hamre, 'Reflections: Improving the Interagency Process', Center for Strategic and International Studies (CSIS), accessed 5 May 2017 at http://defense360.csis.org/improving-the-interagency-process/; see also Alan G Whittaker et al, *The National Security Process: The National Security Council and Interagency System*, University of Virginia National Security Law (2011), accessed 11 May 2017 at http://issat.dcaf.ch/download/17619/205945/icaf-nsc-policy-process-report-08-2011.pdf (This provides an overview of the National Security Council interagency process).

have called for a pre-emptive ban of LAWS.[195] This interagency process would enable the President to make a decision about whether the development and fielding of LAWS are in the national security and foreign policy interests of the United States.

Second, similar to UAVs, once LAWS are developed and successfully used, the technology will spread as other states will seek to develop and use them.[196] We have seen UAVs and its related technology proliferate to over 90 states as well as non-state groups.[197] Although the United States' use of UAVs can serve as a model for how other countries have used and will continue to use these systems, states, in accordance with their national security interests, will likely use UAVs in ways that are different from the United States.[198] It would likely be no different with LAWS. This proliferation certainly increases the potential for 'misperception, miscalculation, and—if improperly managed—conflict'.[199] To address some of these concerns, in February 2015, the United States issued a new export policy for military UAVs to address the proliferation and use of military UAVs.[200]

Decisions by the United States to transfer or export conventional weapons are guided by the Conventional Arms Transfer Policy (CAT Policy), which was revised in January 2014.[201] It was revised to reflect 21st century national security and foreign policy objectives.[202] The CAT Policy sets forth its goals; the criteria that guide any arms transfer decision; and explains how such transfers will support arms control objectives and how this policy supports responsible arms transfers around

---

[195] Future of Life Institute, *Autonomous Weapons: An Open Letter from AI and Robotics Researchers*, Future of Life (18 July 2015), accessed 5 May 2017 at http://futureoflife.org/open-letter-autonomous-weapons/.

[196] Sayler et al (n 52) 1.

[197] Ibid.

[198] Sayler et al (n 52) 12.

[199] Ibid.

[200] Fact Sheet, *US Export Policy for Military Unmanned Aerial Systems*, US Department of State, Office of the Spokesperson (17 February 2015), [hereinafter Fact Sheet, US UAV Export Policy], accessed 11 May 2017 at https://2009-2017.state.gov/r/pa/prs/ps/2015/02/237541.htm.

[201] Presidential Policy Directive 27, United States Conventional Arms Transfer Policy, 15 January 2014 [hereinafter PPD 27].

[202] Rachel Stohl, 'Promoting Restraint: Updated Rules for US Arms Transfer Policy' Arms Control Today (4 March 2014), accessed 5 May 2017 at https://www.armscontrol.org/act/2013_03/Promoting-Restraint-Updated-Rules-for-US-Arms-Transfer-Policy.

the world by the United States.[203] Although the United States released the revised CAT Policy in 2014, a year later the US Export Policy for Military Unmanned Aerial Systems (UAV Export Policy) was released in recognition of the special nature and concern that the proliferation of advanced UAVs raise around the world.[204]

The UAV Export Policy places stringent standards for the sale or transfer of US origin military UAVs, including armed systems. Under this policy, the United States will consider, on a case-by-case basis, pursuant to US statutes and the CAT Policy, sales or exports of military UAVs under the following conditions: (1) sales and transfers of military UAVs, including armed systems will be made only under the government-to-government Foreign Military Sales program; (2) review of any potential transfers will be made through the Department of Defense Technology Security and Foreign Disclosure processes; (3) each recipient state must agree to end-use assurances as part of the sale or transfer; (4) end-use monitoring and possible additional security conditions; and (5) the recipient state must agree to certain principles of proper use.[205] The policy reaffirms the United States' long-standing commitments under the Missile Technology Control Regime (MTCR), including the 'strong presumption of denial' for exports of Category I systems.[206]

The inclusion of principles of use is recognition by the United States that it has an interest in ensuring that military UAVs are used lawfully and a responsibility to protect US national security and foreign policy interests.[207] Recipients must agree to these principles prior to any authorization of a sale or transfer. The principles require recipients to: (1) use these systems in accordance with international law, including IHL

---

[203] Ibid.

[204] Fact Sheet, US UAV Export Policy (n 200).

[205] Ibid.

[206] Ibid; see also Fact Sheets, 'The Missile Technology Control Regime at a Glance' Arms Control Association (6 November 2015), accessed 5 May 2017 at https://www.armscontrol.org/factsheets/mtcr (Established in April 1987, the voluntary Missile Technology Control Regime (MTCR) aims to limit the spread of ballistic missiles and other unmanned delivery systems that could be used for chemical, biological, and nuclear attacks. The annex is divided into two separate groupings of items, Category I and Category II. Category I includes complete missiles and rockets, major sub-systems, and production facilities. Specialized materials, technologies, propellants, and sub-components for missiles and rockets comprise Category II. Potential exports of Category I and II items are to be evaluated on a case-by-case basis. Approval for Category I exports is supposed to be rare).

[207] Fact Sheet, US UAV Export Policy (n 200).

and international human rights law, as applicable; (2) armed and other advanced UAVs are only to be used in operations involving the use of force when there is a lawful basis for the use of force under international law, such as national self-defense; (3) not use military UAVs to conduct unlawful surveillance or use unlawful force against their domestic populations; and (4) as appropriate, operators shall receive technical and doctrinal training on the use of these systems to reduce the risk of unintended injury or damage.[208] Importantly, the UAV Export Policy is part of a broader US effort, which 'include[s] plans to work with other countries to shape international standards for the sale, transfer, and subsequent use of military [UAVs]'.[209]

Many of the aspects of the UAV Export Policy can provide a basis for an export policy with respect to LAWS and its related technology. Turning first to the export or transfer of LAWS or related technology, a strict policy that improves criteria that promotes restraint and would provide this technology to only the closest allies of the United States, that share US values and international goals and objectives, will help ensure that this technology does not fall into the hands of adversaries or non-state actors.[210] It certainly will not eliminate the risk of this future technology proliferating, but it can mitigate this risk.

If the United States develops principles of use that it applies to its own operations, just like with UAVs, the United States' use can serve as a model for other countries. Therefore, it will be incumbent upon policy-makers to articulate clear standards for US use.[211] Once the United States has these standards in place, the United States, either through the CCW process or with its closest allies, should begin to shape international standards with respect to the use of LAWS. The warning with respect to UAVs holds true for any potential use of LAWS in that failure to articulate clear standards for use 'could hold significant implications for crisis stability, escalation dynamics, and norms regarding state sovereignty—with particular consequences for the international system'.[212]

---

208 Ibid.
209 Ibid.
210 See PPD (n 201) 27.
211 Sayler et al (n 52) 12.
212 Ibid.

5. CONCLUSION

In conclusion, the debate surrounding the potential development of LAWS raises many issues, including whether it will have an impact on global and regional security concerns, which will need to be addressed moving forward. Although opinions may differ about whether LAWS may lower the threshold for engaging in armed conflict, increase the risk of unintended engagements, result in proliferation and asymmetric warfare, these are legitimate concerns that the United States and the international community will need to address when considering the development of such systems.

The United States will need to take steps both internationally, through the CCW, and on a national basis, through an overarching policy with respect to LAWS, to help mitigate the risks to global and regional security concerns. By working with like-minded states in CCW, the United States should focus the LAWS discussion on a compliance-based approach using existing IHL. The United States should also continue work on a comprehensive weapons review process outcome document. Finally, the United States should encourage other states to develop their own national policies on LAWS. On a national basis, the United States should adopt an overarching policy on LAWS that would include an export policy for LAWS and related technology and establishing principles of use for LAWS that could provide a framework for other states to follow. These approaches would provide a framework for states to carefully consider the complex issues surrounding LAWS and help address many of strategic implication concerns.

Index